Computational Intelligence in Medical Decision Making and Diagnosis

Computational intelligence (CI) paradigms, including artificial neural networks, fuzzy systems, evolutionary computing techniques, and intelligent agents, form the basis of making clinical decisions. This book explains different aspects of the current research on CI technologies applied in the field of medical diagnosis. It discusses critical issues related to medical diagnosis, like uncertainties in the medical domain, problems in the medical data, especially dealing with time-stamped data, and knowledge acquisition.

Features:

- Introduces recent applications of new computational intelligence technologies focusing on medical diagnosis issues.
- Reviews multidisciplinary research in health care, like data mining, medical imaging, pattern recognition, and so forth.
- Explores intelligent systems and applications of learning in health-care challenges, along with the representation and reasoning of clinical uncertainty.
- Addresses problems resulting from automated data collection in modern hospitals, with possible solutions to support medical decision-making systems.
- Discusses current and emerging intelligent systems with respect to evolutionary computation and its applications in the medical domain.

This book is aimed at researchers, professionals, and graduate students in computational intelligence, signal processing, imaging, artificial intelligence, and data analytics.

Computational Intelligence Techniques

Series Editor: Vishal Jain

The objective of this series is to provide researchers a platform to present state of the art innovations, research, and design and implement methodological and algorithmic solutions to data processing problems, designing and analyzing evolving trends in health informatics and computer-aided diagnosis. This series provides support and aid to researchers involved in designing decision support systems that will permit societal acceptance of ambient intelligence. The overall goal of this series is to present the latest snapshot of ongoing research as well as to shed further light on future directions in this space. The series presents novel technical studies as well as position and vision papers comprising hypothetical/speculative scenarios. The book series seeks to compile all aspects of computational intelligence techniques from fundamental principles to current advanced concepts. For this series, we invite researchers, academicians and professionals to contribute, expressing their ideas and research in the application of intelligent techniques to the field of engineering in handbook, reference, or monograph volumes.

Computational Intelligence Techniques and Their Applications to Software Engineering Problems
Ankita Bansal, Abha Jain, Sarika Jain, Vishal Jain and Ankur Choudhary

Smart Computational Intelligence in Biomedical and Health Informatics
Amit Kumar Manocha, Mandeep Singh, Shruti Jain and Vishal Jain

Data Driven Decision Making using Analytics
Parul Gandhi, Surbhi Bhatia and Kapal Dev

Smart Computing and Self-Adaptive Systems
Simar Preet Singh, Arun Solanki, Anju Sharma, Zdzislaw Polkowski and Rajesh Kumar

Advancing Computational Intelligence Techniques for Security Systems Design
Uzzal Sharma, Parmanand Astya, Anupam Baliyan, Salah-ddine Krit, Vishal Jain and Mohammad Zubair Kha

Graph Learning and Network Science for Natural Language Processing
Edited by Muskan Garg, Amit Kumar Gupta and Rajesh Prasad

Computational Intelligence in Medical Decision Making and Diagnosis
Techniques and Applications
Edited by Sitendra Tamrakar, Shruti Bhargava Choubey, and Abhishek Choubey

Applications of 5G and Beyond in Smart Cities
Edited by Ambar Bajpai and Arun Balodi

For more information about this series, please visit: *www.routledge.com/Computational-Intelligence-Techniques/book-series/CIT*

Computational Intelligence in Medical Decision Making and Diagnosis

Techniques and Applications

Edited by Sitendra Tamrakar,
Shruti Bhargava Choubey, and
Abhishek Choubey

CRC Press
Taylor & Francis Group
Boca Raton London New York

CRC Press is an imprint of the
Taylor & Francis Group, an informa business

Designed cover image: Shutterstock

First edition published 2023
by CRC Press
6000 Broken Sound Parkway NW, Suite 300, Boca Raton, FL 33487–2742

and by CRC Press
4 Park Square, Milton Park, Abingdon, Oxon, OX14 4RN

CRC Press is an imprint of Taylor & Francis Group, LLC

ISBN: 978-1-032-31377-1 (hbk)
ISBN: 978-1-032-31378-8 (pbk)
ISBN: 978-1-003-30945-1 (ebk)

DOI: 10.1201/9781003309451

Typeset in Times
by Apex CoVantage, LLC

This book is dedicated in fond memories of our loving elder brother Late Dr.Sitendra Tamrakar ,to his wife Dr. Shambhavi Tamrakar and his son Master Atharava Tamrakar

Contents

About the Editors

Dr. Sitendra Tamrakar is working as an associate professor and research coordinator in the Department of Computer Science and Engineering at Nalla Malla Reddy Engineering College, Hyderabad, Telangana, India. He has more than 17 years of experience in the field of teaching and research. He has guided 5 PhD and 19 MTech dissertations. He has authored a total of 92 publications which include books, research papers, and book chapters which have been published nationally and internationally. He has 5 patents published and granted with IP Australia and IP India. He had delivered 15 invited talks in various national and international conferences and seminars. He has been appointed as reviewer in various journals and conferences. He has attended 35 FDP/workshops and organized 7 conferences, FDPs, and workshops. His research interests are focused on the area of artificial intelligence, cloud computing, and computer networks. He is an active member of the Computer Society of India (CSI), Hyderabad Chapter, and ACM CSTA.

Dr. Shruti Bhargava Choubey has received her BE with honors (2007) from RGPV Bhopal and her MTech degree in Digital Communication Engineering (2010) from RGPV Bhopal; subsequently, she carried out her research from Dr. K. N. Modi University Banasthali Rajasthan and was awarded PhD in 2015. Presently, she is working as an associate professor and dean of innovation and research in the Department of Electronics and Communication at Sreenidhi Institute of Science and Technology, Hyderabad. She has published more than 100 papers (5 SCI, 18 Scopus) of national and international repute. She has been a member of many selection committees for recruitment of staff and faculty. Her research areas include signal processing, image processing, and biomedical engineering. She has produced 17 MTech degrees and guided more than 70 BTech projects. She is a senior member of IEEE and a member of IETE, New Delhi, and International Association of Engineers (IAENG). She worked in different positions, like dean of academics and HOD, with numerous capacities. She was awarded MP Young Scientist fellowship in 2015 and received MP Council fellowship in 2014 for her contribution to research.

Dr. Abhishek Choubey has received his PhD degree in the field of VLSI for digital signal processing from Jayppe University of Engineering and Technology, Guna MP, in 2017. He is currently associated with Sreenidhi Institute of Science and Technology, Hyderabad, as an associate professor. He has published nearly 70 technical articles. His research interest includes reconfigurable architectures, approximate computating, algorithm design, and implementation of high-performance VLSI systems for signal processing applications. He was a recipient of the Sydney R. Parker and M. N. S. Swamy Best Paper Award for Circuits, Systems, and Signal Processing in 2018.

About the authors

Contributors

Dr. Rajanikanth Aluvalu,
Noble International University

Pradeep Kumar Bheemavarapu,
Capgemini Technology Services India Limited, Bangalore, Karnataka, India

Dr. Jay A. Dave,
Sankalchand Patel University

R. Deebalakshmi,
SRM Institute of Science and Technology, Tiruchirappalli, Tamilnadu, India

Dr. Nagaraju Devarakonda,
VIT-AP University

Abhishek Gandhar,
Bharati Vidyapeeth's College of Engineering, New Delhi

Shashi Gandhar,
Bharati Vidyapeeth's College of Engineering, New Delhi

Ms. Pallavi Gholap,
Sanjay Ghodawat University

P. Susan Lalitha Grace,
Raghu Engineering College, Visakhapatnam

Drishti Hans,
Bharati Vidyapeeth's College of Engineering, New Delhi

Umang Hans,
University of Florida

P. Sharon Priya Harika,
Raghu Engineering College, Visakhapatnam

Sudhir Hebbar,
Saptagiri Institute of Medical Sciences, Bengaluru

Muhammad Fazal Ijaz,
Sejong University Republic of Korea

Dr. Panchanand Jha,
Raghu Engineering College, Visakhapatnam

Inderpreet Kaur,
Chandigarh Group of Colleges, Landran

Dr. Utku Kose,
Suleyman Demirel University, Turkey

Sujith Kumar,
Vishnu Institute of Technology

Yogesh Kumar,
Indus University, Ahmedabad, Gujarat

R. Shantha Selva Kumari,
Mepco Schlenk Engineering College, Sivakasi, India

Dr. Rongxing Lu,
University of New Brunswick (UNB)

Ch. Madhu,
Acutus Technology, Georgia, USA

Dr. Shivaprasad More,
Sanjay Ghodawat University

Gaurav Narula,
Bharati Vidyapeeth's College of Engineering, New Delhi

Dr. Nimisha Patel,
Indus University

B. Pavitra
Anurag University

Subba Rao Polamuri,
Kakinada Institute of Engineering and Technology, Korangi, AP, India

R. Meena Prakash,
P.S.R. Engineering College, Sivakasi, India

Indrakanti Raghu,
Anurag University

K. Ramalakshmi,
PSR Engineering College, Sivakasi, India

Ch. Chakradhara Rao,
Raghu Engineering College, Visakhapatnam

P Srinivasa Rao,
MVGR College of Engineering (A), Vizianagaram, AP, India

Dr. B. Thirumala Rao,
Vishnu Institute of Technology

A. Jayasimha Reddy,
North Dakota State University

Dr. Ravi Kumar Saidala,
Vishnu Institute of Technology

Dr. Mohammad Salim,
Malaviya National Institute of Technology

Amanpreet Kaur Sandhu,
Chandigarh University Gharuan Mohali, India

Henry Selvaraj,
University of Nevada, Las Vegas, United States

Ankur Narendra Bhai Shah,
C. U. Shah University

Dr. Girraj Sharma,
JECRC Foundation, India

Dr. Ashish Singh,
Anurag University

D. Narendar Singh,
Anurag University

Dr. M. Madhusudhana Subramanyam,
King Khalid University, Daharan Al Janoob, Kingdom of Saudi Arabia

B. Sundaravadivazhagan,
University of Technology and Applied Science–AL Mussanah

Chaudhuri Manoj Kumar Swain,
Anurag University

D Swapna,
BVRIT Hyderabad College of Engineering for Women, Telangana, India

Jabez Syam,
University of Technology and Applied Sciences, Sultanate of Oman

Dr. Sitendra Tamrakar,
NallaMalla Reddy Engineering College

S. Thayammal,
PSR Engineering College, Sivakasi, India

R. Thilagavathy,
College of Engineering and Technology, SRM Institute of Science and Technology (Deemed-to-Be University)

Dr. Tien Anh Tran,
Vietnam Maritime University, Vietnam

Dr. Abinash Tripathy,
Raghu Engineering College, Visakhapatnam

Dr. Nitin Jaglal Untwal,
MIT, Aurangabad

Vivek Upadhyaya,
Poornima University, Jaipur

B. Vandana,
Visvesvaraya Technological University

T. Veeramani,
Chennai Institute of Technology

Preface

Medical diagnosis, being the very first step in practice, is very critical for making clinical decisions. Computational intelligence paradigms, including artificial neural networks, fuzzy systems, evolutionary computing techniques, intelligent agents, and so on, provide a basis for humanlike reasoning in medical systems. Also, computational intelligence (or CI) techniques are fast-growing and promising research topics that have drawn a great deal of attention from researchers for many years. This book brings together many different aspects of the current research on computational intelligence technologies and also investigates contemporary computational intelligence (CI) techniques applied in the field of medical diagnosis. The book attempts to present the performance of these techniques in diagnosing different diseases, along with a detailed description of the data used.

The present book includes basic as well as hybrid CI techniques that have been used in recent years so as to know the current trends in medical diagnosis domain. It also presents the merits and demerits of different techniques in general as well as application-specific context. The presented book discusses some critical issues related to medical diagnosis, like uncertainties in the medical domain, problems in medical data, especially dealing with time-stamped (temporal) data, and knowledge acquisition. In addition, this book also discusses the features of good CI techniques in medical diagnosis. It is a collection of chapters that covers a rich and diverse variety of computer-based CI techniques, all involving some aspect of computational intelligence, but each one taking a somewhat-pragmatic view.

Overall, this book provides new insight for future research requirements in the medical diagnosis domain.

Preface

1 Prediction of Diseases Using Machine Learning Techniques

Abinash Tripathy, Panchanand Jha,
Ch. Chakradhara Rao, P. Susan Lalitha Grace,
P. Sharon Priya Harika, and Ch. Madhu

CONTENTS

1.1 INTRODUCTION

A human body can be comparable to a machine. Like a machine, the human body also needs rest, servicing, and proper maintenance. Due to mishandling of operators or some technical issues, a machine stops working, and again, in order to be in running condition, it needs the help of a mechanic. Likewise, in the case of a human being, many factors make the system be affected with issues that, ultimately, affect performance. These factors may include weather changes, wrong habits, mixing with people who have some symptoms of any disease. What a mechanic does for a machine, similarly, for the human being, the doctor does the same work. It is told that "prevention is better than cure"; thus, before getting affected by any issue, precaution must be taken. Some symptoms in a machine, or the human body, give us an idea whether in the future the symptom may cause a serious problem and damage to the system. Studying the symptoms might help us predict the future, which may or may not be correct. Thus, in this work, a review paper is prepared where manuscripts present are studied to analyze the prediction mechanism adopted to predict diseases based on symptoms collected from patients with machine learning approaches.

DOI: 10.1201/9781003309451-1

For human beings, or any animal, there is a chance they will be affected by any disease. In the present scenario, there are medicines and other medical facilities available that help cure the issue or try to reduce the ill effects of a problem. However, the process of development of these medical facilities has grown in different phases. The process of development can be informed as follows, informally:

- When the human civilization starts, they used to stay in forests, and whatever they collect from the forest, they feed on that. Therefore, there is a chance of being infected with diseases, and other issues like death and health issues come into the picture. In such cases, they have to seek the help of medicinal plants, knowledge of which comes in the course of time. In such cases, success rate is very less.
- Later, when civilization starts, a set of people known as "Baidya" (ancient Indian doctor) helps others take care of their health by using medicinal plants available in their locality. However, if a person from another place comes with some diseases, it is difficult to find the reason behind the issue and, hence, the proper cure. Again, most of the medicinal plants the Baidya used for cure are seasonal, so without the proper mechanism to store medicine, these are not available throughout the year.
- After the improvement of technology, medicines provided by the Baidya come in the form of tablets and syrup, and now it is available to each one based on a prescription provided by doctors. Again, different versions and doses, along with strength, are available in the market. Still, an issue comes that every five to ten years, it is observed that the original virus for which medicine is prepared mutates, and another version of the disease comes.
- In the next level of analysis, vaccination comes into the market. These vaccines are given to kids and other people who may have a risk of suffering from the same type of disease. For the preparation of these vaccines, a good number of researchers sit together and plan. They find out the symptoms of the diseases from the affected people and analyze how the virus present in the body works, and then work on that. However, it is always difficult to predict that. Thus, the concept of machine learning techniques comes into play, helping to predict not only future symptoms but also the age group who may be affected and if there is any chance that the virus will mutate.

In this work, the approaches adopted by the researcher to predict different diseases are collected, and a review paper is prepared based on the categories of diseases. For this chapter review, four different diseases are considered for analysis, that is, COVID-19, cancer, diabetes, and heart diseases.

1.1.1 Motivation for Current Approach

For any research work, whether it is a research paper or a review paper, motivation in needed. This expresses the purpose for which the document is prepared and for which audience the research work is meant. The present work can be discussed under two categories:

1. **For machine learning research.** This work gives an idea about how machine learning techniques are used for research work. Each machine learning process consists of steps, like collecting data, preprocessing, using techniques, obtaining the results, and finally, analyzing the results. Thus, the scholar doing research in the field of ML can go through the paper and get the required information.
2. **For disease-based research.** In the present work, four different diseases are considered, and the machine learning approaches adopted to predict them are analyzed. Thus, it gives an idea to the set of researchers who are not that much concerned with machine learning techniques on how to predict a disease based on the dataset collected.

So the motivation for the paper can be discussed as follows:

- It is observed that many researchers have worked in the area of disease prediction using ML techniques. However, in most of the cases, they considered a particular disease for analysis. But for a research person who wants to start his/her research work, they have to go through a number of papers for collecting the required information about the disease. The present work, however, gives them a chance to go through papers where machine learning techniques are carried out on a particular disease.
- For any research work, it is most important to know about the dataset and how the dataset is processed to get a proper result. It is observed that new researchers do not get the dataset for their work. The present work suggests datasets that are used by authors for their research work. Thus, they can start their work and even compare their obtained results with existing ones.
- Many ML techniques are presented for use of research personnels. The scope of these techniques is also fixed, that is, some techniques work in some domain and provide a better analysis result, and in other areas, their performance is not up to the mark. Thus, choosing a correct approach and machine learning technique for work is quite important for any analysis. This problem is also somewhat solved by the current chapter.

The chapter is organized as follows: Section 2 discusses information collected from manuscripts having prediction of diseases using ML techniques. Section 3 discusses the concluding remarks and future work.

1.2 LITERATURE SURVEY ON PREDICTION OF DISEASES USING MACHINE LEARNING

This section discusses the use of ML techniques for prediction of diseases. This is subdivided into four sections, namely, COVID-19, cancer, diabetes, and heart diseases. The details of the analysis follow.

1.2.1 Prediction of COVID-19 Using Machine Learning

Mohimont *et al.* concentrate on numerous CNN-based approaches for the prediction of COVID-19 estimate [1]. The precision of the local forecasts was improved by utilizing various leveled pre-preparing plan, and a proficient equal execution takes into account fast preparation of numerous local models. They show how data-driven models can deliver amazing forecasts. Since deep learning, for the most part, relies upon enormous data, it was imagined that it could not proficiently display the COVID-19 pandemic because the available data during the pestilence change was diminished and divided. They utilized four datasets of various data sources, namely, worldwide dataset, French dataset, French dataset with moving or ever-changing dataset, and worldwide dataset with mobility data. Their proposed model was effectively tested on five COVID-19 markers: the patient gets a confirmation report, patient is hospitalized, ventilation service, recovery, and demise. Additionally, their proposed approach was effective in different scales: provincial predictions may be produced using public data available, and public expectations were carried out on the whole data. Temporal convolutional network (TCN) appears to accomplish great exactness, yet this correlation is restricted because the blunder rates were determined for various periods. Generally, in the best-case scenario, where all information is present, their proposed approach has achieved an error rate of 1% or less and kept on introducing great outcomes.

Annavarapu has used a deep learning–based snapshot ensemble procedure for proficient COVID-19 chest X-ray characterization, and this proposed strategy exploits the exchange learning method utilizing a pretrained model, namely, ResNet-50 model [2]. He has proposed the ensemble algorithm to upgrade model training and exactness. The proposed model is contrasted with benchmark techniques that show the productivity of COVID-19 CXR arrangement. The proposed strategy utilizes information expansion to deal with and address class unevenness, which misleadingly adds pictures to fewer classifications to approach those of the biggest class. Their proposed method arbitrarily picked and replicated the pictures, having a place with the class with fewer examples to make copy pictures while resampling. Nonetheless, in light of the fact that DNN performs better with a lot of information, information increase makes pictures that portray its class highlights at each conceivable point. Their proposed approach additionally utilizes ADAM (adaptive moment estimation) enhancer with weight to lessen overfitting and get the best approval precision after preparing the data. The proposed model can accomplish a general precision of 95% and particularity of 97% for the dataset having multiclasses.

Perumal *et al.* have proposed the exchange learning strategy that is used for clinical pictures of various sorts of aspiratory infections, including COVID-19 [3]. Transfer learning model to stimulate the expectation cycle and help the clinical experts and this model beats other existing models. Transfer learning is a technique where the information acquired by a model for a given data is moved to assess one more issue of comparable errand. In this, underlying preparation is done utilizing a lot of datasets to perform characterization. Haralick surface highlights are acquired from the improved pictures, and these altered pictures are then taken care of into different pre-characterized CNN models. The COVID-19 dataset is acclimatized from different assets accessible in Google Images, RSNA, and GitHub open repository. VGG16 modes give high exactness of 93.8% in contrast with different modes, like Resnet50, Inception v3. Their suggested approach produces accuracy of 91%, review of 90%, and accuracy of 93% by VGG-16 utilizing transfer realizing, which beats other existing models during the same period.

Sinha and Rathi proposed an artificial intelligence–based factual way to deal with foreseen endurance probability of corona-contaminated individuals in the region of South Korea with the investigation effect on factors like gender, age group, and temporal evolutions [4]. They incorporate all conceivable condition of craftsmanship philosophies from ML and DL spaces in their experimentation to come up with a convincing comment in regard to display execution with further developed exactness while making the forecasts of endurance probability of the patients suffer from coronavirus in South Korea. The forecast of endurance chances of the coronavirus-contaminated populace is directed in two stages, including the model investigation and model expectation utilizing ML and deep neural networks. They have considered the coronavirus disease dataset from Kaggle that contains information about patients in South Korea. The preparation and testing of put together artificial intelligence (AI) models were led with respect to 5,165 coronavirus examples and approved more than 1,533 isolated patients. During the time-spent structure, a productive learning model, hyperparameter tuning, is considered as quite possibly the main aspect. Logistic regression (LR) is utilized to play out the expectation with 4,000 cycles. Their proposed approach gave a normal full-scale precision of 91%. Further, SVM is applied with boundary streamlining utilizing the grid search cross-approval method, and the exactness was obtained as 97%.

Shi *et al.* have made an effective knowledge strategy for the analysis of COVID-19 according to the point of view of biochemical lists [5]. They proposed a system consisting of a stochastic approach called colony prediction algorithm (CPA) with the combination of kernel extreme learning machine (KELM) to generate a model called ECPA-KELM. The core part of the approach is an ECPA technique that combines two fundamental administrators, which works without the dim wolf-streamlining agent and moth-fire enhancer to improve and re-establish the CPA research works and, at the same time, is used to upgrade the boundaries and to choose highlights for KELM. The effectiveness of the proposed enhanced CPA (ECPA) is broadly checked and examined with different algorithms on IEEE CEC 2017 benchmark. Moreover, it has been suggested to synchronize the boundary value and component choice in KELM; the subsequent ECPA-KELM was utilized effectively for early distinguishing proof and victimization of COVID-19. For future work, various issues can be additionally examined. More factors and coefficients are added, and equal handling can likewise diminish the figuring trouble in the application stage. Furthermore, the proposed ECPA-KELM can likewise be utilized to anticipate other assortment of conditions, like clustering of the different aspects and dividing the images used into CTs to grow the utilization of the created framework.

Zhijin Wang and Bing Cai proposed the multivariate shapelet learning (MSL) model to take in shapelets from chronicled perceptions in numerous spaces [6]. A trial assessment was done to think about the expectation execution of 11 algorithms, utilizing the information gathered from 50 US areas/states. The model should handle the issues as follows: (1) the assurance of COVID-19's hatching period among geologically associated regions, like regions/states in a nation, and (2) the acquisition of key transmission patterns in different associated regions. The shapelet has high interpretability and great clarifications. Nevertheless, it is yet a test effectively to find great shapelets. Besides, three learned shapelets portray the developing pattern and plummeting pattern of the infection. All models are prepared utilizing the ADAM-streamlining agent. The mean squared error (MSE) is picked as the misfortune capacity of the relative multitude of models. The cluster size is set to 32. For RNN, LSTM, ED, and LSTNet, the quantity of stowed-away neurons is in {32, 64}. Their learning rates are set to 0.001. In future, the multi-skyline COVID-19 forecast would be additionally explored, which would give further dreams to infection avoidance and control.

Nayak *et al.* recommended the relevance of canny frameworks, for example, ML, DL, and others, in tackling COVID-19-related flare-up issue [7]. The primary expectation behind this review is:

1. To comprehend the significance of savvy approaches, for example, ML and DL, for COVID-19 pandemic.
2. To talk about the effectiveness and effect of these techniques in the forecast of COVID-19.
3. To develop improvements in the sort of ML and progressed ML strategies for COVID-19 prognosis.
4. To examine the effect of data types and the idea of information alongside challenges in handling the information for COVID-19.
5. To zero in on some future difficulties in COVID-19 anticipation to rouse the specialists for advancing and upgrading their insight and exploration on other affected areas because of COVID-19.

In light of the experimentation, multinomial Naive Bayes and logistic regression have outflanked better outcomes with 94%, 96%, 95%, and 96.2% of accuracy, review, F1-score, and exactness rate, respectively. Based on this exploration work, we emphatically prescribe that numerous other investigations should be directed with ML just as DL on COVID-19 information with those techniques not applied at this point. Additionally, there is a need to address the difficulties of absence of information on COVID-19 to direct additionally progressed explorations.

1.2.2 Prediction of Cancer Using Machine Learning

Breast cancer, which starts and spreads from breast tissue, is mostly widespread among women [8]. Survival rate increases if it is identified in the initial stage by using microarray technology, which makes convincing benefaction for both diagnosis and treatment. In their paper, for the prediction of breast cancer, many ML algorithms are used, and their categorization presentations are differentiated among them. The genes that are responsible for breast cancer were spotted using the approach of selection of attribute, and this conducted analysis gave 90.72% success rate with 139 features.

TABLE 1.1
Consolidated Information about the Prediction of COVID-19

Authors	Techniques used	Dataset used for analysis	Result obtained	References
Mohimont *et al.*	Convolutional neural network, temporal neural network	Worldwide dataset, French dataset, French dataset with mobility data, worldwide dataset with mobility data	The normalized root mean square error is 1% or less than 1%	[1]
Annavarapu	CNN architecture with ResNet50	COVID-19 CXR dataset with 2905 X-ray images	Accuracy, 96.18%; precision, 95.23%; recall, 95.63%; F1-score, 95.42	[2]
Perumal *et al.*	Transfer learning method along with Haralick features	Chest X-ray-14 dataset collected from NIH	Accuracy obtained, 93%	[3]
Sinha and Rathi	Machine learning techniques with hyperparameter tuning, deep learning with encoder-based approach	Dataset taken from Kaggle, which contain information about health data of South Korea from Dec. 19–Mar. 20	Support vector machine + convolution neural network (CNN): 97%	[4]
Shi *et al.*	Colony predation algorithm with kernel extreme learning machine (ECPA-KELM)	51 COVID-19 patients at Whenzhou Medical University between Jan. 21 and Mar. 21, 2020	Accuracy obtained: 92.129%	[5]
Wang and Cai	Multivariate shapelet learning method	Dataset collected from GitHub website having COVID-19 cases from 50 provinces in the US	Authors provide information about the place where cases are in increasing tone and where it is in decreasing trend	[6]
Nayak *et al.*	Combination of the ML and DL	Review paper, thus information about various methods discussed	Review paper, thus result of various papers compared	[7]

The rapid growth and improvement in medical sciences and technology determined many new techniques for both diagnosis and treatment of metastatic tumors; as a result, the cancer patient's survival rate has been increased. However, even after good treatment, there is a chance of getting affected again [9]. Exploring the mechanism of tumor recurrence and metastasis to predict recurrence and metastasis of cancer is a major clinical issue. At the same time, the rapid development of the Human Genome Project and gene microarray technology has enabled the activity of many genes in the patient's body to be intuitively measured through the chip. The rapid development of machine learning has contributed to data mining and medical science of this DNA microarray technology.

Firstly, they perform simple data cleaning and normalization processing on clinical data and genetic data. Second, they perform differential gene screening. Next, they select PCA, sparse PCA (SPCA), nuclear PCA (NPCA), and multidimensional scaling algorithms to reduce the dimension of the data. Finally, the genetic data uses random forest (RF), SVM, SVM using linear kernel, guided aggregation algorithm, gradient boosting algorithm, and ensemble learning for machine learning and then finds the best parameters and methods through grid search and selects the appropriate model evaluation method. The clinical data is manually selected and classified using machine learning. Finally, the results of clinical data and genetic data are combined to predict the site of recurrence. Using the abovementioned method to predict the recurrence and the location of the recurrence, a good effect was achieved. Taking likelihood of recurrence as an example, the accuracy rate of the verification set achieved was 0.825; the recall and the F1 values are 0.801 and 0.800, respectively. Through the retrospective study and prediction of gastric cancer recurrence, the model proposed in this paper has potential clinical value. This paper provides a novel method for bioinformatics data mining. Their proposed approach achieved a comparatively better result in judging the recurrence of gastric cancer and has potential clinical application value.

In view of the multitude of reports, this disease is causing an extremely huge issue in human existence [10]. In their paper, they have conjectured reoccurrence of breast cancer growth by staggered perceptron with two individual yields. For highlight extraction, deep neural organization, and as a classifier, multifacet perceptron, ANN with two nonindistinguishable yields, and afterwards, the SVM. Then, at that point, they have analyzed the outcome information acquired by every technique; it tends to be perceived that ANN with two yields prompts the most elevated exactness and the least change among different constructions. For ANN analysis, the matrices of the internal values are considered, and those analyses help the result of the proposed approach obtain better accuracy. In any case, the fluctuation of results by SVM was significantly more than those of the other different strategies. Their proposed paper to arrange the examples having breast cancer depends on the symptoms, to check whether or not their malignant growth has backslid or vice versa. In order to perform the task, various types of classifiers are appealed to various designs. The reproduction results communicated ANN with two yields in the secret layer prompted the most elevated exactness. Additionally, largely rough neural network brought about the least change, contrasted with different techniques.

Quick and exact determination of normal infections, for example, breast cancer growth that is normal among women, is vital [11]. While expert specialists make this assurance, studies are completed with ML algorithms to support them. ML algorithms make derivations from existing information and anticipate what is obscure. Administered ML algorithms utilized in arrangement of clear-cut information and of new information are utilized as often as possible. They used the dataset of 357 malignant and 212 benign, utilizing the elements of the University of Wisconsin, by making a few estimations in the mass pictures on the breast, and the element techniques separated from these elements were utilized. Subsequent to applying the fundamental prehandling and standardization ventures for the dataset, information was isolated as training and test information, and training was performed for six directed ML algorithms (k-nearest neighbors' algorithm, random forest algorithm, decision tree, naive Bayes algorithm, support vector machines, and logistic regression). In addition, similar tasks were completed by applying principal parts examination and linear discriminant analysis to the dataset. In their review, the exactness esteems for every one of their models were expanded in the wake of applying LDA, and a triumph pace of 96.49% was accomplished with logistic regression. It was intended to choose the appropriate algorithm and to acquire perception that will be the wellspring of the following examinations with the outcome to be gotten in this review.

In view of Breast Cancer Institute (BCI), the malignant growth is the most hazardous kind of disease that is exceptionally compelling for the women on the planet [12]. According to clinical masters, recognizing this malignant growth in its primary stage helps in saving life. According to cancer.net, information aids for in excess of 120 sorts of cancer and its related genetic conditions. For distinguishing breast cancer, for the most part, ML methods are utilized. In their paper, they proposed versatile gathering casting a ballot strategy for analyzed breast cancer utilizing Wisconsin breast cancer growth information base. The proposed discussion is to analyze and clarify how the combination of ANN and logistic algorithm furnishes proper arrangement when the method is collaborated with gathering ML methods for breast cancer diagnosis, regardless of whether the factors are diminished. Their proposed approach utilized the Wisconsin determination breast cancer dataset. Their proposed approach, which is a combination of ANN and LR, accomplished 98.50% precision compare to other ML methods.

They suggested ensemble ML method with ANN and linear regression (LR) for the conclusion and identification of breast cancer growth. They utilized normalization approach as prehandling for breast cancer dataset, then they have applied features selection algorithm (FSA) using univariate. UFSA utilized chi-square technique for finding out the features from UCI dataset, that is, 16 features. Upon gathering the last 16 highlights from UFSA, they executed logistic and ANN on 16 elements obtained and, lastly, applied democratic calculation on outcome and accomplished 98.50% precision. The Wisconsin breast cancer dataset contains 699 rows which highlights classes of 30 elements. After UFSA is applied, the top 16 highlights are chosen from conclusive model execution, since huge highlights have an impact on the cost of model execution. Accomplished precision is great from individual accomplished exactness from both ML algorithms. This work proposed a gathering ML strategy for analysis of breast cancer growth, a proposed technique that appears to be with 98.50% exactness. They utilized just 16 highlights for cancer prediction. Later, they will take a stab at all elements of UCI to accomplish best exactness. Their work demonstrated that ANN is additionally powerful for indispensable human information examination, and they can do preanalysis with next to no extraordinary clinical information.

In their proposed review work, it is expected to arrange breast cancer information achieved from UCI (University of California–Irvine) ML laboratory for certain machine learning techniques [13]. With this point, grouping execution of some distance gauges in MATLAB has been analyzed, utilizing breast cancer information. Later, without utilizing any prehandling, a portion of the ML strategies is utilized for grouping breast cancer information, utilizing WEKA data mining programming. Therefore, it has been seen that distance estimates impacts the bunching execution almost at a rate of 12, and the achievement of the order shifts from 45% to 79%, as per the techniques. The values of the configuration boundaries of CSO utilized in tests are thus: count of aspects is 2, count of chickens is hundred, count of cycles is 100. The frequency for swarm reorganization is considered as 10, least flight length is 0.5, greatest flight length is considered as 0.9, epsilon value is considered to be 0.000000001, and roosters, hens, and chicks are considered to be 20%, 20%, and 60%, respectively. The qualities gamma and C got for every grouping issue. Those qualities are utilized in the preparation of an alternate SVM for every order issue. The finding of cervical cancer drawn in utilizing an ML strategy is based on SVM, where the gamma and the C boundaries of the SVM classifiers were determined utilizing an adjusted form of the CSO algorithm. The exactness of the results accounts for the Hinselmann (0.953), and the best outcomes as far as AUC were acquired because of the Citology (0.66). As future examination work, the accompanying bearings are proposed: (1) use of oversampling strategies, (2) use of different classification techniques, (3) the correlation with known techniques, and (4) adaption of the approach for different demonstrative datasets.

1.2.3 Prediction of Diabetes Using Machine Learning

Perhaps the most basic persistent medical care problem is diabetes. If the problem is not handled properly, they may affect various parts of the human body, like the kidney and the eyes, and furthermore, if the process of medication is not proper, the issue may cause the patient to die [14]. The authors, in their paper, tested various machine learning techniques and tried to find out an ML technique that provides a result based on certain criteria, including accuracy, F-measure, specificity, etc. The authors carried out a detailed analysis with different machine learning techniques, namely, random forest, CART, LDA, SVM, and KNN, to test and obtain the result of the analysis. Their proposed system has shown that RF has produced a better accuracy result compared to other ML techniques they have used for analysis. For the purpose of analysis, they have considered the CSV diabetes dataset from R-Studio, which is a free dataset source. In the wake of performing preprocessing of information, they applied distinctive ML techniques for analysis of the dataset and to check for different categories of diabetes. They have categorized the dataset into different classes, namely, type 1, type 2, and gestational, having percentage values of 19.03%,

TABLE 1.2
Consolidated Information about the Prediction of Cancer

Authors	Techniques used	Dataset used for analysis	Result obtained	References
Bektaş and Babur	Attribute selection approach	Kaggle cancer dataset	90.72% with 139 features	[8]
Gao et al.	PCA, SPCA, NPCA to reduce the dimension of the dataset, RF, SVM, gradient boosting for classification	Microarray dataset from GSE62254 dataset downloaded directly from gene expression collection	Accuracy, 82.5; recall, 80.1; F-measure, 0.800	[9]
Jafarpisheh et al.	For classification, the multilayered perceptron with ANN and SVM	Dataset contains information about 1983 patients having best cancer	SVM, 92.97; multilayer perceptron with 1 output, 94.53	[10]
Kaya and Yağanoğlu	Six machine learning techniques, namely, KNN, RF, DT, NB, SVM, LR	Dataset collected from the University of Wisconsin having the images of the breast	LR, 96.49; KNN, 95.9; RF, 94.7; NB, 95.9; DT, 95.9; SVM, 94.7	[11]
Khuriwal and Mishra	Wisconsin determination of breast cancer dataset	Combination of ANN and logistics regression	Accuracy obtained, 98.5%	[12]
Kolay and Erdoğmuş	Breast cancer dataset from the University of California–Irvine	K means, random forest, naive Bayes	K-means, 79; random forest, 68.88; naive Bayes, 71.68	[13]

56.73%, and 24.03%, respectively. Subsequent to preprocessing of dataset, ML algorithms were applied and investigated. The assessed results are compared, and kappa metrics is plotted on a dot plot platform. RF gives better exactness compared to that of SVM, CART, LDA, and k-NN. The obtained accuracy and kappa value of RF is found out to be 0.99, which is best among all. The accuracy and review values of this algorithm are 1 and 1 separately, which likewise show to be most minimal in case of RF.

Diabetes is a most normally spread illness which is not anticipated previously, so a proficient strategy for expectation will assist patients in self-determination [15]. Nonetheless, the regular strategy is to perform blood glucose tests for predicting diabetes by a specialist, and clinical assets need to be restricted. In this manner, most patients cannot get the analysis right away. Since early indications of diabetes are not self-evident and the connection among side effects and the disease is intricate, the patient performs the test on their own, which may or may not be correct, which may cause a big issue in the later part of their life. The main purpose of using ML technique is to suggest a proper mechanism to test for diabetes, which will work on various datasets—that is, the method must be accepted by all, not confined to or works on a single dataset. The method should also work on the datasets irrespective of both the size and time; accuracy is a very vital factor for the result also, and thus, all factors need to be handled properly. As mentioned, they have performed the test using different ML techniques; in their paper, they have considered six different ML techniques, namely, linear regression, decision tree, boosting, neural network, SVM, and RF. Among these techniques, the results obtained by RF, boosting, and NN are found out to be better compared to those of the other three approaches. The predicted result for NN is 95.59%. They have used six different ML techniques and observed that for three ML approaches, the system shows a better result; for others, it does not perform that good. Thus, according to them, a more detailed analysis must be carried out to obtain a better result which will work on almost every dataset.

Diabetes is a non-transmittable illness related to an increase in the level of glucose in the body and can be categorized into two main different classes, namely, type 1, where the patient's body is not able to use insulin properly, and type 2, where the production of insulin is not done properly in the patient's body [16]. However, the specific reason for type 1 diabetes is obscure; the likely explanations are hereditary qualities and ecological elements (for example, susceptibility to infections). Then again, type 2 diabetes is largely connected to undesirable ways or life decisions. The authors have considered patient details in a dataset from Nigeria. According to the authors, the disease is found out to be most common in that area and lots of people suffer from that, and it is observed that the number of patients in that area is in a higher rank in Africa. In order to find out the reason behind the increase in the number of diabetes patients, an online investigation was done by the media there. The investigation was carried out upon billions of people considered as random, which included patients suffering from diabetes and the medical staff taking care of them, and utilized web-based social media platforms to unreservedly share their encounters and examine numerous well-being-related subjects. None of the current examination focuses on the African crowd, who are, additionally, significant clients of social media. Subsequently, their work found out an approach to fill the gap by using social media platforms present in Africa, situated particularly at Nigeria, to obtain diabetes information and then, using ML, to find out the main reason for the spread of diabetes in that area. The authors have tried to collect data on the habits and lifestyles of the patients and, based on that, predict the chances of having diabetes in the future. They used their model to predict the classes for the posts collected from social media. The predicted accuracy obtained by the system is 87.09%, which gives information that the process of analysis is going in right direction. Their proposed approach has obtained a true classification for 209 cases from a list of 240 cases as per the confusion matrix obtained by them. Their work further discloses how to enable people to assume responsibility for their well-being through changes in their ways of life that are quantifiable and compelling, accordingly comprising a reasonable methodology for abridging the predominance of diabetes in Nigeria.

Diabetes is quite possibly the most widely recognized disease that can influence anybody at any ages and is affected when sugar content or glucose level in the blood is increased [17]. Determining it early is required. There are around a few methods applied on Indian Pima dataset. In their review, they have applied some exceptionally well-known methods, including, ANN, SVM, RF, and so on. They have used those techniques in many ways. Primarily, they have implemented different algorithms in the first dataset. Afterwards, at that point, they utilized a few preprocessing strategies to recognize diabetes. Finally, they applied those methods to think about and get the best exactness. Neural network has given the best exactness (80.4%) when compared to some other procedures. There are a few examinations made in this, which are implementation environment, dataset description, accuracy examination of various preprocessor procedures, connection between features, feature precision as per correlation, comparison with different strategies, and execution time correlation. They contrasted many algorithms with different preprocessing procedures and found out the algorithms with the best execution preprocessing strategy. They observed that ANN provides a better result, that is, 80.5%, than some other strategies. They have observed the execution time of a few techniques and noted that NB has taken comparatively less time than the other strategies. They likewise tested the co-connection between the features. As for future work, they attempt to apply further developed strategies to ANN, like adding more multiple hidden layers, optimizing the algorithm for better results.

In everyday life, the utilization of huge information in medical care is expanding. Social networks, banking systems, utilization of sensors and smart gadgets lead to a lot of information [18]. Thus, Swarna *et al.* have suggested a model and prepared a gadget that can help in the prediction of the disease. They tried to predict the disease using different ML techniques, including NB, RF, KNN, and LR, on a particular dataset. Their approach consists of the combination of ML along with big data, and in order to test their process, they take the help of different matrices present. In order to test their proposed approach, they have considered four types of ML, performed testing on the dataset, and predicted the results using parameters that include accuracy, precision, and exactness. The amount of data available for analysis is large, and thus, they prefer to use the concept of big data. They have used the supervised technique to perform the work and, with the help of confusion matrix, check the result. In case of data present in the medical field, it is difficult to find out a proper working model and test the patient data on that model; thus, the combination of both models is carried out. The machine learning techniques work differently on different datasets, and thus, the result obtained after the analysis is also different. Among the four ML techniques they have used, the LR has shown the best result when compared to the other three approaches.

Data on diabetes mellitus, commonly called diabetes, entails a process of collecting "problems" from multiple persons and finding out the related issues to suggest a proper solution before any mishap happens [19]. PIMA Indian Diabetes dataset is available for researchers to perform research work in the field of diabetes, but as it includes personal information of patients, proper authentication is needed before any processing is made. The design of the dataset started in 1965, when it was observed that the number of cases of the disease increased in an alarming rate. The reason for the increase in cases is improper testing process, and actions needed to be taken afterwards to reduce the affect. The authors concluded that the testing process and the steps needed to be taken afterwards were not done properly, and that, finally, affected many persons. In order to test their approach, they have combined DNN and SVM to find out the reason behind the disease and perform processing of the dataset before it was given to ML. They not only performed the preprocessing activities

TABLE 1.3
Consolidated Information about the Prediction of Diabetes

Authors	Techniques used	Dataset used for analysis	Results obtained	References
Kumar and Pranavi	Random forest, CART, LDA, SVM, KNN	CSV diabetes dataset from R-studio	RF, 0.99; LDA, 9435; SVM, 96.9; CART, 93.2; KNN, 59.69	[14]
Ma	Logistics regression, SVM, decision tree, random forest, boosting, and neural network	UCI machine learning repository collected by direct question-naire from the patients of Sythet Diabetes Hospital	LR, 87.2; SVM, 90.4; DT, 93.6; RF, 94.9; boost-ing, 95.5; NN, 96.2	[15]
Oyebode and Orji	Naive Bayes with three variations, namely, multinomial, binarized, Bernoulli	Dataset contains 371,996 posts from 74,224 topics within the health forum using web scraping technique	Accuracy obtained, 87.08	[16]
Saha *et al.*	Neural network, SVM, random forest, LR	Pima Indian women diabetes dataset	NN, 80.4; RF, 77; LR, 76.9; SVM, 75	[17]
Swarna *et al.*	NB, KNN, RF, LR	Pima Indian women diabetes dataset	NB, 69; RF, 60; LR, 74; KNN, 0.88	[18]
Wei *et al.*	LR, deep NN, SVM, DT, NB	Pima Indian women diabetes dataset	LR, 77; DNN, 77.8; SVM, 77.6; DT, 76.3; NB, 75.7	[19]

but also updated the boundary of the dataset to obtain the results. Among the ML approaches used by them for analysis, DNN is found to provide an accuracy level of 77.8%, and there is no separation between training and testing of the data. They then performed a tenfold cross-validation. In addition, a practical method can be utilized straightforwardly on any new information collection, and the information ought to be preprocessed with "scale." As per the outcome, the three most significant elements in this informational index are the level of plasma glucose two hours after food intakes; for women, frequency of pregnancy; and lastly, patient's present age. The most unsignificant component is two-hour serum insulin. The most reliable classifier, DNN, in any case, actually can possibly work in the future. The suggested approach by the authors is to add a greater number of hidden layers compared to that in the traditional approach. As suggested by them, the increase in the number of hidden layers may show an improvement in the result obtained but can also cause issues, like dropouts—that is, some cases may not be considered for analysis. Along with that, an increase in hidden layer may also increase regularization.

1.2.4 Prediction of Heart Diseases Using Machine Learning

Yadav *et al.* has proposed an automated system for heartbeat classification [20]. They have categorized recorded heart auscultations between normal and abnormal heartbeats. To record the heartbeats and murmurs, they have used a phonocardiogram (PCG) machine. The recorded heartbeats are fed through band pass filter to remove the noise and unwanted signals and attain better accuracy for the proposed ML algorithm. These heartbeat samples are later divided into multiple non-overlapping frames to extract important features from the data. These important properties are based on the P-value; distinguished features are used to train the classification model. Once the data is preprocessed, they adopt multiple classification models, including, naive Bayes, SVM, random forest, and K-nearest neighbor. Based on the training and testing of preprocessed data, SVM, K-NN, and naive Bayes classifiers yield the best result, with accuracy greater than 95%.

Narayan and Sathiyamoorthy proposed a system which is based on FFT with ML also known as Fourier transformation–based heart disease prediction system (FTH-DPS) [21]. In this work, they adopted Tunstall Healthcare real-time data of patients. The dataset is a time series data obtained from day-to-day monitoring of normal and abnormal heart conditions of patients. The dataset consists of various features, such as patient ID, measurement questions, date, value and corresponding detail features, such as mean arterial pressure, diastolic blood pressure, blood glucose, etc. Further, the dataset is decomposed by FFT and used to extract important features. Further, these processed data are fed to various machine learning models. The adopted models are LS-SVM, naive Bayes, NN, and ensemble learning. Comparison has been made on accuracies of each algorithm. The claim has been made on the accuracy of the proposed algorithm, which outperforms the traditional machine learning algorithms.

Doppala *et al.* have proposed a hybrid machine learning model which is a fusion of genetic algorithm (GA) and radial basis function (RBF) network [22]. In this work, an open-source coronary diseases dataset from Cleveland repository has been taken. The dataset consists of 14 features with 303 entries. Dataset preprocessing, including, missing value treatment, data cleansing, and analysis, has been made. In this work, GA is adopted for the selection of optimum features from the dataset. The selected features are further used to train various machine learning models. The proposed GA-RBFNN models are compared with NN, KNN, naive Bayes, SVM, random forest, and logistic regression. Based on various features extracted from GA, the target label is classified. Confusion matrix with accuracy, sensitivity, specificity has been considered for the comparison of the state-of-the-art models with proposed architecture. The proposed model, GA-RBFNN with tenfold cross-validation, yields maximum accuracy of 94% when compared to other state-of-the-art machine learning models.

Meng et al. have proposed ensemble learning models like AdaBoost, gradient boosting, regression tree, and random forest model with tenfold grid search cross-validation to get optimum tuning parameters for all models [23]. The data is collected through activity tracker Fitbit Charge 2. A total of 200 patients with stable ischemic heart diseases (SIHD) used the activity tracker for 12 weeks to gather personal physical activity report. Further, these patients have been inquired about their

TABLE 1.4
Consolidated Information about the Prediction of Heart Diseases

Authors	Techniques used	Dataset used for analysis	Results obtained	References
Yadav *et al.*	Tenfold cross-validation with feature selection, SVM, NB, RF, KNN	PCG recorded heart sound signals database created by the National Institute of Health (NIH)	SVM, 97.78; NB, 96.67; RF, 97.78; KNN, 95	[20]
Narayan and Sathiyamoorthy	Fourier transformation–based heart disease prediction system (FTHDPS) (i.e., combination of FT and ML techniques)	Tunstall dataset gathered from pilot research	Accuracy obtained, 93	[21]
Doppala *et al.*	Feature selection with GA, then classification using radial basis function (GA-RBF)	Coronary disease dataset from Cleveland repository available from the University of California–Irvine	GA-RBF, 94	[22]
Meng *et al.*	Patient-reported outcome using activity tracker, data with stable ischemic heart diseases; 182 patients with SIHD monitored over a period of 12 weeks	Hidden Markov model (HMM)	Highest mean area under curve (AUC), 79	[23]
Pal et *al.*	PPG signal used to identify ischemic heart disease; PPG signal collected from patients visiting outdoor units of cardiology departments in medical colleges, hospitals in Kolkata	Decision Tree, Discriminant analysis, LR, SVM, KNN, Boosted Tree	Boosted tree, 94	[24]
Hemanth *et al.*	Heart sounds recorded 3 times in 10-second interval of 10 hospital staff working at Suleiman Demiral University, Turkey	Augmented reality and virtual reality	Normal sound, 87.5; murmured sound, 92.5	[25]

health status through some predefined questionnaires, which we call patient-reported outcome measurement information system (PROMIS). Further, these responses and activities data are classified using machine learning models. The comparison has been made with independent weeks classifications and hidden Markov model with forward algorithm (HMM). T-tests are considered to validate various p-values to get statistically significant features. Further sensitivity analysis has been made to investigate the performance of various models in the independent weeks; RF classifier attends an average area under curve of 0.76 for the classification output, while the HMM model yields area under curve of 0.79.

Pal et *al.* have proposed a machine learning model for the prediction of ischemic heart disease on multiple subjects [24]. They have carried out the experimentation and data collection using fingertip photoplethysmography (PPG) sensor in the cardiology departments of medical colleges and hospitals in Kolkata. In this work, time domain analysis and features extraction of signal have been made. In total, 12 different features are extracted to train the various supervised learning models. They adopted decision tree, discriminant analysis, logistic regression, support vector machine, KNN, and boosted tree algorithms. Further, ten different evaluation metrics have been used to compare the performance of adopted ML models. Comparison has been made on accuracy, true positive rate (TPR), precision, specificity, MCC, F1 score, false positive rate, informed, markedness, and critical success index. Based on these metrics, the highest accuracy of 94%, specificity 95%, sensitivity 95%, and precision 97% were achieved by the boosted tree model.

Hemanth *et al.* have proposed an augmented and virtual reality–based mobile application for the detection of heart diseases [25]. In this work, they recorded the heart sounds for every ten-second interval using mobile phones with suction cup microphone attached to it. These heart sounds are categorized in systole and diastole by virtual graphs. These heart patterns can also be collected from patients' heart condition report. Finally, they have deployed the augmented reality and virtual reality–based detection of heart diseases in the form of mobile application. This assists the doctors to know the condition of heart for given sound and ECG reports.

1.3 CONCLUSION AND FUTURE WORK

In the present time, many researchers have worked in the field of prediction of diseases using machine learning techniques. However, the result mostly depends on the dataset they are working on. In a machine learning approach, choosing the correct dataset is the most important part. If researchers get the correct size of dataset, then the prediction result also comes out correct, but if the dataset is not a correct one, the prediction result may indicate the accuracy obtained to be more than 90%, but it is useless. Therefore, the most important part in the prediction of diseases is the dataset. The dataset is mainly for the patients and is personal and may not be available for analysis; in such cases, it is always difficult to predict the result. In the questionnaire given by doctors to their patients, the latter should give correct information about their habits and other issues. It is observed, however, that patients try to hide significant information from their doctors, and that affects the dataset prepared, which makes prediction difficult.

It has been found that a new virus has now come to the world, and many people have gotten affected by the virus; sometimes, that is the reason for their sad demise. When machine learning techniques are used to predict any disease and find out the solution of the problem, the virus changes its structure, and other sets of people are also affected by that. Thus, the process of analysis or prediction must be continued until all phases are complete. However, in real life, it is always difficult to predict the shape and effect of the virus; thus, the system must be quite flexible enough to handle the ever-changing situation.

It is observed from the literature survey that ML approaches are adopted as a single source of analysis, but very few papers combine multiple ML or even the pre-processing done before the process has started. Thus, the obtained result is found to be better. Again, in the future, if someone has a health issue and prediction is done for a particular issue at that time, whatever other issues that come out because of the present symptoms also need to be analyzed.

REFERENCES

[1] L. Mohimont, A. Chemchem, F. Alin, M. Krajecki and L. A. Steffenel, "Convolutional neural networks and temporal CNNs for COVID-19 forecasting in France," *Applied Intelligence*, pp. 1–26, 2021.

[2] C. S. R. Annavarapu, "Deep learning-based improved snapshot ensemble technique for COVID-19 chest X-ray classification," *Applied Intelligence*, pp. 3104–3120, 2021.

[3] V. Perumal, V. Narayanan and S. J. S. Rajasekar, "Detection of COVID-19 using CXR and CT images using transfer learning and Haralick features," *Applied Intelligence*, pp. 341–358, 2021.

[4] A. Sinha and M. Rathi, "COVID-19 prediction using AI analytics for South Korea," *Applied Intelligence*, pp. 1–19, 2021.

[5] B. Shi, H. Ye, L. Zheng, J. Lyu, C. Chen, A. A. Heidari and P. Wu, "Evolutionary warning system for COVID-19 severity: Colony predation algorithm enhanced extreme learning machine," *Computers in Biology and Medicine*, pp. 136–150, 2021.

[6] Z. Wang and B. Cai, "COVID-19 cases prediction in multiple areas via shapelet learning," *Applied Intelligence*, pp. 1–12, 2021.

[7] J. Nayak, B. Naik, P. Dinesh, K. Vakula, B. K. Rao, W. Ding and D. Pelusi, "Intelligent system for COVID-19 prognosis: A state-of-the-art survey," *Applied Intelligence*, pp. 2908–2938, 2021.

[8] B. Bektaş and S. Babur, "Machine learning based performance development for diagnosis of breast cancer," in *2016 Medical Technologies National Congress, volume: 1 (TIPTEKNO)*, pp. 1–4. IEEE, October 2016.

[9] Y. Gao, H. Wang, M. Guo and Y. Li, "An adaptive machine learning pipeline for predicting the recurrence of gastric cancer," in *2020 5th International Conference on Information Science, Computer Technology and Transportation (ISCTT)*, pp. 408–411. IEEE, 2020.

[10] N. Jafarpisheh, N. Nafisi and M. Teshnehlab, "Breast cancer relapse prognosis by classic and modern structures of machine learning algorithms," in *2018 6th Iranian Joint Congress on Fuzzy and Intelligent Systems (CFIS)*, pp. 120–122. IEEE, 2018.

[11] S. Kaya and M. Yağanoğlu, "An example of performance comparison of supervised machine learning algorithms before and after PCA and LDA application: Breast cancer detection," in *2020 Innovations in Intelligent Systems and Applications Conference (ASYU)*, pp. 1–6. IEEE, October 2020.

[12] N. Khuriwal and N. Mishra, "Breast cancer diagnosis using adaptive voting ensemble machine learning algorithm," in *2018 IEEMA Engineer Infinite Conference (eTechNxT)*, pp. 1–5. IEEE, 2018.

[13] N. Kolay and P. Erdoğmuş, "The classification of breast cancer with machine learning techniques," in *2016 Electric Electronics, Computer Science, Biomedical Engineerings' Meeting (EBBT)*, pp. 1–4. IEEE, April 2016.

[14] P. S. Kumar and S. Pranavi, "Performance analysis of machine learning algorithms on diabetes dataset using big data analytics," in *2017 International Conference on Infocom Technologies and Unmanned Systems, (trends and future directions)(ICTUS)*, pp. 508–513. IEEE, December 2017.

[15] J. Ma, "Machine learning in predicting diabetes in the early stage," in *2020 2nd International Conference on Machine Learning, Big Data and Business Intelligence (MLBDBI)*, pp. 167–172. IEEE, October 2020.

[16] O. Oyebode and R. Orji, "Detecting factors responsible for diabetes prevalence in Nigeria using social media and machine learning," in *2019 15th International Conference on Network and Service Management CNSM)*, pp. 1–4. IEEE, December 2019.

[17] P. K. Saha, N. S. Patwary and I. Ahmed, "A widespread study of diabetes prediction using several machine learning techniques," in *2019 22nd International Conference on Computer and Information Technology (ICCIT)*, pp. 1–5. IEEE, December 2019.

[18] S. R. Swarna, S. Boyapati, P. Dixit and R. Agrawal, "Diabetes prediction by using big data tool and machine learning approaches," in *2020 3rd International Conference on Intelligent Sustainable Systems, (ICISS)*, pp. 750–755. IEEE, December 2020.

[19] S. Wei, X. Zhao and C. Miao, "A comprehensive exploration to the machine learning techniques for diabetes identification," in *2018 IEEE 4th World Forum on Internet of Things (WF-IoT)*, pp. 291–295. IEEE, February 2018.

[20] A. Yadav, A. Singh, M. K. Dutta and C. M. Travieso, "Machine learning-based classification of cardiac diseases from PCG recorded heart sounds," *Neural Computing and Applications*, pp. 17843–17856, 2020.

[21] S. Narayan and E. Sathiyamoorthy, "A novel recommender system based on FFT with machine learning for predicting and identifying heart diseases," *Neural Computing and Applications*, pp. 93–102, 2019.

[22] B. P. Doppala, D. Bhattacharyya, M. Chakkravarthy and T. H. Kim, "A hybrid machine learning approach to identify coronary diseases using feature selection mechanism on heart disease dataset," *Distributed and Parallel Databases*, pp. 1–20, 2021.

[23] Y. Meng, W. Speier, C. Shufelt, S. Joung, J. E. Van Eyk, C. N. B. Merz and C. W. Arnold, "A machine learning approach to classifying self-reported health status in a cohort of patients with heart disease using activity tracker data," *Journal of Biomedical and Health Informatics*, pp. 878–884, 2019.

[24] P. Pal, S. Ghosh, B. P. Chattopadhyay, K. K. Saha and M. Mahadevappa, "Screening of ischemic heart disease based on PPG signals using machine learning techniques," in *2020 42nd Annual International Conference of the IEEE Engineering in Medicine & Biology Society (EMBC)*, pp. 5980–5983. IEEE, July 2020.

[25] J. D. Hemanth, U. Kose, O. Deperlioglu and V. H. C. de Albuquerque, "An augmented reality-supported mobile application for diagnosis of heart diseases," *The Journal of Supercomputing*, pp. 1242–1267, 2020

[illegible] P. K. [illegible] and [illegible], "Predictive analysis [illegible] heart disease using big data analytics," [illegible] *Advances in Computing, Communication [illegible] Information Technology*, [illegible], IEEE [illegible], 2017.

[illegible] [illegible] "Heart disease prediction using machine learning techniques," [illegible], IEEE, 2018.

[illegible] [illegible], "Predicting [illegible] disease using data mining [illegible]," [illegible], IEEE, [illegible].

[illegible] [illegible] and [illegible], "[illegible] machine learning [illegible]," [illegible].

[illegible] [illegible], "[illegible] prediction [illegible]," in [illegible], [illegible].

[illegible] [illegible], "[illegible] machine learning [illegible]," [illegible].

[illegible] [illegible], "[illegible]," [illegible], IEEE, [illegible].

[illegible] [illegible], "A [illegible] approach [illegible]," *[illegible] Computational [illegible]*, [illegible], 2019.

[illegible] [illegible], "[illegible] approach [illegible] structure [illegible]," [illegible].

[illegible] [illegible], "A [illegible] machine learning approach [illegible]," [illegible], 2019.

[illegible] [illegible], "[illegible]," [illegible], [illegible], 2017.

[illegible] [illegible], "[illegible] heart disease [illegible]," *[illegible]*, [illegible].

2 A Novel Virtual Medicinal Care Model for Remote Treatments

Drishti Hans, Shashi Gandhar, Gaurav Narula, Abhishek Gandhar, and Umang Hans

CONTENTS

2.1 INTRODUCTION

Telemedicine refers to the procedure of remotely worrying for sufferers where both the doctor and the patient are not actually present. Doctors and sick people can share data from one computer screen to another in real time. And even in a remote location, they can view and record readings from medical equipment. Patients can use telemedicine software to look to a health practitioner for diagnosis and remedy while not having to watch for an appointment. Patients should meet with a doctor at their own convenience. Furthermore, patients who live in rural areas who had previously had trouble finding a physician can now easily get in touch with them. Various researchers have found benefits of using telemedicine, like price effectiveness, portability, government funding, expert exposure, flexible working hours, less exposure to

DOI: 10.1201/9781003309451-2

illness of patients, attracting more patients and managing them efficiently. However, there are more advantages of this system: In numerous remote places or postcatastrophe circumstances, reliable medical services are inaccessible. Telemedicine can be applied in such places or circumstances to give crisis medical care. It saves lives in an emergency situation, while there is no ideal opportunity to take the patient to an emergency clinic. This framework is valuable for patients dwelling in distant regions or confined districts. Patients can get clinical medical services from their homes without the challenging travel to a clinic. This framework likewise works with health-care education, as beginners in the health-care profession can notice the functioning system of medical services specialists in their particular fields and the specialists can supervise the work. Present-day developments of data innovation, for example, mobile phones, have made simple data sharing and conversation about basic clinical cases among medical care experts from numerous areas. [1–14]

While cost-cutting incentives may stimulate more telemedicine testing for clinical, educational, and administrative objectives, health-care providers should be careful about how telemedicine may impair the real and perceived quality of their services. The continual creation and implementation of complex clinical, research, management, and other information systems are directly tied to the quality testing and development of telemedicine, as it is elsewhere.

In this chapter, the authors have tried to solve the major drawbacks of the telemedicine system. They try to reduce the gap between the doctors and patients by using a proper, feasible database, along with a new algorithm for modified and better prediction of disease. The database that the authors use stores every piece of relevant information, along with providing security from various issues, by using end-to-end data transfer encryption, and also makes system handling easier as all relevant information of the patient has already been preserved in the database. The chapter describes the entire system interface so created and the proposed new algorithm. The system proposed helps in making the appointment with the specialist for the disease that has been so predicted. In this way, it is easier to contact the doctor than to visit different doctors. These services can help an individual user on many levels, as during this pandemic. The act of telemedicine sets out open doors for greater access to clinical consideration, further develops access to clinically trained professionals, and offers more prominent accommodation for patients, all at a conceivably lower monetary expense. With the worldwide extension of internet connectivity through cables, fiber optics, satellites, drones, and surprisingly high elevation balloons, telemedicine administrations can reach each person's doorstep on Earth with a mobile phone.

2.1.1 History and Evolution of Telemedicine

Telemedicine in its initial structure started with the approach of the telegraph in 1844 and the telephone in 1876, which permitted patients to bring doctors rapidly and expanded the accessibility of doctor-to-doctor conferences. During the American Civil War, the telegraph was utilized to send loss/death records and requests for clinical supplies. After telegraphy, phones were created and turned into the essential method for far-off clinical communication. The National Aeronautics and Space Administration (NASA) reformed long-distance communication in the mid- to late

1900s by creating frameworks to both speak with and analyze the strength of astronauts. NASA played a key position in the improvement of telehealth. NASA has a defined obligation to evaluate astronauts' fitness whilst in orbit for the reason of Mercury programs within the early sixties. NASA and Lockheed worked on a project to provide clinical services to the Papago Indians in Arizona as part of the process of testing and improving tactics employing satellite TV for PC-based telemedicine. STARPAHC (Space Technology Applied to Rural Papago Advanced Health Care) was a program that lasted until the late 1970s. Thanks to the satellite era, many rural regions now have access to telemedicine.

2.1.1.1 Georgia

Georgia adopted the rules in 1992 to create a state-supervised network with resources from the Medical College of Georgia and the area's administrative services department and has been praised for its foresight in the discipline of telemedicine. The Georgian School and Operational and Medical Council helped plan and implement several programs in the 1990s.

2.1.1.2 Texas

The University of Texas (UTMB) Medical Branch of Galveston, in partnership with Texas Tech Health Sciences, devised a hub-and-spoke telemedicine technique that provided a wide range of special treatments to inmates. As one of the first people in telemedicine in the 1990s, the UTMB overcame challenges, such as the development of luxury devices and telecommunications infrastructure, to become a pioneer in correctional/penitentiary telemedicine.

2.1.1.3 Alaska

In 1996, the National Library of Medicine started an observation on telemedicine to deal with the overabundance of otitis media. Two years later, a large coalition of federal corporations, the Office for the Advancement of Telemedicine inside the Department of Defense, the Departments of Health and Human Services, Indian Health Services, and Veterans Affairs, has given the Alaska Federal Health Care Access Network, a big telemedicine community, free range to set up (AFHCAN). Alaska Native establishments, public fitness centers, and Navy and Veteran facilities are among the 235 websites maintained through AFHCAN.

2.1.1.4 Arizona

Based on early telemedicine experience, Arizona leaders founded the Arizona Rural Telemedicine Network, in the end renamed the Arizona Telemedicine Network, in the early 1990s (as described earlier on this bankruptcy, which includes STARPAHC, NASA, and revels in at Massachusetts General Hospital). This substantial initiative has incorporated studies with teaching and coaching and has consolidated several revenue sources, inclusive of membership costs, country and university aid, and grants, given its inception within the early days of telehealth.

Tele-ICU has advanced into some other telemedicine specialty that has surged in prominence in the closing decade. Even though far-flung consultations for deep assist devices have existed in preceding many years, a brand-new, complete version has emerged with a command center staffed with the aid of resuscitators, nurses crucial, and different employee bodies, in addition to the capacity to conduct protocol-based remote surveillance with "smart alarms" and interactive videoconferencing. Sentara Healthcare in Virginia became the first to call this manner in the year 2000. Despite contradictory statistics on mortality consequences and fee effectiveness, a famous research has advocated this technique. In 2010, Tele-ICU safeguarded approximately 5,000 long-term care unit beds in almost 250 institutions.

Fitness care became impacted by the ARRA Act of the late 2000s, which touched an extensive variety of firms. Fitness information era (HIT) and electronic health report growth had been deployed at a fee of $19 billion (EHR). Despite the reality that telehealth changed into no longer right now addressed in the EHR adoption incentives, the underlying expertise of HIT and EHR aided the development and implementation of era-enabled care fashions. The ARRA investment allowed telehealth initiatives to increase and develop in a spread of ways. During this decade, the National Telecommunications and Information Administration's Broadband (NTIAB-USA) aimed to leverage ARRA monies to build telehealth catastrophe training programs across the United States (formerly the Broadband Technology Opportunities Program, BTOP).

By the late 2000s, California and some other states had authorized rules mandating health insurers to cover telemedicine. Since then, five states have applied telemedicine guidelines which might be just like California's and limit compensation based on face-to-face interplay in a few ways. Georgia, Louisiana, Maine, New Hampshire, and Oregon have stricter rules, with three of them (Louisiana, New Hampshire, and Oregon) mandating telemedicine insurance for any in-character treatment. This decade saw a shift within the region, indicating that telehealth has an area inside the shipping of health care in the United States. During this period, the maximum commonplace forms of care fashions found had been those who were modeled after conventional in-individual care but added through technology. The most famous styles of care fashions found for this term have been those that had been modeled after conventional in-person care but brought through generations. As these times display, there was a developing drive closer to new care fashions that did not "look" like in-character care. In the final decade, however, most of that innovation changed, nevertheless.

2.1.2 Literature Review

Recent improvements in wireless and community technology, in addition to nanotechnologies and ubiquitous computer structures, make it possible to set up a healthcare device in the past decade. *Telemedicine* [15] is the use of telecommunication generation for scientific analysis, treatment, and affected person care. The motive of online medicine is to deliver professional-based help to understaffed rural regions using the modern-day telecommunication and statistics era. Information is less expensive to deliver than individuals. In many countries, advances in the clinical era have led to a great boom inside the older population, prompting a bigger name for domestic fitness tracking to assure that aged human beings can stay independent

[16]. During ordinary sports, several physiological signals may be captured from human beings in their houses and utilized to choose early signs of fitness problems or automatically alert paramedics in an emergency [17]. All the research that has been finished and is being used in this area, in particular, for far-off tracking of physiological markers, can be classified into the following categories: sensor sorts, facts communique kinds, monitoring tool kinds, and sign processing/medium sorts [18]. As a result, this segment will gain popularity on those subjects further to fashionable-day research. The biosignal sensors capture and transmit physiological data to the signal processing unit. Several types of research were undertaken to develop the sensors of the one to be compact in length [19], hold affected person mobility [20], and utilize low jogging strength to minimize battery duration and allow for longer battery existence [21]. A personal place community, more known as a body network [22], might be used to link a set of wearable medical sensors which may be placed within the client's clothes [23]. Before being sent to the verbal exchange layer, every remote monitoring gadget's sensor layer is usually associated with the processing device for signal amassing, processing, evaluation, and records formatting. The affected individual's present-day fame, in addition to dispositions in his or her clinical condition, may be assessed via the processing unit. Processing devices encompass PCs [24], mobile telephones [25], and embedded systems [26]. Many clinical algorithms had been developed in cutting-edge telemedicine research to help in affected person evaluation [27] and early identification of cardiovascular illnesses [28]. Pulse evaluation has long been a place of interest within the physiology zone for the motive that pulse symbolizes someone's kingdom of fitness. Much research has encouraged tracking systems that could choose and classify unique biosignals, consisting of QRS detection and arrhythmia type [28], actual-time ECG grouping set of regulations [27], and coronary heart fee variability size [29]. Recent enhancements in Wi-Fi and network technology have also authorized the improvement of a Wi-Fi telemedicine system that offers health care to patients at a low fee. Telemedicine systems are divided into classes: actual-time and save-and-ahead. Patient facts are available on the server in actual-time mode at once after the seizure, while it's far to be had later in hold-and-ahead mode. Computer networks [30], mobile networks [31], public phone networks [32], and cable TV networks [33] are all used to offer critical signals to the server.

To grow the medical physician's mobility, the global machine for cell (GSM) communique mobile telephony network became used to hyperlink the server [34]. In [35], Hung and Zhang created a Wi-Fi software program protocol (WAP)–based completely telemonitoring device. It made use of WAP devices as mobile entry to terminals and allowed clinical doctors to read monitored statistics on WAP gadgets in store-and-ahead mode [36]. In comparison to the doctor, the patient's motion is drastically reduced with such systems. In many previous telemedicine structures, the given sensor unit protected an ECG records amassing circuit, an A/D converter, and a storage unit. This device covered an interior wireless transmitter that sent the monitored facts to a community-linked PC, allowing the patient with limited mobility to use it [31, 35]. [37] utilized a PC and a GSM modem to broadcast actual-time ECG information from a moving ambulance automobile. Rasid and Woodward provided a cell telemonitoring device primarily based on a Bluetooth-enabled processing unit that sends monitored records to a Bluetooth cell phone, which then sends it to a server over the GSM/GPRS community. Seeing from the alternative hand, Engin et al. [38] used a cell mobile phone to speak the determined ECG check in real-time mode. In the given designs, the affected person's efficiency is greater [37, 38]. ECG assessment, as a substitute, isn't always done in which the ECG is received; instead, its miles completed on the server cease, for instance. In reality, using the GSM/GPRS community is wasteful due to the fact that every day ECGs are also dispatched, incurring a massive expense. Lin et al. [39] made a cellular affected person tracking device that combines PDA generation with WLAN technology to relay a patient's critical signs in actual time to a distant relevant manipulate unit. That tool is primarily based totally on a tiny mobile ECG recording tool that transmits dimension records to the cell phone through Bluetooth [40]. Any anomalies detected among the sections of the scale records are transmitted to a server once the obtained statistics are processed inside the cellular telephone. The common overall performance became, however, far from last due to the bounds of the processing devices inside the cell phone [41].

In this chapter, a telemedicine system was proposed with a centralized database with the data controller in patients' hands, integrated with state-of-the-art disease prediction algorithm which lets patients enter their symptoms and provides them with real-time analysis of probable diseases, with the option to consult an expert immediately via online conference mode. The system also allows users to find the drugs prescribed by the physician/expert at the lowest cost and the most appropriate alternative in case of nonavailability of any particular drug. The system also comes with a reminder system for the appointments/consultations and can provide the available schedule of any expert to discuss any issue. The conference held is end-to-end encrypted, and the patients' data stored in the database also will be encrypted to prevent any personal data leak, thus providing the comfort of visiting a doctor while staying at home.

2.2 MATERIALS AND METHODS

The system consists of the following components:

1. The main system
2. Drug lookup and distribution system
3. System interface
4. Disease prediction algorithm
5. Database

2.2.1 The Main System

The key structure is a mixture of all the present subsystems as shown in Figure 1.1. First, the user interacts with the app, after signing in to the account; the user may choose to communicate with the doctor to check the schedule or help with disease prediction.

Users can check the schedule, book an appointment, or predict the disease according to their choice. For predicting the disease, the user enters the symptoms, and the

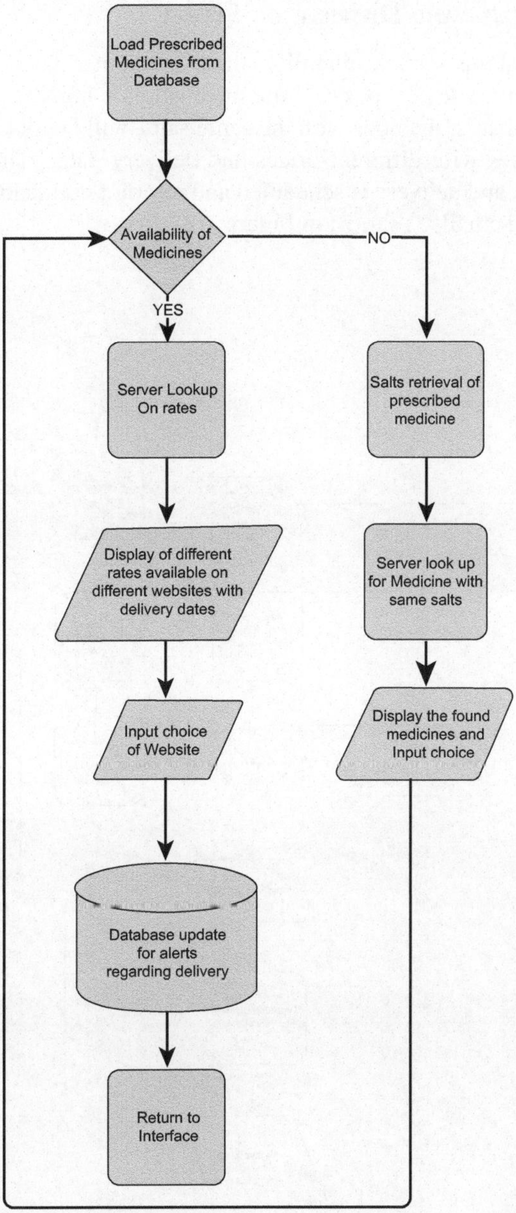

FIGURE 2.1 The layout and workings of the main system, along with all subsystems.

predicted disease is displayed along with a list of specialists in the same field. Whenever a doctor prescribes a new medicine to a patient, the database is updated side by side to maintain proper history. The appointments generally take place through video-conferencing software which is readily available. Any further appointment made by the doctor is updated for alerts so no appointment is missed by the patient.

2.2.2 Drug Lookup and Distribution System

This subsystem, along with availability, finds medicines from the database and searches online for the lowest price. If the medication is not available, then alternative medications, that is, the ones with the same salts, will be looked for. It gives the user various choices with different prices and delivery dates. The user selects one with their consent, and delivery is scheduled and relevant notifications are sent to the user via phone and email, as shown in Figure 2.2.

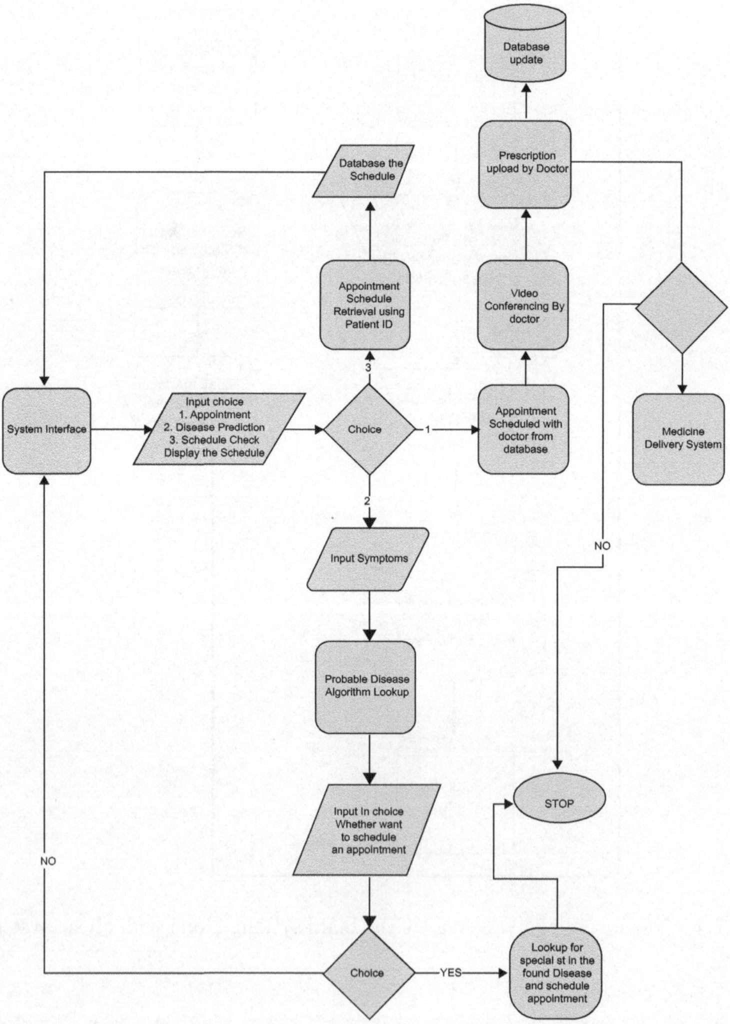

FIGURE 2.2 Flowchart depicting the workflow of medicine lookup and delivery subsystem.

2.2.3 System Interface

The program has a user-friendly baseline GUI, as shown in Figure 2.3. Either a previous user must log in or a new user must register. The user enters the required information for the database during the signup process. When logging in, the user's ID is encrypted with a salt created for security purposes from the doctor's ID. Then, the doctor's ID is retrieved from the device before being entered into the system to perform the requisite decryption to access the system. A pseudo-ID is given for a new patient before the doctor ID in the database is modified.

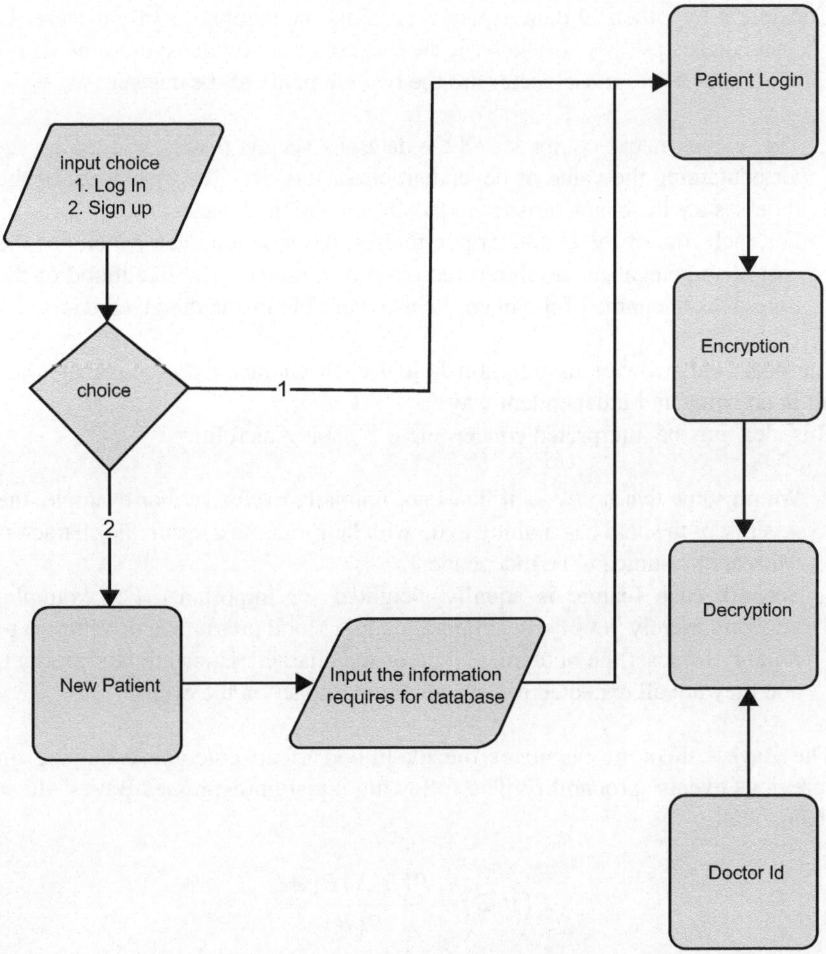

FIGURE 2.3 Flowchart depicting the design and flow of the system's interface.

2.2.4 Disease Prediction Algorithm

The majority of the method is based on the naive Bayes algorithm, which is based on probability prediction P(c|x) from P(x), P(x|c), and P(c):

$$P(c \mid x) = \left(P(x \mid c) * P(c)\right)/P(x) \tag{1}$$

- *P(c|x)* is the occurrence probability (*c*, target) given (*x*, attributes) that occurred.
- *P(c)* is the likelihood of class *c* incidence.
- *P(x|c)* is the *x* probability provided that *c* has already occurred.
- *P(x)* is the likelihood of event *x*.

Consider a hypothetical dataset that represents the symptoms of a disease. Each row is categorized as "yes" or "no" for the disease based on the symptoms. The feature matrix and the response vector are the two elements of the dataset.

- The feature matrix comprises all the dataset's vectors (rows), with each vector containing the value of dependent characteristics. The symptoms of the illnesses are the characteristics in the aforementioned dataset.
- For each row of the characteristic matrix, the reaction vector includes the cost of the elegance variable (prediction or output). "The likelihood of the illness" is the name of the magnificence variable inside the given dataset.

The basic naive Bayes assumption is that each characteristic contributes to the result in an equal and independent way.

This idea may be interpreted concerning our dataset as follows:

- We presume that no two attributes are mutually exclusive. For example, the severity of the cold has nothing to do with hair loss. As a result, the characteristics are presumed to be independent.
- Second, each feature is equally weighted (or important). For example, knowing merely your body temperature and blood pressure is insufficient to reliably forecast the outcome. None of the characteristics are unimportant, and they are all expected to have an equal impact on the event.

The Bayes' theorem calculates the likelihood of an outcome occurring given the previous events' probability. The following equation expresses Bayes' theorem mathematically:

$$P(A/B) = \frac{P(B/A)P(A)}{P(B)} \tag{2}$$

- P(B)! = 0, where *A* and *B* are events.
- Essentially, we're looking for the likelihood of *A* event if *B* event is true. Evidence is also referred to as event B.

- The priory of A is P(A) (which means the prior probability, that is, Probability of event earlier than evidence is seen). The proof is a value assigned to an unknown instance's attribute (here, it is event B).
- P(A|B) is the probability of happening of A when B has already occurred or the likelihood of an occurrence after seeing the evidence.

Now, we can use Bayes' theorem to our dataset in the following manner:

$$P(y/X) = \frac{P(X/y)P(y)}{P(X)} \quad (3)$$

Where *X* is an n-dimensional dependent feature vector and *y* is a class variable.

$$X = (x_1, x_2, x_{3,\ldots\ldots} x_n) \quad (4)$$

Now it is set to apply a basic assumption to Bayes' theorem, particularly feature independence. So now we've separated the evidence into its component elements. If any two occurrences A and B are independent of one another, then:

$$P(A, B) = P(A) \times P(B)$$

As a consequence, we arrive at the following conclusion:

$$P(y/x_1,\ldots x_n) = \frac{P(x_1/y)P(x_2/y)\ldots P(x_n/y)P(y)}{P(x_1)P(x_2)\ldots P(x_n)} \quad (5)$$

Which may be represented in the following way:

$$P(y/x_1,\ldots x_n) = \frac{P(y)\prod_{i=1}^{n} P(x_i/y)}{P(x_1)P(x_2)\ldots P(x_n)} \quad (6)$$

We can now eliminate that term because the denominator remains constant for every given input:

$$P(y/x_1,\ldots x_n) \infty P(y)\prod_{i=1}^{n} P(x_i/y) \quad (7)$$

We must now construct a classifier model. To do so, we calculate the probability of a given set of inputs for all possible values of the class variable *y*, then select the output with the highest probability. This can be expressed numerically as:

$$y = argmax_y P(y)\prod_{i=1}^{n} P(x_i/y) \quad (8)$$

The duty of computing P (y) and P (x_i|y) is eventually left to us.
P(y) stands for class probability, whereas *P(x_i|y)* stands for conditional probability. The assumptions that various naive Bayes classifiers make about the distribution of P(x_i|y) are what distinguish them.

The probabilities to be computed are then constructed into a training set, which is subsequently supplied to an ANN that has been trained using the prior dataset to provide accurate illness prediction. The functions of the human brain are used to create artificial neural networks. When the relationship between the underlying data is complicated or uncertain, this method is applied. There are three layers in this image: an input layer, an output layer, and a concealed layer. The input layer is made up of neurons that are used to compare input and output, both of which are independent variables. The input function and the activation function are the two major functions that play a significant role in neural networks. The weighted sum of the neuron output connected with a node is its input. Neti = Vij × x_j; in production of Jth unit, the unit's activation value is $g(w_{ji} \times x_i)$, *g(.)* is the activation verb, and *xi* is the unit's output associated with unit *j*.

$$g\left(\text{input}\right) = 1 / \left(1 + \left(e^{\text{Input}}\right)\right) \tag{9}$$

The activation function has the function of mapping the output between gaps [j1, j2].

The vector of weights for the jth processing unit immediately (t + 1) in terms of the weight vector in a moment (t) is as follows:

$$Y\left(t + 1\right) = y_1\left(t\right) + y_2\left(t\right) \tag{10}$$

The weight vector difference (10) is in $y_1(t)$. The network adjusts in the following way: to distinguish between the desired output and the actual output, the weight is modified proportionally. Its formula is as follows:

$$Y = \left(d - y\right) * i \tag{11}$$

In (11), *d* represents the intended output, *y* represents the actual output, and *Ii* represents the jth input. The perceptron learning rule is what it's called. Figure 2.4 shows the certified network that will incorporate the right input matrix and forecast the illness in the prediction phase.

2.2.5 Database

The database is maintained centrally so that if the user chooses a new doctor, the doctor will be able to access that particular user's past medical history, but only if the user gives access to the respective doctor.

Central maintenance helps remove the paper trails and make record updation simpler. The database contains all the things that are important not only for the diagnosis of disease but also for ease of the doctor seeing the signs and history.

Also, consultation of appointments and the timing of any doctor's availability can be found in the database, along with the area of expertise. Even the appointment schedules are kept in the database so that no two appointments overlap and reminders about the schedule can be sent to the patients so they don't miss their appointments.

The database is represented by the diagram of entity relationship given in Figure 2.5. The database administrator can easily translate this entity relationship

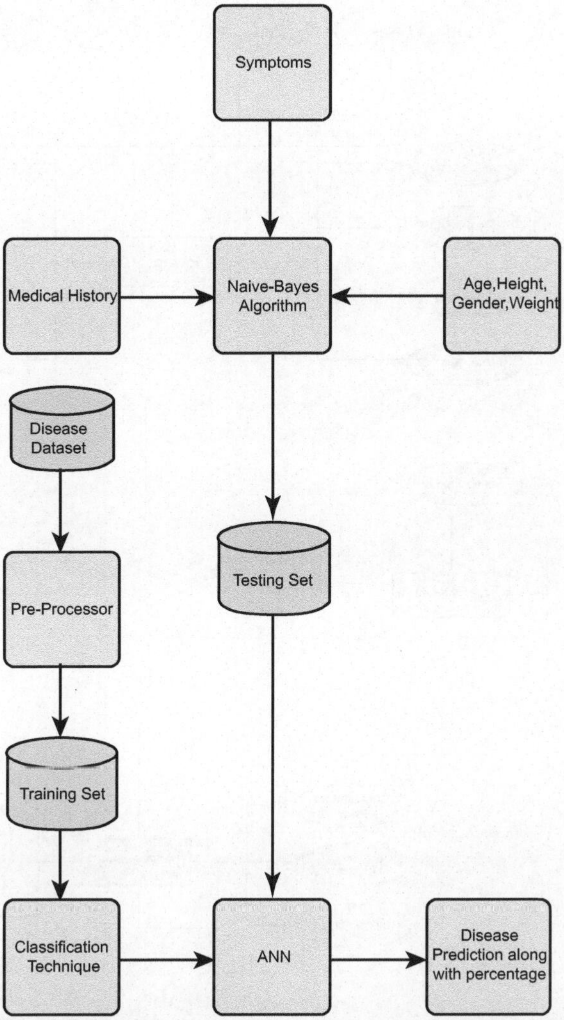

FIGURE 2.4 Flowchart explaining the working of disease prediction algorithm depicting necessary inputs and the output.

diagram into the said database with appropriate rules so that who can access what information is determined, and the encryption in the framework provides the security layer to secure the database.

2.3 RESULTS

What follows is an estimated study that was conducted when a patient visits a private medical facility every week equivalent to that of telemedicine. This means that telemedicine can not only be applied in rural sectors but can also be encouraged by inculcating some basic knowledge in urban sectors.

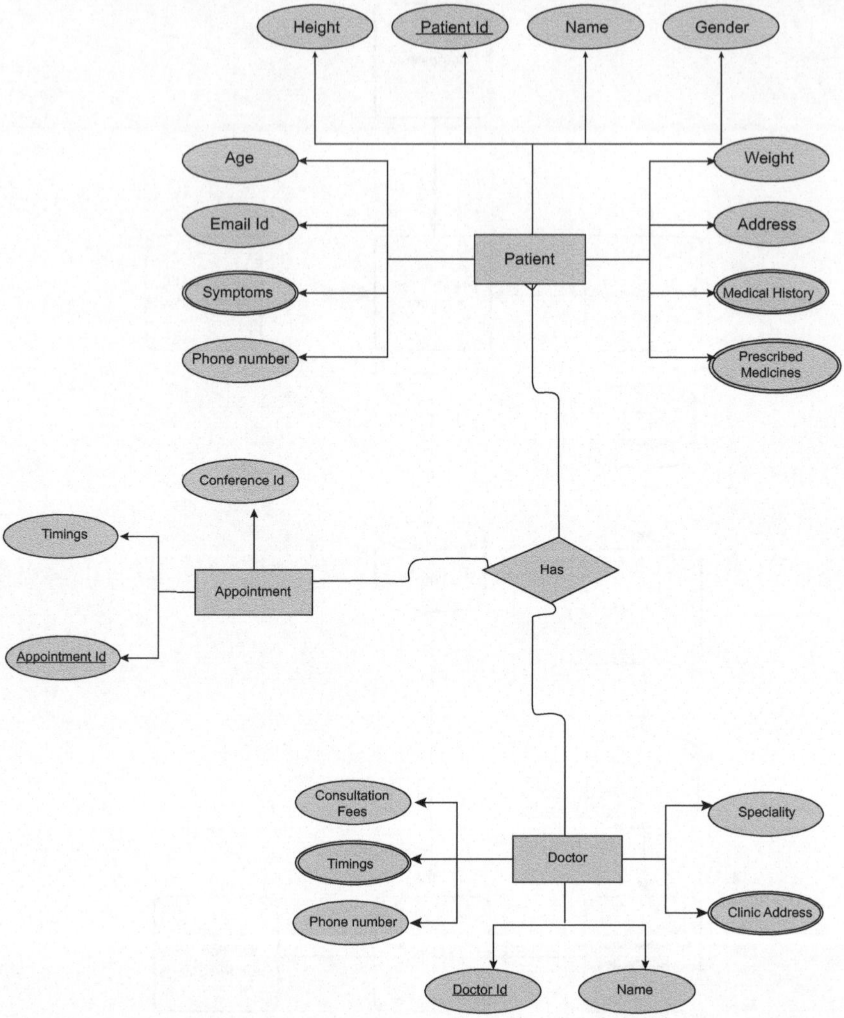

FIGURE 2.5 ER diagram of the database.

This 43-year-old patient had leg pain and went through the process as follows: the first appointment with the doctor and clarification of the symptoms that indicated a few of the tests and the X-ray. Then it went on to physiotherapy for a couple of weeks and then continued the medications and then calmed the problem for a few months.

On the other side, if the same goes via telemedicine, the travel costs would firstly be minimized for each visit. Second, some of the assessments can only be carried out via the meeting. Physiotherapy may also be performed by the doctor, who uploads the videos to their own sites. The biggest expense is that of pharmaceutical products. It cannot be assumed because every service can be based on telemedicine, but

TABLE 2.1
Overall Cost, Including Transportation, Consultation Fee, and Tests for Offline Visits

	Transportation (rupees)	Consultation fee (rupees)	Tests
Week 1	200–300	800–1,000	• X-rays (whole lower limb AP) (2,500) • Blood tests (2,000)
Week 2	200–300		• Physiotherapy (continues for 3 weeks) (600 per day)
Week 5	200–300	800–1,000	• Medicines (4,000)
Total cost (in rupees)	23,500–24,000		

TABLE 2.2
Overall Cost, Including Transportation, Consultation Fee, and Tests, Using the Proposed System

	Transportation (rupees)	Consultation fee (rupees)	Tests
Week 1	0	300–500	• X-rays (whole lower limb AP) (2,000) • Blood tests (1,800)
Week 2	0		• Physiotherapy (70% at home through professional videos) (180 per day)
Week 5	0	300–500	• Medicines (3,600)
Total cost (in rupees)	11,500–12,000		

indeed, it can be used for patients by the period it functions well, as it is only used for community welfare. Telemedicine will reduce costs by almost 50%.

2.4 DISCUSSION

The authors conclude that the system required to overcome numerous problems present in existing systems was successfully designed. The proposed system has many more advantages which make it more preferable from any existing system in use.

1. Since the proposed system uses a combination of the naive Bayes model with neural network, disease prediction is more accurate than in existing models.

2. It provides the facility for a central database that not only allows easy access but also saves time and ensures both patient and doctor security.
3. Quick access to licensed substitutes where prescribed medical products are not available, and compared to general pharmacies, medicines are available at lower prices.
4. It is a seamless process between the patient and the doctor, as each procedure is organized with both sides agreeing. There will be no delays in taking care as the patient can select the timing according to their choice.
5. The cost to each consumer is much lower since the device is more affordable.
6. Privacy and confidentiality are no longer a problem because end-to-end encryption is used to ensure that there are no compromises.
7. The inability to assess patients has been solved, as it offers all the information required to ensure proper safety due to a large database.

REFERENCES

1. E. F. Grange, V. Warren, S. Khemka, A. N. Theodorou and A. J. Vallance-Owen, "The role of telemedicine in helping to manage hypertension—a review of health outcomes and cost-benefit," *2009 International Conference on eHealth, Telemedicine, and Social Medicine*, Cancun, 2009, pp. 204–208, doi: 10.1109/eTELEMED.2009.32.
2. A. Pal, V. W. A. Mbarika, F. Cobb-Payton, P. Datta and S. McCoy, "Telemedicine diffusion in a developing country: The case of India (march 2004)," in *IEEE Transactions on Information Technology in Biomedicine*, vol. 9, no. 1, pp. 59–65, March 2005, doi: 10.1109/TITB.2004.842410.
3. B. Woodward, R. S. H. Istepanian and C. I. Richards, "Design of a telemedicine system using a mobile telephone," in *IEEE Transactions on Information Technology in Biomedicine*, vol. 5, no. 1, pp. 13–15, March 2001, doi: 10.1109/4233.908361.
4. R. D. Chand, A. Kumar, A. Kumar, P. Tiwari, R. Rajnish and S. K. Mishra, "Advanced communication technologies for collaborative learning in telemedicine and tele-care," *2019 9th International Conference on Cloud Computing, Data Science & Engineering (Confluence)*, Noida, 2019, pp. 601–605, doi: 10.1109/CONFLUENCE.2019.8776970.
5. L. Jiang, H. Zhang and Z. Cai, "A novel Bayes model: Hidden Naive Bayes," in *IEEE Transactions on Knowledge and Data Engineering*, vol. 21, no. 10, pp. 1361–1371, October 2009, doi: 10.1109/TKDE.2008.234.
6. D. Jarrett, J. Yoon and M. van der Schaar, "Dynamic prediction in clinical survival analysis using temporal convolutional networks," in *IEEE Journal of Biomedical and Health Informatics*, vol. 24, no. 2, pp. 424–436, February 2020, doi: 10.1109/JBHI.2019.2929264.
7. S. Junnila et al., "Wireless, multipurpose in-home health monitoring platform: Two case trials," in *IEEE Transactions on Information Technology in Biomedicine*, vol. 14, no. 2, pp. 447–455, March 2010, doi: 10.1109/TITB.2009.2037615.
8. C. A. P. de Lucena, C. R. Mont'Alvão, F. Pierantoni and L. Frajhof, "Telemedicine and design: Relationships that create opportunities," in Kurosu M (ed.), *Human-Computer Interaction. Applications and Services. HCI 2013. Lecture Notes in Computer Science*, vol. 8005, 2013, Springer, Berlin, Heidelberg, doi: 10.1007/978-3-642-39262-7_15.
9. Q. Xue, L. Zhou, J. Hao and M. Liu, "The research on the benefit of telemedicine to human based on evolutionary game theory," in Stephanidis C (ed.), *HCI International 2018—Posters' Extended Abstracts. HCI 2018. Communications in Computer and Information Science*, vol. 852, 2018, Springer, Cham, doi: 10.1007/978-3-319-92285-0_45.

10. Yunzhou Zhang, Huiyu Liu, Xiaolin Su, Pei Jiang and Dongfei Wei, "Remote mobile health monitoring system based on smart phone and browser server structure," *Journal of Healthcare Engineering*, vol. 6, no. 4, pp. 717–738, 2015.
11. Jessica F. Robb, Megan H. Hyland and Andrew D. Goodman, "Comparison of telemedicine versus in-person visits for persons with multiple sclerosis: A randomized crossover study of feasibility, cost, and satisfaction," *Multiple Sclerosis and Related Disorders*, vol. 36, 2019, 101258, doi: 10.1016/j.msard.2019.05.001.
12. www.ajmc.com/journals/ajac/2017/2017-vol5-n1/telemedicine-and-its-role-in-revolutionizing-healthcare-delivery
13. https://relymd.com/blog-using-telemedicine-to-improve-patient-loyalty-and-satisfaction/
14. Jarrett Evans, Amy Papadopoulos, Christine Tsien Silvers, Neil Charness, Walter R. Boot, Loretta Schlachta-Fairchild, Cindy Crump, Michele Martinez and Carrie Beth Ent, "Remote health monitoring for older adults and those with heart failure: Adherence and system usability," *Telemedicine and E-health*, vol. 22, no. 6, pp. 480–488, 2016.
15. S. Pavlopoulos, R. H. Istepanian, S. Laxminarayan and C. S. Pattichis, "Emergency health care systems and services: Section overview," in *Proceedings of the 5th International IEEE EMBS Special Topic Conference on Information Technology Applications in Biomedicine*, pp. 371–374, 2006.
16. R. Anta, S. El-Wahab and A. Giuffrida, *Mobile Health: The Potential of Mobile Telephony to Bring Health Care to the Majority*, Inter-American Development Bank, USA, 2009.
17. M. F. Wyne, V. K. Vitla, P. R. Raougari and A. G. Syed, "Remote patient monitoring using GSM and GPS technologies," *Journal of Computing Sciences in Colleges*, vol. 24, no. 4, pp. 189–195, 2009.
18. M.-K. Suh, C.-A. Chen, J. Woodbridge, M. K. Tu, J. I. Kim, A. Nahapetian, L. S. Evangelista and M. A. Sarrafzadeh, "A remote patient monitoring system for congestive heart failure," *Journal of Medical Systems*, vol. 35, no. 5, pp. 1165–1179, 2011.
19. A. Hande, T. Polk, W. Walker and D. Bhatia, "Self-powered wireless sensor networks for remote patient monitoring in hospitals," *Sensors*, vol. 6, no. 9, pp. 1102–1117, 2006.
20. E. Sardini and M. Serpelloni, "Instrumented wearable belt for wireless health monitoring," *Procedia Engineering*, vol. 5, pp. 580–583, 2010.
21. F. E. H. Tay, D. G. Guo, L. Xu, M. N. Nyan and K. L. Yap, "MEMSWear-biomonitoring system for remote vital signs monitoring," *Journal of the Franklin Institute*, vol. 346, no. 6, pp. 531–542, 2009.
22. H. Alemdar and C. Ersoy, "Wireless sensor networks for healthcare: A survey," *Computer Networks*, vol. 54, no. 15, pp. 2688–2710, 2010.
23. P. S. Pandian, K. Mohanavelu, K. P. Safeer, T. M. Kotresh, D. T. Shakunthala, P. Gopal and V. C. Padaki, "Smart vest: Wearable multi-parameter remote physiological monitoring system," *Medical Engineering & Physics*, vol. 30, no. 4, pp. 466–477, 2008.
24. B. Mehta, D. Rengarajan and A. Prasad, "Real time patient tele-monitoring system using LabVIEW," *International Journal of Scientific and Engineering Research*, vol. 3, no. 4, pp. 435–445, 2012.
25. A. Loutfi, G. Akner and P. Dahl, *An android based monitoring and alarm system for patients with chronic obtrusive disease* [M.S. thesis], Department of Technology at Orebro University, 2011.
26. V. K. Sambaraju, *Design of a Wireless Cardiogram System for Acute and Long-Term Health Care Monitoring*, ProQuest, Michigan, 2011.
27. C. Wen, M.-F. Yeh, K.-C. Chang and R.-G. Lee, "Real-time ECG telemonitoring system design with mobile phone platform," *Measurement*, vol. 41, no. 4, pp. 463–470, 2008.
28. G. Tartarisco, G. Baldus, D. Corda, R. Raso, A. Arnao, M. Ferro, A. Gaggiol and G. Pioggia, "Personal health system architecture for stress monitoring and support to clinical decisions," *Computer Communications*, vol. 35, no. 11, pp. 1296–1305, 2012.

29. C.-M. Chen, "Web-based remote human pulse monitoring system with intelligent data analysis for home health care," *Expert Systems with Applications*, vol. 38, no. 3, pp. 2011–2019, 2011.
30. E. Dolatabadi and S. Primak, "Ubiquitous WBAN-based electrocardiogram monitoring system," in *Proceedings of the 13th IEEE International Conference on e-Health Networking, Applications and Services (HEALTHCOM'11)*, Columbia, MO, June 2011, pp. 110–113.
31. R. Sukanesh, S. P. Rajan, S. Vijayprasath, S. J. Prabhu and P. Subathra, "GSM based ECG tele-alert system," *International Journal of Computer Science and Application*, pp. 112–116, 2010.
32. J. Bai, Y. Zhang, D. Shen, L. Wen, C. Ding, Z. Cui, F. Tian, B. Yu, B. Dai and J. Dang, "A portable ECG and blood pressure telemonitoring system," *IEEE Engineering in Medicine and Biology Magazine*, vol. 18, no. 4, pp. 63–70, 1999.
33. R.-G. Lee, H.-S. Chen, C.-C. Lin, K.-C. Chang, and J.-H. Chen, "Home telecare system using cable television plants—an experimental field trial," *IEEE Transactions on Information Technology in Biomedicine*, vol. 4, no. 1, pp. 37–44, 2000.
34. R. Sukanesh, P. Gautham, P. T. Arunmozhivarman, S. P. Rajan, and S. Vijayprasath, "Cellular phone based biomedical system for health care," in *Proceedings of the IEEE International Conference on Communication Control and Computing Technologies (ICCCCT'10)*, Ramanathapuram, October 2010, pp. 550–553.
35. K. Hung and Y.-T. Zhang, "Implementation of a WAP-based telemedicine system for patient monitoring," *IEEE Transactions on Information Technology in Biomedicine*, vol. 7, no. 2, pp. 101–107, 2003.
36. M. V. M. Figueredo and J. S. Dias, "Mobile telemedicine system for home care and patient monitoring," in *Proceedings of the 26th Annual International Conference of the IEEE Engineering in Medicine and Biology Society (EMBC'04)*, vol. 2, pp. 3387–3390, September 2004.
37. E. Kyriacou, S. Pavlopoulos and D. Koutsouris, "An emergency telemedicine system based on wireless communication technology: A case study," in *M-Health: Emerging Mobile Health Systems*, Springer, New York, NY, 2006, pp. 401–416.
38. M. F. A. Rasid and B. Woodward, "Bluetooth telemedicine processor for multichannel biomedical signal transmission via mobile cellular networks," *IEEE Transactions on Information Technology in Biomedicine*, vol. 9, no. 1, pp. 35–43, 2005.
39. M. Engin, E. Çağlav and E. Z. Engin, "Real-time ECG signal transmission via telephone network," *Measurement*, vol. 37, no. 2, pp. 167–171, 2005.
40. Y.-H. Lin, I.-C. Jan, P. C.-I. Ko, Y.-Y. Chen, J.-M. Wong, and G.-J. Jan, "A wireless PDA-based physiological monitoring system for patient transport," *IEEE Transactions on Information Technology in Biomedicine*, vol. 8, no. 4, pp. 439–447, 2004.
41. S. Khoór, J. Nieberl, K. Fügedi and E. Kail, "Internet-based, GPRS, long-term ECG monitoring and non-linear heart-rate analysis for cardiovascular telemedicine management," in *Proceedings of the Computers in Cardiology*, vol. 28, Thessaloniki, September 2003, pp. 209–212.

3 Artificial Intelligence in Future Telepsychiatry and Psychotherapy for E-Mental Health Revolution

Sudhir Hebbar and Vandana B

CONTENTS

DOI: 10.1201/9781003309451-3

3.1 INTRODUCTION

Psychological problems and mental illnesses contribute significantly to the burden of illness. Depression is a major killer in young adults. In younger generations, the incidence of depression and anxiety is increasing. As socioeconomic conditions improve, people increasingly start utilizing mental health services. Unfortunately, especially in developing countries, there is a dearth of skilled psychotherapists and psychiatrists. They are available mainly in the cities. Hence, the availability of services is a major concern for rural areas.

In contrast to the aforesaid situation, there is major progress in the field of treatment of mental problems. There is a significant progress in the use of effective medications for psychological disorders and a revolution in the psychotherapy/counseling techniques, like cognitive behavior therapies, acceptance and commitment therapy, dialectical behavior therapy, psychodynamic therapy, and other evidence-based therapies. The major challenge lies in the dissemination of these psychotherapy skills and further accessibility of therapists to the patients. Fortunately, there is a silver lining in the darkness.

Digital technology is bridging this gap. The internet is very useful in the dissemination of psychotherapeutic skills. Many prestigious institutions of psychotherapy in the world are offering online training courses, which were not possible even in the last decade because of the low penetration of broadband. High penetration of broadband has revolutionized this area. Digital recording of therapy sessions provides a good opportunity for supervision of psychotherapy skills. The internet has revolutionized the delivery of psychiatric consultation (telepsychiatry) and psychotherapy (telepsychotherapy). Aiding all these advances is artificial intelligence.

Artificial intelligence is the intelligence exhibited by machines in a similar manner demonstrated by humans. Some of the tasks it is intended to perform are learning, thinking, and problem-solving. It aims to design computer programs to perform tasks that would otherwise need natural intelligence. The goal of artificial intelligence is to develop intelligent agents to find solutions to real-world problems in a more human-like manner. It is used to generate new insights from data. An agent solves complex problems by utilizing a huge amount of knowledge and heuristic techniques. Limitations of humans, like the ability to analyze the minimal number of options at a time, constraints related to time, and memory which is not affected by stress, can be addressed by intelligent agents. But these agents lack common sense and the ability to handle new situations. Humans and intelligent agents can take the benefit of their complementary strength to execute the task in a better approach.[1]

Artificial intelligence offers various benefits to the mental health field. Individuals can access this when they are in crisis or need support. Even mental health professionals can monitor patient progress and suggest possible treatment or support. When AI is combined with a good therapist, it can offer greater support that is affordable and able to reach more people. Data analysis techniques based on AI help practitioners identify diseases precisely, which can be used to provide treatment quickly. AI allows the therapist to observe their patients from remote locations and offer help in case of mental distress and keeps them safe. Mental health is different from other areas of health. People suffering from mental illness face discrimination. Because of the sensitivity of data, consent for data sharing in mental health cannot always be guaranteed. Special care has to be taken to ensure technology safeguards people's privacy. Artificial intelligence tools provide the solution for mental health services. It streamlines repetitive activities. The practitioner can spend more time on personalized patient care. These tools should meet the requirement of patients, health-care professionals, and the general community. Accessing the right treatment at the right time is a big problem for people in rural areas. This is a significant concern for those living with severe mental health illness and their families.

Telepsychiatry and telepsychotherapy have changed the health-seeking behavior in the young generation. People are using the internet for seeking help for health-related issues. There is an increasing trend in using social media platforms and other resources for finding mental health–related information online than in person. The present generation is increasingly pursuing online support for health-related issues than does the previous generation. In the coming years, a large number of people is turning to e-mental health than conventional practice. Urban people can easily access these types of services when compared to the rural population. But considering the increased technology penetration and smartphone usage in rural areas, the percentage of people using E-mental health services in rural sectors will increase substantially. Developing countries like India, where most of the population are from rural areas, would benefit more from such initiatives. Different types of platforms are used to assist mental health treatment delivery. It ranges from internet-based video psychotherapy to virtual reality and artificial intelligence–based programs. Therapist time is progressively limited, and clinicians are overloaded with increased documentation. These problems are cumbersome, with psychiatrists who must rely on their skills to poster therapeutic rapport with their patients.[2]

Artificial intelligence techniques can be used to diagnose mental illness, monitor treatment progress, identify medication adherence, determine the severity of a mental illness, design personalized treatment, and focus on the clinician–patient relationship. In the case of mental health, there are few tests and techniques in finding diagnosis, which are lengthy and need extensive assessment. Mental illness is intrinsically complex, which involves biological and psychological factors. Active involvement by each member of the community is very important in promoting mental health. Review is conducted to identify the opportunities and future challenges associated with this new technology in treatment intervention and outcome prediction.

3.2 ARTIFICIAL INTELLIGENCE

Artificial intelligence is the science and engineering branch that is associated with system development that imitates the features associated with intelligence in humans, like language processing, learning and adaptation, and responding to the external environment. Artificial intelligence provides access to all sectors of people in the field of quality medical care, personalized medical assistance programs, prevention

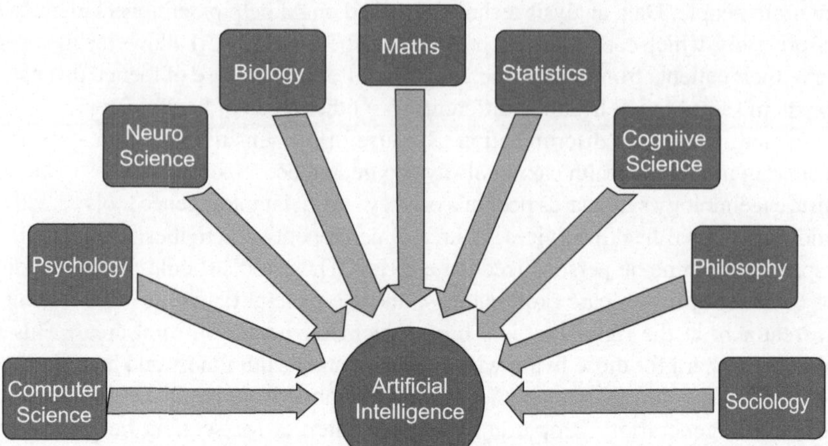

FIGURE 3.1 Interdisciplinary areas of artificial intelligence.

of illness, and discovery of new treatment approaches. From self-driving vehicles to IBM Watson, Siri, and Alexa, AI techniques are creating a major influence on the daily lives of an ordinary person.

Figure 3.1 describes different disciplines associated with artificial intelligence. Artificial intelligence is an interdisciplinary field that has roots in computer science, mathematics, statistics, cognitive science, psychology, and neuroscience.[3]

3.2.1 Domains of Artificial Intelligence

Prospering research areas in the field of artificial intelligence include diverse fields.

Figure 3.2 shows the different domains of AI, which include machine learning, deep learning, gaming, expert systems, computer vision, natural language processing, neural network, fuzzy logic, and robotics.

3.2.1.1 Machine Learning

Machine learning originated on the idea that machines can acquire and adapt over experience. A huge volume of data growth, increased computational processing speed, recent advances in data storage technology have contributed to the growth of machine learning algorithms.[4] In supervised learning, data samples are trained from a known data source with assigned class labels. *Unsupervised learning* refers to learning from the unlabeled data to distinguish the given input data.[5] *Reinforcement learning* extends the machine learning domain to handle control and decision problems that cannot be solved by supervised or unsupervised techniques. It is a method of programming agents using reward and punishment techniques without specifying how a task should be achieved. It acquires knowledge about behaving fruitfully to attain the objective while intermingling with an environment.[6] Machine learning allows identifying the patterns which can be used to predict treatment outcome.

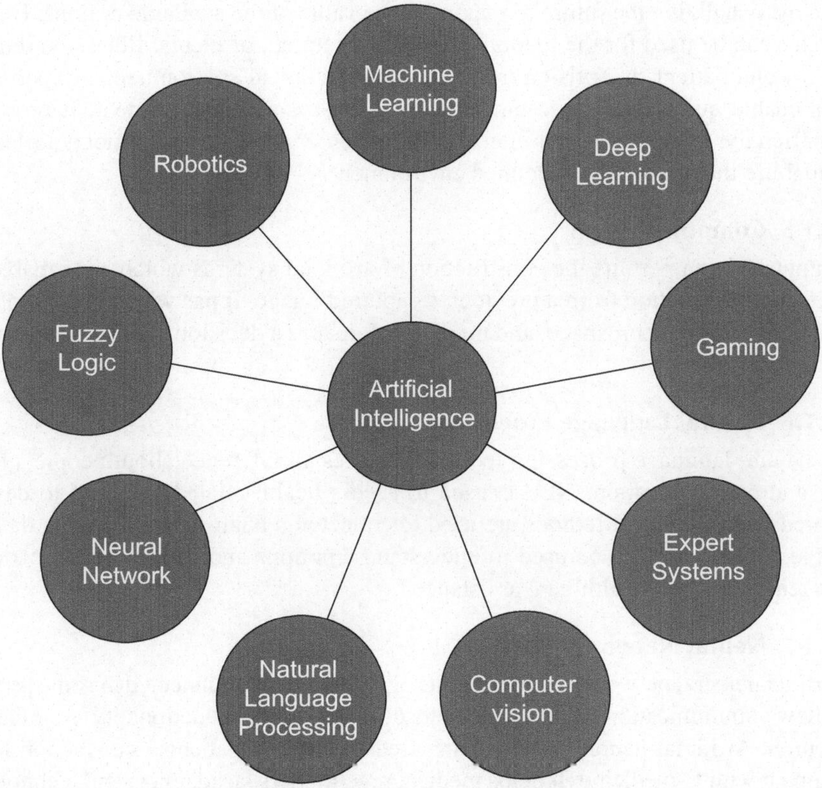

FIGURE 3.2 Domains of artificial intelligence.

3.2.1.2 Deep Learning

Deep learning is a multiple-layer neural network which is based on an unsupervised pertained network, convolutional network, recurrent network, recursive network. Automatic feature extraction is the major advantage of deep learning.[7]

3.2.1.3 Gaming

Artificial intelligence–based chatbots in virtual reality games help clinicians treat patients with depression symptoms. It provides a platform to improve the concentration of patients. Games support building positive emotions and reduce anxiety level.[8] Virtual reality can also be used to deal with posttraumatic stress disorder.

3.2.1.4 Expert System

These are the decision-making systems used to solve problems using facts and heuristics. It is based on the information gathered from an expert. Expert systems are used in the interpretation of data, diagnosis of malfunctions, predicting future, such as weather data, etc. In this, intelligent agents must be able to set goals and achieve them by visualizing the future to maximize the value of the available options. Expert systems can be used for the identification and treatment of mental illness on remote therapy and patient supervision in intra- and interclinical environments. Expert systems enable quick diagnosis and suggest the best available treatment. It provides user-friendly, computer-based intervention which enables mental patients to lead a normal life than living in a sidelined environment.[9]

3.2.1.5 Computer Vision

Computer vision permits the construction of artificial systems which are capable of retrieving information from a previously captured image. It has widespread application in the military, aerospace, and medicine. It makes a decision about objects based on captured images.[10]

3.2.1.6 Natural Language Processing

The natural language processing technique makes an inference about people's feelings and communication. NLP is used to access health-related data and to design tailored interventions. Methods are used to collect and analyze social media data of people, which can then be used to understand emotions and determine the patients who require mental health-care assistance.[11]

3.2.1.7 Neural Network

In artificial neural network assembly, units or nodes are interconnected in some pattern to allow communication between the units or nodes, whose functionality is similar to a neuron. Artificial neural networks are used to find out patients' satisfaction level judgment with E-medicine. It helps medical practitioners, academics, and technology experts to design quality techniques in subsequent applications of E-medicine.[12]

3.2.1.8 Fuzzy Logic

It is a way of decision-making from enormous ambiguous, unclear, and imprecise data. It is similar to human decision-making, which deals with drawing output from approximate data, unlike traditional logic, which requires precise numeric values.

3.2.1.9 Robotics

Robotics is a subdomain of AI which deals with artificial agents called robots acting in real-world environments. It can be used in industry, medicine, and the military to avoid repetitive tasks. Robots have autonomy which ranges from completely teleoperated, in which the operator makes all the decisions for the robots, to completely autonomous, in which the robot is completely independent. In recent years, robotics has focused on the problem associated with interactions in a real-time environment. Human-centered robotics is an emerging area that has potential use in the future.[13] The pandemic has severely affected mental health globally. Older adults faced problems because of lockdowns and social distancing. Social robots are used to support mental health. Natural conversations similar to those between humans are needed in a less-expensive, implementable way. The robot can be used to offer cognitive behavior therapy for adults and patients having dementia. Facial expressions and speech can be handled efficiently by multimodal robots to improve human–robot relations.[14]

3.3 E-MENTAL HEALTH

E-mental health is a promising upcoming technique in the domain of mental health service used to fill the gap in treatment levels. It is based on the internet and mobile technologies. It uses smartphone applications and web applications to deliver health-care services. The usage of online interventions for the prevention, diagnosis, treatment, and monitoring of mental illness is a major benefit of E-mental health-care service. It has the capability to reach people living in remote locations and rural areas. It can also be used as an effective mental health service delivery method to people facing various barriers, such as transportation, physical disability, and time scheduling constraints.[15]

Mental health services in rural areas are neglected a lot, which requires instant attention with respect to the treatment gap. Identification of different types of barriers, like geographical, social, and economic, while utilizing existing services would strengthen E-mental health service delivery in rural areas.[16]

Synchronous (real-time live interaction) and asynchronous (store-forward) are commonly used in telepsychiatry. Synchronous offers bidirectional interactive communication between patient and therapist. Examples are telephony, chat forms, audio-conference, and videoconference. In this mode, a patient will get real-time response that is nearer to face-to-face interaction. In the store-and-forward technique, data is collected by the patient and later it is transferred through email or any other web application to the clinician for detailed analysis. In this mode, patients and therapists can select their convenient time to interchange information.

Primary care physicians can be offered training and supervision by specialists through telepsychiatry. Mental health care can be combined into primary care to offer psychiatric services to the doorstep of patients to fill the mental health gap also, which strengthens human resource and existing infrastructure, which is a promising solution to the rural–urban division in accessing mental health services.[17]

Telemental health services are offered to patients with symptoms of depression and trauma-related problems in psychotherapy. Cognitive behavior therapy is also offered in remote practice. Exercises during the session and between sessions (home assignments), which include problem-solving, mindfulness, relaxation, addressing core beliefs, are given to patients through email, message, or fax technique.[18]

3.3.1 Advantages of E-Mental Health

The exponential growth in the domain of information and communication technologies and the effective intervention of digital tools to address psychiatric disorders are creating a promising impression in the E-mental health sector.

Telemental health aids in the effective management of psychiatric disorders and includes observing, health promotion activities, disease prevention techniques, and biopsychosocial treatments. Cognitive behavior therapy was the typical model in telemental health interventions when handling symptoms of depression, anxiety, and mood disorder. Different sectors of people, including children and adults, with various types of health problems, like migraine, depression, etc., have accepted telemental health services in the form of various technologies like the internet and the telephone.[19]

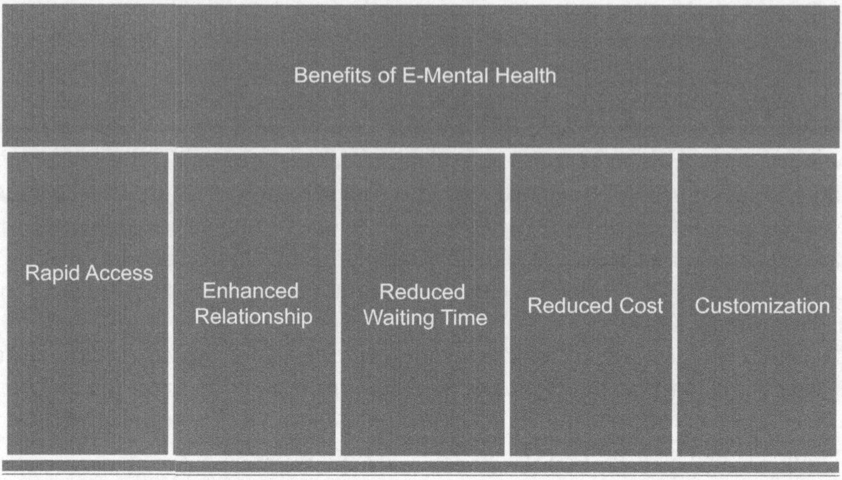

FIGURE 3.3 Benefits of E-mental health.

Figure 3.3 describes various advantages of E-mental health.
Some of the major benefits are explained as follows.

3.3.1.1 Rapid Access

- Can be accessed anytime, anywhere.
- Information can be reviewed.
- Can provide treatment for the patients who are not able to attend a face-to-face session.
- Can be used for people who are not satisfied with conventional services.
- Available in many languages.
- Used for people staying in remote locations.
- Can be used for populations having other types of barriers, like transportation and physical disability.
- Useful to address the stigma that surrounds mental health which discourages people from seeking proper care.
- Provides better-quality patient care.
- Enables the practitioner to follow up with their patients in a more effective way to ensure the recovery status.
- Hospitals can expand their access to medical specialists.

3.3.1.2 Enhanced Relationship

- Enhances the relationship between patient and practitioner.
- A standardized way to exchange information across the health-care organization.
- Can enhance continuity of care.
- Can improve patient engagement in therapy through psychoeducation and multimedia.
- Builds the research network.

3.3.1.3 Reduced Waiting Time

- Patients and practitioners can schedule the appointment in their flexible, convenient timing.
- The therapist can focus on patients.
- Therapist time can be optimized for the treatment process.

3.3.1.4 Reduced Cost

- Many interventions used for the treatment are of low cost and can be accessed many times.
- Operating cost is low.
- Can be used to reach a large number of people simultaneously.

3.3.1.5 Customization

- More personalized treatment can be given based on individual requirements.
- Customized health-care training to professionals.
- Research on ICBT.
- Can be developed according to the best and recent research techniques.
- Virtual techniques can be used to improve patient involvement in the therapy process.

3.3.2 Telepsychiatry and Telepsychotherapy

Access to mental health-care services is limited by physical disability, social constraints, or residency in underserved locations. For this group of people, interventions via remote communication technology like telephone and the internet may be more suitable. Telephone-based intervention is more popular because of its extensive availability and ease of operation. Technology-assisted psychotherapy has the potential to reach all sectors of the people. The location of the patient and the therapist does not affect the treatment quality. With the intensification of digital devices and mobile phones, more people can access treatment for mental illness according to their requirements. Mental health services provided through the internet, videoconferencing, and smartphone-related technologies could have impacts on population health. Data can be collected by taking input from the patient or with the help of motion detection. In the field of support, smartphone-based interventions can be used to monitor mood variations or harm tendencies that may occur in patients.

Large-scale research has to be conducted in this domain to analyze the impact of technology in telemental health.[20] The pandemic has changed the field of psychotherapy overnight from face-to-face to virtual, remote teletherapy. Regular hospital visits are expensive in rural sectors in the era of the pandemic due to travel costs and the social distancing concept. Videoconferencing can be used to suggest telemedicine, which avoids physical interaction. Telemedicine is effective in saving both patient and practitioners time for the treatment. It can also be used to manage discharged patients' recovery status. It aids patients to schedule follow-up, which reduces the probability of missing appointments and optimizes patient recovery.

Telemedicine supplements physical consultation and is not a substitute for it. Patients who cannot go to the doctor can consider this as an alternate option.[21] During the period of the pandemic, telepsychiatry and telepsychotherapy are the most commonly used treatment method to offer mental health services. There are issues in using this model both from the patient's and the therapist's viewpoint. There are some issues with respect to rapport establishment, reconsultation management, confidentiality, fee structure, prescribing and monitoring of medicine. We must include the advantages of virtual reality in telemental health to assist patients effectively.[22]

3.3.2.1 Modes of Telepsychiatry

Teletherapy uses traditional telephones, smartphones, apps, internet video calls, or computer-based treatment programs. The great strength of telepsychotherapy and telepsychiatry is that it expands access. Some patients with social anxiety disorders or posttraumatic stress disorder have requested telephone rather than video sessions. A therapist offers patients different choices of treatment methods, like traditional telephone or video calls, based on their request. Research has to be done on whether video therapy is preferable to traditional telephone therapy. Video has advantages over the telephone for group therapy, but it provides more distraction for some patients in individual therapy.[23]

3.4 APPLICATIONS OF AI TECHNOLOGY IN MENTAL HEALTH CARE

The requirement of social distancing and the deficiency of therapists during the pandemic have catalyzed the implementation of digital technologies in the field of telepsychiatry. An E-health service provides accessible and cost-effective services for the management of patients. Advanced technologies may become a safer treatment option in the future. AI tools relieve psychiatrists from repeated tasks, allowing them to establish a supportive doctor–patient relationship. The issues associated with the fruitful implementation of digital technology in the field of psychiatry have to be given importance.[24]

Figure 3.4 describes applications of artificial intelligence in various domains of E-mental health.

3.4.1 Health Service Management

- Management of inpatient and outpatient scheduling
- Appointment scheduling
- Alerts to practitioners
- Electronic medical record maintenance
- Medical billing
- Intradepartmental and interdepartmental communication

In the present system, most counseling sessions occur behind closed doors, without empirically supported outcome metrics. In contrast to the present, a predominant system of outpatient mental health care, many digital health platforms collect patient data on all interactions, which allow continuous evaluation of the treatment and patient response to medication.[25]

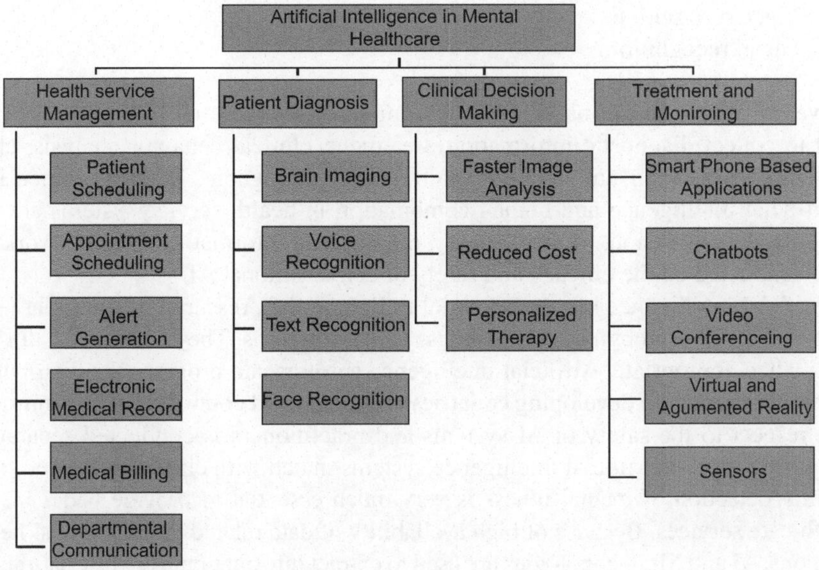

FIGURE 3.4 Artificial intelligence in E-mental health.

The prospect of AI in mental health care is encouraging. An active role has to be taken in incorporating AI in clinical care and patient monitoring. Clinical expatiation of psychiatrists has to be collaborated with those of computational scientists to assist in the transformation of traditional mental health-care practices to E-mental health service.[26]

In developing nations, variations between rural and urban health-care systems are a major issue. The unavailability of qualified health service sources is an important factor for the reduced quality of health service in rural areas. The application of AI based services could improve mental health-care service in rural sectors of developing countries. Medical AI systems can also be used to train other health workers, which helps improve rural health services. A multilevel AI service network model is proposed.[27]

Computer intelligence can enhance manual procedure; it can discover a best practice for the medical procedure, and the concept of big data technology is used to retrieve health-care data. Cloud computing can be used to manage infrastructure issues. The versatility of artificial intelligence and telemedicine has generated endless possibilities in the sector of patient handling, health-care technology, and intelligent assistance identification and information retrieval.[28] Patient movement in mental care entities can be improved with artificial intelligence techniques. It is the capability of health-care units to manage the patients efficiently with minimum delay, quality, and patient satisfaction.[29]

3.4.2 Patient Diagnosis

- Artificial intelligence is used to understand the structure of the brain, which aids clinicians to make decisions on time.
- Text recognition.

- Voice recognition.
- Facial recognition.

Conversational artificial intelligence is augmenting the work of psychotherapy. It is used to collect diagnostic information, treatment, clinical behavior analysis, clinician–AI collaboration, and migration path remains ambiguous. Different dimensions of artificial intelligence and human combination in health service systems are discussed in the dimensions of care, quality, rapport, and information sharing. Work has to be conducted on the efficacy and safety of conversational AI.[30]

Artificial intelligence has the potential to lighten the pressure on health-care systems. AI can be used by patients to assess their symptoms. They can take health services when appropriate. Artificial intelligence systems can progress the distribution of health care in both developing countries and developed countries. Validation study with respect to the safety of AI systems and practitioners is conducted separately. Diagnostic facts of artificial intelligence systems match with clinical precision.[31]

Early detection of mental illness is very much essential to provide better mental health-care services. Because of high availability of data related to individuals' health positions, AI and ML technologies are used to extract information from raw data and to understand the mental health condition, which assists mental health service providers.[32]

3.4.3 Clinical Decision-Making

- Faster image analysis
- Reduced cost
- More personalized therapies

Internet-based interventions in the youth community may provide cost-effectiveness. Until now, online programs depend on human mediators to provide therapeutic content. More refined models are required to deliver tailored online therapy. Moderated online social therapy (MOST) delivers social media–based intervention for mental health service. Computational intelligence methods are used to enhance the delivery of therapeutic content. Horizon site is analyzed, which is an online social therapy for the first episode of psychosis. Natural language examination and chatbot tools are used to tailor the interventions, with the aim of enhancing user engagement.[33]

Data that has been generated from biomedical research, which includes electronic health records, images, text, and sensor data, belong to an unstructured format. Deep learning is used to generate learning models from heterogeneous data.[34]

3.4.4 Treatment and Monitoring

- Smartphone-based applications to track patient moods
- Chatbots for online support
- Videoconferencing to conduct exposure and response prevention therapy
- Virtual reality and augmented reality to conduct imaginary exposure
- Usage of sensors in smartphones to count the number of repetitive activities and to focus on reducing them

Artificial intelligence–based systems augment medical staff inpatient treatment process for different types of diseases. AI systems also make an impact on nursing and hospital management activities. AI applications provide both new opportunities and challenges for health-care providers. Applications of AI in the medical field require planning and methods to change the medical services to acquire the benefits of the technology. Most of the factors related to individual health correlate to lifestyle factors, like stress reduction, sleep, exercise, and diet. Different types of digital technologies are used to provide lifestyle modification suggestions tailored to the individual requirement based on artificial intelligence.[35]

Precision psychiatry is the practice of providing accurate dosage and type of medication at an accurate time. Advanced AI and ML algorithms are used to analyze neuroimaging to trace the biomarkers and genetic loci associated with psychiatric disorders.[36]

Ecological momentary programs based on self-supervision and tailored feedback help individuals deal with depressive symptoms. Few participants reported behavioral improvement over time. It aids in enhancing self-awareness and self-management.[37]

Globally, 4%–8% of children are facing the problem of attention-deficit/hyperactivity disorder. Serious video games with multimodal interventions are suggested for handling this problem. Virtual reality–based therapy, which includes *The Secret Trail of Moon* game, is used to offer cognitive training for ADHD patients.[38] Cognitive behavior therapy using internet-based techniques achieved noteworthy changes in self-rated OCD symptoms and behavior.[39]

A computer program is used for OCD patients which includes animations to demonstrate exposure with response prevention, which give rise to a decreased washing habits.[40]

The single-session virtual reality technology is used in exposure therapy for spider phobia, which has a negative role on the lifestyle of a patient. A new generation of virtual reality devices has the potential for treatment effectiveness. Game elements allow patients to experience situations impossible in real life. These VR games provide a promising improvement in exposure therapy. It is also used to offer maintenance therapy after the completion of regular treatment.[41]

Information and communication technology (ICT)–based intervention program tools can help patients with gambling disorders. It assists in cost reduction when compared to the traditional approach. These intervention programs are designed on motivational interviews, cognitive behavioral therapy, and other tools. Outcomes are measured using gambling severity, urges to gamble, managing with urge frequencies. In handling depression and anxiety symptoms, therapeutic support is combined with ICT tools.[42]

Many clinical trials have used virtual reality as an interactive tool for patients having the symptoms of depression, anxiety disorder, schizophrenia, addiction disorder, etc. Common characteristics like anxiety and avoidance can be handled with virtual reality tools, which provide virtual environments with well-controlled sensors. It can also be used to produce behavior change in patients with illnesses like autism and dementia. VR tools can also be used to address intensive stress. In the future, VR tools have to manage hurdles like motion sickness, dry eyes, and addiction.[43]

Cognitive behavioral therapy based on internet (ICBT) is used in patients having rheumatoid arthritis and osteoarthritis. This therapy provides hopeful results in improving the psychological impact on daily life activities. It also provides intervention for people having adjustment problems along with medication.[44]

ICBT is also useful in handling illnesses like diabetes mellitus with psychiatric comorbidities. ICBT has been found less expensive for patients and their families. It can also be useful in rural and urban sectors because of the penetration of technology that covers various languages, regions, and cultures across the globe. Larger randomized control trials are needed to focus on the effectiveness of ICBT in clinical practice.[45]

Internet interventions for the treatment of OCD are promising. Different modules, like psychoeducation, cognition, mindfulness, exposure and reaction prevention, etc., are introduced to OCD patients. Self-confidence and improvement of life quality and depression symptoms are considered in detail. Patients who are facing treatment barriers can select internet-based CBT for OCD. Further research is required on the usage of ICBT in the wider public.[46]

Patients with agoraphobia, social phobia, are treated with 8 to 12 sessions of virtual reality exposure therapy with head-mounted displays. Including drug therapy with VRET is effective in shortening the duration of treatment. The impact of virtual reality exposure therapy in phobia treatment is high when there is no other psychiatric comorbidity. Long-term research shows the continued outcome of VRET in the treatment of agoraphobia and social phobia.[47]

Smartphone-based intervention is found to be cheaper in handling generalized anxiety disorder. College students with self-identified generalized anxiety disorder are assisted with a smartphone-based guided self-help program. These students are compared from pre- to posttreatment and six months' follow-up programs. Participants notice a greater reduction in the DASS stress and greater remission from GAD when compared to students facing GAD with no treatment.[48]

AI-based interventions can be used in the diagnosis, prediction, and treatment of mental illness. A computerized platform can be used to provide the groundwork for the development of reliable clinical tests in psychiatry. Techniques like computer-assisted therapy and MOST already proved effective in treating symptoms of depression, anxiety, and psychosis through online support and providing customized therapy while facilitating privacy and autonomy. AI brings a new dimension in handling psychiatric disorders, which plays a crucial role in improving the standard of life of patients who suffer from mental health disorders.[49]

Machine learning techniques can be used to predict the probability of psychosis in at-risk individuals. Speech pattern analysis is used to analyze the syntactic complexity and other patterns to predict the future occurrence of psychosis with 82% accuracy. AI in mental health care is highly transformative, leading to a vast improvement in inpatient treatment and monitoring.[50]

Smartphone-based chatbots for mental health problems assist psychiatrists in substituting human-based interaction. It provides a chance to fill the self-care gap. It expands the availability of mental health-care services to patients with depression and anxiety disorders. It also provides CBT and mindfulness interventions. Low-cost assistance can be given to patients to reduce the psychological impact caused by anxiety disorders.[51]

There is an increasing trend in the use of smartphones and other consumer technologies in the field of psychiatry. Due to complex issues like commercialization, technological, legal, and medical issues, more productive direction is required in the usage of consumer technology in the field of psychiatry. Domain knowledge of consumer devices is required to combine apps in the treatment of psychiatric disorders. App accuracy, efficacy, privacy, and security issues are taken into consideration while using the apps for treatment and patient monitoring purposes. Along with smartphones, a variety of technology platforms like desktops, tablets, and laptops can be considered to provide mental health services to maximize patient participation.[52]

Smartphone apps are becoming popular in delivering psychological interventions to patients having disorders with high prevalence, like depression, anxiety disorders, trauma, stress-related, addiction and substance-related, schizophrenia, psychotic disorders. There is a requirement for developing programs for the large coverage of psychiatric disorders rather than focusing on the most prevalent disorders. Systematic focus, including randomized control trials, is required to enhance the robustness and trustworthiness of the application. The innovative capabilities of smartphones have to be explored to understand their potential in handling psychiatric disorders.[53]

The mental state of a patient can be estimated without any load on the mental monitoring service using smartphones based on machine learning algorithms. These algorithms are used to learn complex relationships between mental states reported by patients or clinicians and sensor data collected through a smartphone. Categorization of mental state, regression of score, and correlation analysis are carried out. Assessment metrics and cross-validation techniques are designed to conduct the evaluation.[54]

With the help of virtual reality, patients can be transported into an immersive and interactive virtual environment that is fully controlled by clinicians. VR can be harnessed in the assessment and management of mental health. Clinical valuation is an essential part of mental health service delivery practice. Assessment ranges from diagnosis, treatment plan, and supervision. It enables more precise and customized valuation. The data capture capability of VR can reveal important insights while understanding the patient condition and help design more tailored treatment. Development in VR technology can radically transform the landscape of assessment in psychiatry.[55]

Virtual reality techniques are used to improve the accessibility of exposure and response prevention therapy. The negative impact of anxiety-related disorders shows the importance of clinician and patient operative treatment requirements. VR may help reduce the current problems faced by exposure therapy, like refusal of exposure therapy by patients or dropout from the treatment, difficulties in arranging exposure, difficulties in imaginal exposure and in vivo exposure, lack of appropriate training and experience in exposure and response prevention therapy. Virtual reality therapy can improve the treatment for anxiety disorder by making it acceptable to clinicians.[56]

A computer-based biofeedback treatment system is used for the treatment of stress-related psychological problems. A computer-based biofeedback system is used to classify the patient correctly and to provide training. The decision support system is used to determine the changes in finger temperature. Case-based reasoning (CBR) is used to categorize patients, choose the parameters, and collect feedback. CBR technology, which is based on artificial intelligence, assists in information sharing among practitioners. It can be used effectively in stress treatment.[57]

Panic attacks are reproduced in a controlled atmosphere using the safe carbon dioxide test. The carbon dioxide challenge test helped to judge the antipanic properties of medicines and cognitive behavior therapy.[58] A more tailored approach for panic disorder can be developed with the help of upcoming technologies, like machine learning and wearable devices, which acts as the best intervention for the individual patient. Symptoms during panic attacks and physical health status are considered while designing the treatment procedure.[59]

Technology is used to deliver interventions that can be accessed anonymously and by multiple people at the same instance of time. These services which are provided through smartphone and web applications have the potential to address the stigma and transportation barriers. Computerized CBT shows a remarkable reduction in symptoms of depression and anxiety.[60]

3.5 CHALLENGES OF AI IN PSYCHOTHERAPY

AI or machine learning approach can be suitable in psychopharmacology to select appropriate treatments. Medication administered can be predicted based on the patient's genetic information and other biological data collected at baseline. Unlike pharmacology, psychotherapy involves a combination of psychoeducation, exposure to feared stimuli, progressive muscle relaxation techniques, and other techniques that must be delivered by the therapist, by motivating the patient despite fear and anxiety. The clinician has to create a mixture of difficulty levels while carrying out exposure and response therapy with continuous observations of individual patient's stress-taking ability. The therapist has to adjust the next exposure level to achieve the desired results. Most cognitive behavior techniques focus on behavior and emotional aspects more than biological features. These vary dramatically from person to person. Artificial intelligence can be used for solving the limitations of regression techniques. It can collect data and provide interventions beyond the boundaries of the clinics.[61]

3.5.1 Legal Aspects of Telepsychiatry and Telepsychotherapy in India

A legal and ethical framework is highly essential in telepsychiatry and telepsychotherapy to protect the rights of patients and mental health-care authorities. Guidelines are developed to address the emerging needs of E-mental health services during the period of a pandemic. The Indian Psychiatric Society, Telemedicine Society, and NIMHANS have published guidelines for practicing telepsychiatry, with the title "Telepsychiatry Operational Guidelines," in May 2020.[62]

The Clinical Psychology Department of NIMHANS has released a telepsychotherapy service guidelines for clinical psychologists. (Version 1.0, dated April 14, 2020). These guidelines are not limited to this period. It helps the practitioner deliver telepsychotherapy services in a standard way. Ongoing penetration of technologies in the field of mental health-care service delivery systems has generated new challenges with respect to the practice and legal procedures. Social distancing has demanded the use of technology for addressing tailored mental health-care services. Updated guidelines are designed for the implementation of the E-medicine concept in the field of telemental health. The pros and cons of digital technology usage in the E-mental health delivery system has to be considered to make this transition path well accepted by mental health-care professionals and patients.[63]

3.6 DISCUSSION

Denial of treatment and dropout from cognitive behavior therapy are a foremost problem in patients having anxiety-related disorders in a general hospital environment. Still, patients who incorporate CBT techniques can find remarkable benefits. Mainly, youth who are located nearer to the hospital, with good education background, good economic status, prefer CBT treatment.[64] In the future, internet interventions can be merged with traditional psychotherapy treatment to address the location barrier, transportation problem, stigma associated with mental illness, and economic constraints.

The first author is using teleconsultations and telepsychotherapy. Both are well received by patients. With the improvement in net connectivity, video consultations are increasing. Prescriptions sent to patients over smartphones are honored by pharmacists. Videos on relaxation skills are available on the internet. This partly saves time spent on teaching these skills in session. A few clients have used apps like Woebot with some benefits. Biofeedback-assisted relaxation is used by some clinicians, though it is not superior to traditional relaxation methods. Virtual reality is not used much in psychotherapy in India as of now. Internet-based telehealth providers are also increasing in number. One of the important concerns is confidentiality and data safety. This problem does not arise when the therapist contacts the patient directly.

3.7 CONCLUSION AND FUTURE WORK

This chapter provides an overview about E-mental health and usage of internet and artificial intelligence in delivering mental health services. It also focuses on application of artificial intelligence in different domains of mental health service delivery system. In addition, it discusses challenges associated with its implementation. Since E-mental health is an emerging field, a clear road map has to be defined with significant milestones.

Further research has to be carried out in the direction of E-mental health application design, usage of machine learning techniques in the estimation of mental illness diagnosis, the outcome of the treatment, dropout ratio, and relapse chances. After successful completion of treatment, relapse prevention modules can be incorporated along with therapist feedback to aid the patient's therapy process and well-being. Monitoring engagement of patients in therapy and deterioration of symptoms can be further researched to spread E-mental health services. Still, it is in its nascent nature in terms of wide-scale implementation.

Artificial intelligence is a promising technology in delivering E-mental health services, but they are a long way off from reaching the best coverage. Experience of a client with respect to the amalgamation of AI in the field of psychiatry has the main attention of mental health care with the patient-centered approach. Patient's concern is vital to the large-scale adoption of artificial intelligence in order to uphold autonomy around their emotions, which is highly sensitive information, especially in the domain of E-mental health.

NOTES

1 (Tecuci, 2012)
2 (Graham et al., 2019)
3 (Tecuci, 2012)
4 (Attaran and Deb, 2018)
5 (Sathya and Abraham, 2013)
6 (Attaran and Deb, 2018)
7 (Patterson and Gibson, 2017)
8 (Ren, 2020)
9 (Oguoma et al., 2020)
10 (Lugli and De Melo, 2017)
11 (Calvo et al., 2017)
12 (Zobair et al., 2021)
13 (Riek, 2015)
14 (Lima et al., 2021)
15 (Lal, 2019)
16 (Kallivayalil and Enara, 2018)
17 (Malhotra et al., 2013)
18 (Bunnell et al., 2021)
19 (Bashshur et al., 2016)
20 (Bee et al., 2008)
21 (Haleem et al., 2021)
22 (Sousa et al., 2020)
23 (Markowitz et al., 2021)
24 (Roth et al., 2021)
25 (Hirschtritt and Insel, 2018)
26 (Graham et al., 2019)
27 (Guo and Li, 2018)
28 (Pacis et al., 2018)
29 (Cecula et al., 2021)
30 (Miner et al., 2019)
31 (Baker et al., 2020)
32 (Su et al., 2020)
33 (Alfonso et al., 2017)
34 (Miotto et al., 2018)
35 (Lee and Yoon, 2021)
36 (Lin et al., 2020)
37 (Folkersma et al., 2021)
38 (Yanguas et al., 2021)
39 (Andersson et al., 2011)
40 (Clark et al., 1998)
41 (Miloff et al., 2016)
42 (Sanahuja et al., 2021)
43 (Park et al., 2019)
44 (Terpstra et al., 2021)
45 (Kumar et al., 2017)
46 (Schroder et al., 2020)
47 (Krzystanek et al., 2021)
48 (Newman et al., 2020)
49 (Fakhoury, 2019)

50 (Kalanderian and Nasrallah, 2019)
51 (Ahmed et al., 2021)
52 (Bauer et al., 2020)
53 (Miralles et al., 2020)
54 (Fukazawa et al., 2018)
55 (Bell et al., 2020)
56 (Boeldt et al., 2019)
57 (Ahmed et al., 2011)
58 (Amaral et al., 2013)
59 (Caldirola and Perna, 2019)
60 (Dedert et al., 2013)
61 (Horn and Weisz, 2020)
62 (Math et al., 2020)
63 (Nanjegowda and Munoli, 2020)
64 (Hebbar and Romero, 2018).

BIBLIOGRAPHY

Ahmed, Arfan, Nashva Ali, Sarah Aziz, Alaa A. Abd-Alrazaq, Asmaa Hassan, Mohamed Khalifa, Bushra Elhusein, Maram Ahmed, Mohamed Ali Siddig Ahmed, and Mowafa Househ. "A review of mobile chatbot apps for anxiety and depression and their self-care features." *Computer Methods and Programs in Biomedicine Update* 1 (2021): 100012.

Ahmed, Mobyen Uddin, Shahina Begum, Peter Funk, Ning Xiong, and Bo von Scheele. "A multi-module case-based biofeedback system for stress treatment." *Artificial Intelligence in Medicine* 51 (2011): 107–115.

Amaral, Julio Mario Xerfan Do. "The carbon dioxide challenge test in panic disorder: A systematic review of preclinical and clinical research." *Revista Brasileira de Psiquiatria* 35, no. 3 (2013): 318–331.

Amaral, Julio Mario Xerfan Do, Pedro Tadeu Machado Spadaro, Valeska Martinho Pereira, Adriana Cardoso de Oliveira e Silva, and Antonio Egidio Nardi. "The carbon dioxide challenge test in panic disorder: A systematic review of preclinical and clinical research." *Revista Brasileira de Psiquiatria* 35 (2013): 318–331.

Andersson, Erik, Brjann Ljotsson, Erik Hedman, Viktor Kaldo, Björn Paxling, Gerhard Andersson, Nils Lindefors, and Christian Ruck. "Internet-based cognitive behavior therapy for obsessive compulsive disorder: A pilot study." *BMC Psychiatry* 11, no. 125 (2011).

Attaran, Mohsen, and Promita Deb. "Machine learning: The new 'big thing' for competitive advantage." *International Journal of Knowledge Engineering and Data Mining* 5, no. 4 (2018): 277–305.

Baker, Adam, Yura Perov, Katherine Middleton, Janie Baxter, Daniel Mullarkey, Davinder Sangar, Mobasher Butt, Arnold DoRosario, and Saurabh Johri. "A comparison of artificial intelligence and human doctors for the purpose of triage and diagnosis." *Frontiers in Artificial Intelligence* 3 (2020): 543405.

Bashshur, Rashid L., Gary W. Shannon, Noura Bashshur, and Peter M. Yellowlees. "The empirical evidence for TELEMEDICINE interventions in mental disorders." *Telemedicine and e-Health* 22, no. 2 (2016): 87–113.

Bauer, Michael, Tasha Glenn, John Geddes, Michael Gitlin, Paul Grof, Lars V. Kessing, Scott Monteith, Maria Faurholt-Jepsen, Emanuel Severus, and Peter C. Whybrow. "Smartphones in mental health: A critical review of background issues, current status and future concerns." *International Journal of Bipolar Disorders* 8, no. 2 (2020).

Bee, Penny E., Peter Bower, Karina Lovell, Simon Gilbody, David Richards, Linda Gask, and Pamela Roach. "Psychotherapy mediated by remote communication technologies: A meta-analytic review." *BMC Psychiatry* 8, no. 60 (2008): 1471.

Bell, Imogen H., Jennifer Nicholas, Mario Alvarez Jimenez, Andrew Thompson, and Lucia Valmaggia. "Virtual reality as a clinical tool in mental health research and practice." *Dialogues in Clinical Neuroscience* 22, no. 2 (2020): 169–177.

Boeldt, Debra, Elizabeth McMahon, Mimi McFaul, and Walter Greenleaf. "Using virtual reality exposure therapy to enhance treatment of anxiety disorders: Identifying areas of clinical adoption and potential obstacles." *Frontiers in Psychology* 10 (2019): 773.

Bunnell, Brian E., Nikolaos Kazantzis, Samantha R. Paige, Janelle Barrera, Rajvi N. Thakkar, Dylan Turner, and Brandon M. Welch. "Provision of care by "real world" telemental health providers." *Frontiers in Psychology* 12 (2021): 653652.

Caldirola, Daniela, and Giampaolo Perna. "Toward a personalized therapy for panic disorder: Preliminary considerations from a work in progress." *Neuropsychiatric Disease and Treatment Dovepress* 15 (2019): 1957–1970.

Calvo, Rafael A., David N. Milne, and M. Sazzad Hussain. "Natural language processing in mental health applications using non-clinical texts." *Natural Language Engineering* 23, no. 5 (2017): 649–685.

Cecula, Paulina, Jiakun Yu, Fatema Mustansir Dawoodbhoy, Jack Delaney, Joseph Tan, Iain Peacock, and Benita Cox. "Applications of artificial intelligence to improve patient flow on mental health inpatient units—Narrative literature review." *Heliyon* 7 (2021): e06626.

Clark, Augustino, Kenneth C. Kirkby, Brett A. Daniels, and Isaac M. Marks. "A pilot study of computer-aided vicarious exposure for obsessive-compulsive disorder." *Australian & New Zealand Journal of Psychiatry* 32, no. 2 (1998): 268–275.

D'alfonso, Simon, Olga Santesteban-Echarri, Simon Rice, Greg Wadley, Reeva Lederman, Christopher Miles, John Gleeson, and Mario Alvarez-Jimenez. "Artificial intelligence-assisted online social therapy for youth mental health." *Frontiers in Psychology* 8 (2017): 796.

Dedert, Eric, Jennifer R. McDuffie, Cindy Swinkels, Ryan Shaw, Jessica Fulton, Kelli D. Allen, Santanu Datta, John W. Williams Jr., Avishek Nagi, and Liz Wing. "Computerized cognitive behavioral therapy for adults with depressive or anxiety disorders." (2013).

Fakhoury, Marc. "Artificial Intelligence in Psychiatry." *Frontiers in Psychiatry, Advances in Experimental Medicine and Biology* 1192 (2019).

Folkersma, Wendy, Vera Veerman, Daan A. Ornee, Albertine J. Oldehinkel, Manna A. Alma, and Jojanneke A. Bastiaansen. "Patients' experience of an ecological momentary intervention involving self-monitoring and personalized feedback for depression." *Internet Interventions* 26 (2021): 100436.

Fukazawa, Yusuke, Naoki Yamamoto, Takashi Hamatani, Keiichi Ochiai, Akira Uchiyama, and Ken Ohta. "Smartphone-based mental state estimation: A survey from a machine learning perspective." *Journal of Information Processing* 26 (2018): 1–15.

Graham, Sarah, Colin Depp, Ellen E. Lee, Camille Nebeker, Xin Tu, Ho-Cheol Kim, and Dilip V. Jeste. "Artificial intelligence for mental health and mental illnesses: An overview." *Current Psychiatry Reports* 21, no. 116 (2019).

Guo, Jonathan, and Bin Li. "The application of medical artificial intelligence technology in rural areas of developing countries." *Health Equity* 2, no. 1 (2018): 174–181.

Haleem, Abid, Mohd Javaid, Ravi Pratap Singh, and Rajiv Suman. "Telemedicine for healthcare: Capabilities, features, barriers and applications." *Sensors International* 2 (2021): 100117.

Hebbar, Sudhir, and Sylvester Satish Romero. "The utilization pattern of cognitive behavior therapy for anxiety disorders in adults: A naturalistic study from a Medical College Hospital." *IAIM* 5, no. 10 (2018): 37–43.

Hirschtritt, Mathew E., and Thomas R. Insel. "Digital technologies in psychiatry: Present and future." *Focus* 16, no. 3 (2018): 251–258.

Horn, Rachel L., and John R. Weisz. "Can artificial intelligence improve psychotherapy research and practice." *Administration and Policy in Mental Health and Mental Health Services Research* 47 (2020): 852–855.

Kalanderian, Hripsime, and Henry A. Nasrallah. "Artificial intelligence in psychiatry." *Current Psychiatry* 18, no. 8 (2019): 33–38.

Kallivayalil, Roy Abraham, and Arun Enara. "Prioritizing rural and community mental health in India." *Indian Journal of Social Psychiatry* 34, no. 4 (2018): 285–288.

Krzystanek, Marek, Stanisław Surma, Małgorzata Stokrocka, Monika Romańczyk, Jacek Przybyło, Natalia Krzystanek, and Mariusz Borkowski. "Tips for effective implementation of virtual reality exposure therapy in phobias—A systematic review." *Frontiers in Psychiatry* 12 (2021): 737351.

Kumar, Vikram, Yasar Sattar, Anan Bseiso, Sara Khan, and Ian H. Rutkofsky. "The effectiveness of internet-based cognitive behavioral therapy in treatment of psychiatric disorders." *Cureus* 9, no. 8 (2017): e1626.

Lal, Shalini. "E-mental health: Promising advancements in policy, research, and practice." *Healthcare Management Forum* 32, no. 2 (2019): 56–62.

Lee, DonHee, and Seong No Yoon. "Application of artificial intelligence-based technologies in the healthcare industry: Opportunities and challenges." *International Journal of Environmental Research and Public Health* 18, no. 271 (2021).

Lima, Maria R., Maitreyee Wairagkar, Nirupama Natarajan, Sridhar Vaitheswaran, and Ravi Vaidyanathan. "Robotic telemedicine for mental health: A multimodal approach to improve human-robot engagement." *Frontiers in Robotics and AI* 8 (2021): 618866.

Lin, Eugene, Chieh-Hsin Lin, and Hsien-Yuan Lane. "Precision psychiatry applications with pharmacogenomics: Artificial intelligence and machine learning approaches." *International Journal of Molecular Sciences* 21, no. 969 (2020).

Lugli, Alexandre Baratella, and Mauricio Gomes De Melo. "Computer vision and artificial intelligence techniques applied to robot soccer." *International Journal of Innovative Computing, Information and Control* 13, no. 3 (2017): 991–1005.

Malhotra, Savita, Subho Chakrabarti, and Ruchita Shah. "Telepsychiatry: Promise, potential, and challenges." *Indian Journal of Psychiatry* 55, no. 1 (2013): 3–11.

Markowitz, John C., Barbara Milrod, Timothy G. Heckman, Maja Bergman, Doron Amsalem, Hemrie Zalman, Thomas Ballas, and Yuval Neria. "Psychotherapy at a distance." *The American Journal of Psychiatry* 178, no. 3 (2021): 240–246.

Math, Suresh Bada, Narayana Manjunatha, Naveen C. Kumar, Chethan Basavarajappa, and Gangadhar. *Telepsychiatry Operational Guidelines*. Bengaluru: NIMHANS, 2020.

Miloff, Alexander, Philip Lindner, William Hamilton, Lena Reuterskiold, Gerhard Andersson, and Per Carlbring. "Single-session gamified virtual reality exposure therapy for spider phobia vs. traditional exposure therapy: Study protocol for a randomized controlled non-inferiority trial." *Trials* 17, no. 60 (2016).

Miner, Adam S., Nigam Shah, Kim D. Bullock, Bruce A. Arnow, Jeremy Bailenson, and Jeff Hancock. "Key considerations for incorporating conversational AI in psychotherapy." *Frontiers in Psychiatry* 10 (2019): 746.

Miotto, Riccardo, Fei Wang, Shuang Wang, Xiaoqian Jiang, and Joel T. Dudley. "Deep learning for healthcare: Review, opportunities and challenges." *Briefings in Bioinformatics* 19, no. 6 (2018): 1236–1246.

Miralles, Ignacio, Carlos Granell, Laura Díaz-Sanahuja, William Van Woensel, Juana Bretón-López, Adriana Mira, Diana Castilla, and Sven Casteleyn. "Smartphone apps for the treatment of mental disorders: Systematic review." *JMIR Mhealth and Uhealth* 8, no. 4 (2020): e14897.

Nanjegowda, Raveesh Bevinahalli, and Ravindra Neelakanthappa Munoli. "Ethical and legal aspects of telepsychiatry." *Indian Journal of Psychological Medicine* 42, no. (5S) (2020): 63S–69S.

Newman, Michelle G., Nicholas C. Jacobson, Gavin N. Rackoff, Megan Jones Bell, and Barr C. Taylor. "A randomized controlled trial of a smartphone-based application for the treatment of anxiety." *Psychotherapy Research* 31, no. 4 (2021): 443–454.

Oguoma, Stanley Ikechukwu, Kizito Kanayo Uka, Chekwube Alphonsus Chukwu, and Emeka Christian Nwaoha. "An expert system for diagnosis and treatment of mental ailment." *Open Access Library Journal* 7 (2020): e6166.

Pacis, Danica Mitch M., Edwin D. C. Subido Jr, and Nilo T. Bugtai. "Trends in telemedicine utilizing artificial intelligence." 2nd Biomedical Engineering's Recent Progress in Biomaterials, Drugs Development, and Medical Devices, AIP Conf. Proc. 1933, 040009 2018.

Park, Mi Jin, Dong Jun Kim, Unjoo Lee, Eun Jin Na, and Hong Jin Jeon. "A literature overview of virtual reality (VR) in treatment of psychiatric disorders: Recent advances and limitations." *Frontiers in Psychiatry* 10 (2019): 505.

Patterson, Josh, and Adam Gibson. *Deep Learning a Practitioner's Approach*. Sebastopol, CA: O'Reilly Media, Inc., 2017.

Ren, Xinrui. "Artificial intelligence and depression: How AI powered chatbots in virtual reality games may reduce anxiety and depression levels." *Journal of Artificial Intelligence Practice* 3 (2020): 48–58.

Riek, Laurel D. "Robotics technology in mental health care." *Artificial Intelligence in Behavioral Health and Mental Health Care* (2015): 185–203.

Roth, Carl B., Andreas Papassotiropoulos, Annette B. Brühl, Undine E. Lang, and Christian G. Huber. "Psychiatry in the digital age: A blessing or a curse." *International Journal of Environmental Research and Public Health* 18 (2021): 8302.

Sanahuja, Laura Diaz, Daniel Campos, Adriana Mira, Diana Castilla, Azucena Garcia Palacios, and Juana Maria Breton Lopez. "Efficacy of an internet-based psychological intervention for problem gambling and gambling disorder: Study protocol for a randomized controlled trial." *Internet Interventions* 26 (2021): 100466.

Sathya, R., and Annamma Abraham. "Comparison of supervised and unsupervised learning algorithms for pattern classification." *International Journal of Advanced Research in Artificial Intelligence* 2, no. 2 (2013): 34–38.

Schroder, Johanna, Nathalie Werkle, Barbara Cludius, Lena Jelinek, Steffen Moritz, and Stefan Westermann. "Unguided Internet-based cognitive-behavioral therapy for obsessive-compulsive disorder: A randomized controlled trial." *Depression and Anxiety* 37 (2020): 1208–1220.

Sousa, Avinash De, Amresh Shrivastava, and Bhumika Shah. "Telepsychiatry and telepsychotherapy: Critical issues faced by Indian patients and psychiatrists." *Indian Journal of Psychological Medicine* 42, no. 5S (2020): 74S–80S.

Su, Chang, Zhenxing Xu, Jyotishman Pathak, and Fei Wang. "Deep learning in mental health outcome research: A scoping review." *Translational Psychiatry* 10, no. 116 (2020).

Tecuci, Gheorghe. "Artificial intelligence." *WIREs Computational Statistics* 4, no. 2 (2012): 168–180.

Terpstra, Jessy A., Rosalie van der Vaart, Jie He Ding, Margreet Kloppenburg, and Andrea W. M. Evers. "Guided internet-based cognitive-behavioral therapy for patients with rheumatic conditions: A systematic review." *Internet Interventions* 26 (2021): 100444.

Yanguas, Maria Rodrigo, Marina Martin Moratinos, Angela Menendez Garcia, Carlos Gonzalez Tardon, Ana Royuela, and Hilario Blasco Fontecilla. "A virtual reality game (the secret trail of moon) for treating attention-deficit/hyperactivity disorder: Development and usability study." *JMIR Serious Games* 9, no. 3 (2021): e26824.

Zobair, Khondker Mohammad, Louis Sanzogni, Luke Houghton, and Md Zahidul Islam. "Forecasting care seekers satisfaction with telemedicine using machine learning and structural equation modeling." *PLoS One* 16, no. 9 (2021): e0257300.

4 Optimized Convolutional Neural Network for Classification of Tumors from MR Brain Images

K. Ramalakshmi, R. Meena Prakash, S. Thayammal, R. Shantha Selva Kumari, and Henry Selvaraj

CONTENTS

4.1 INTRODUCTION

Medical image processing is one of the emerging and most challenging fields nowadays. One of the emerging techniques in the medical field is image processing. Medical imaging techniques help medical practitioners detect the type of disease and its location in the human body. Various imaging modalities, like X-ray, magnetic resonance imaging (MRI), computerized tomography (CT), magnetic resonance spectroscopy (MRS), positron emission tomography (PET), etc., assist doctors in predicting the well-being and overall survival of patients. MRI has attained great attention, and it is a noninvasive and 3D imaging technique that uses nonionizing

DOI: 10.1201/9781003309451-4

and harmless radiation and could locate abnormality in soft tissues in a very effective manner. Magnetic resonance imaging machine can take several images of the location of the human body under observation from different views, as shown in Figure 4.1, with diverse contrast and physical properties, and because of this, it is known as multiple-modality imaging.

The human brain has three important parts in terms of tissue structure, that is, gray matter, white matter, and cerebrospinal fluid (CSF), as shown in Figure 4.2. These

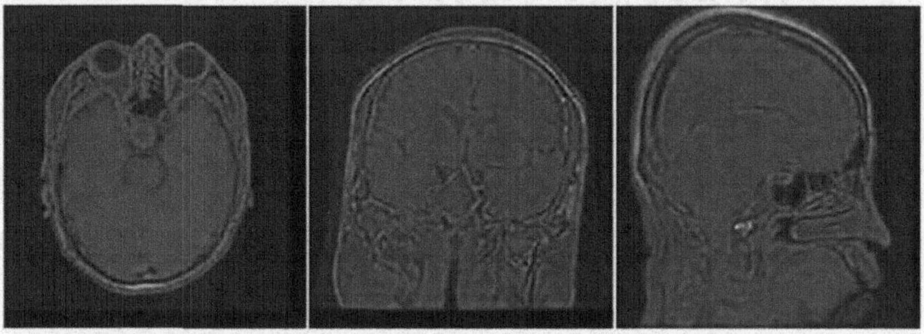

FIGURE 4.1 Three views of an MRI image (axial, sagittal, and coronal).

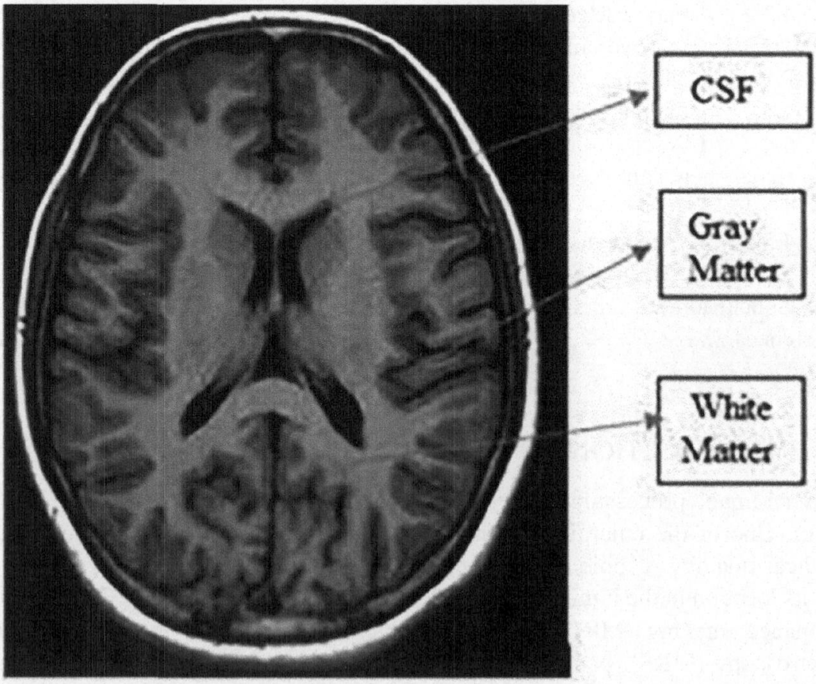

FIGURE 4.2 Image taken from Kaggle dataset showing three parts of the human brain.

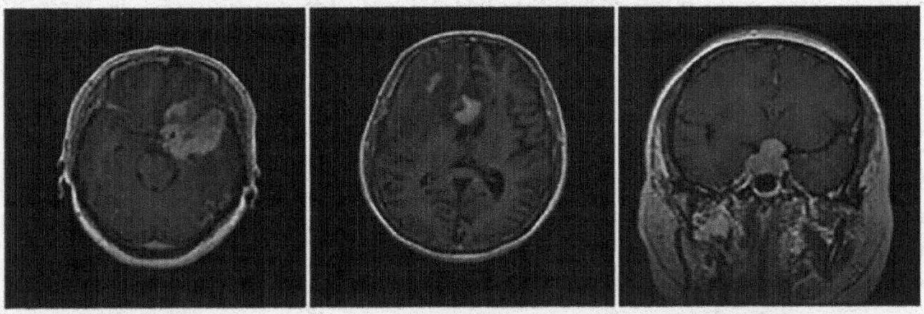

(a) Meningioma (b) Glioma (c) Pituitary

FIGURE 4.3 Three types of brain tumor images taken from CE-MRI dataset.

three parts, along with the tumor, show different contrast when imaged under different physical characteristics and play a vital role in MR imaging for the detection of brain tumor as they consist of soft tissues (Nazir et al., 2021).

Brain tumor is an abnormal cell growth in the brain. Broadly, brain tumors are classified into benign and malignant. Benign tumors are noncancerous, while malignant tumors are cancerous. Tumors that originate in the brain are called primary tumors, such as gliomas, meningiomas, and medulloblastomas. Tumors that originate in other parts of the body and spread to the brain are called secondary tumors, such as breast cancer, colon cancer, and kidney cancer. Abnormal cells within the brain cause brain tumor, and it can be treated if it is found in an earlier stage. It is mainly classified as cancerous or malignant tumor and benign or non-cancerous tumor. Further, these can be classified as primary tumors, within the brain, and secondary tumors, located outside the brain. According to the development of tumor cells, it can be classified as grade I, pilocytic astrocytoma; grade II, low-grade astrocytoma; grade III, anaplastic astrocytoma; and grade IV, glioblastoma (GBM). Symptoms for these brain tumors vary based on the size of the tumor and the part of the brain in which the tumor exists. The proposed method is validated using the contrast-enhanced magnetic resonance images (CE-MRI) benchmark brain tumor dataset. It comprises 3,064 T1-weighted contrast enhanced-MR images from 233 patients with three types of brain tumor, which consists of 708 slices of meningioma, 1,426 slices of glioma, and 930 slices of pituitary brain tumors. Examples of those three types of brain tumors are shown in Figure 4.3.

The succeeding parts of the chapter consist of Section 2, which reviews various works; Section 3, which gives information about the materials and methods used; and Section 4, which deliberates the proposed method. The results are discussed in Section 5, and finally, Section 6 concludes the paper.

4.2 RELATED WORKS

Over the past decade, deep learning algorithms are successfully employed for the classification of medical images. In this work, an optimized convolutional neural network (CNN) is proposed for the classification of brain tumors into benign and malignant tumors. CNN consists of convolution layers, pooling layers, and a fully connected layer or dense layer. The convolution layers and pooling layers are used for feature extraction, while the dense layer with softmax activation is used for classification.

Harish et al. (2020) implemented enhanced, faster region-based convolutional neural network (R-CNN) and AlexNet model for brain tumor detection and classification. Initially, image resizing and contrast-limited adaptive histogram equalization for MRI brain image enhancement are carried out as preprocessing; after this, enhanced faster R-CNN model is used to segment the tumor region from the non-affected brain tumor region. R-CNN is enhanced by adding a new feature extractor network model, ResNet50. Finally, transfer learning, that is, the AlexNet model, is used to classify the brain tumor regions with high speed by using stochastic gradient descent with momentum optimization in the AlexNet train model. Classification accuracy of 99.25% is achieved with this proposed method.

Mehrotra et al. (2020) proposed artificial intelligence in the form of deep learning algorithms for classifying the type of brain tumors. During preprocessing, the raw MR images are downscaled from 512 × 512 × 1 pixels into 225 × 225 × 1 pixels to reduce the dimensionality calculations and help produce support to the system to show a greater outcome in less time. Feature extraction is done using deep learning networks by utilizing its kernels or convolutional filters. The pretrained CNN models are used to perform transfer learning to extract essential features. Finally, feature classification is done with the help of the softmax layer. They used the Cancer Imaging Archive (TCIA) public access repository datasets, which comprise of 696 T1-weighted MR images, with 224 benign images and 472 malignant images, and the image size of 225 × 225 in JPG/JPEG format.

Sajjad et al. (2018) presented a novel convolutional neural network (CNN)–based multigrade brain tumor classification system to segment and classify brain tumor into four different grades using a fine-tuned CNN model in which, initially, the tumor regions from a magnetic resonance image are segmented using a deep learning technique called input cascade CNN. This unique CNN architecture differs from other traditional CNNs due to its two-way processing of image. Input cascade CNN uses a final layer that is a convolutional implementation of a fully connected layer which is 40-fold faster than the other CNN models. The architecture of input cascade CNN consists of two streams: one with 7 × 7 accessible fields for extracting local features, and another with 13 × 13 accessible fields for extracting global features. First, by means of extracting the local features, the highest and lowest intensities are suppressed using bias correction algorithm N4ITK, and then the information of each input channel is normalized by subtracting mean channel and dividing it by standard deviation of that channel to obtain the global features. After the completion of these two steps, a postprocessing step is applied to get rid of the noise from the segmented image, in which connected components-labeling algorithm is applied to get rid of the flat blobs that may appear as a tumor region due to bright corners of the brains near the skull. As a result, the input cascade CNN model gives a segmented brain tumor region. Next, to achieve greater accuracy, the data is augmented using eight different augmentation techniques with different transformational and noise invariance techniques by changing different parameters, which effectively trains the system and avoids the lack of data problem when dealing with MRI for multigrade brain tumor classification. The augmentation techniques, called rotation, flipping, skewness, and shears, are employed for geometric transformations invariance, and the other four techniques, namely, Gaussian blur, sharpening, edge detection, and emboss, are used for noise invariance. Finally, a pretrained CNN model is fine-tuned using augmented data for the final prediction of brain tumor grades. Here, the experiments are carried out using Radiopaedia and brain tumor datasets. These datasets are divided into 50%, 25%, and 25% for training, cross validation, and testing sets, respectively. The Radiopaedia dataset consists of 121 MR images, and the brain tumor dataset consists of 3,064 T1-weighted contrast-enhanced MR images collected from 233 patients. Resolution of these brain tumor images is 512 × 512, with pixel size of 0.49 × 0.49 mm^2. The suggested method achieved the best results, with the highest value of 88.41%, 96.12%, and 94.58% for sensitivity, specificity, and accuracy, respectively, over the brain tumor dataset.

Kabir Anaraki et al. (2018) proposed a CNN-based method to classify three gliomas with MR images. In this proposed method, genetic algorithm is employed to find the CNN structure that produces greater results. The proposed method grades the glioma tumors with high precision, and also, it successfully classified the images of various types of brain tumors. The architecture of CNN is developed using genetic algorithm. In addition, to reduce the variance of prediction error, bagging as an ensemble algorithm is utilized on the best model developed by genetic algorithm, which produces the classification accuracy of 90.9% for classifying three glioma grades in one case study, and in another case study, 94.2% accuracy is achieved for classifying glioma, meningioma, and pituitary tumor types. Due to the effectiveness and flexible nature of this method, it can be greatly used for assisting the doctor to diagnose brain tumors in an early stage. The proposed method is tested on some four databases called normal brain MR images from the IXI dataset, MR images of glioma tumors from the cancer imaging archive datasets, presurgical magnetic resonance multisequence images from REMBRANDT dataset, TCGA-GBM data collection, and low-grade gliomas data from TCGA-LGG.

Gopal S. Tandel et al. (2021) proposed a majority voting (MajVot)–based ensemble algorithm to optimize the overall classification performance of five deep learning (CNN), AlexNet, VGG16, ResNet18, GoogleNet, and ResNet50, and five machine learning-based models, support vector machine, K-nearest neighbors, naive Bayes, decision tree, and linear discrimination using fivefold cross-validation. The proposed method optimized the average accuracy of four clinically relevant brain tumors datasets by 2.02%, 1.11%, 1.04%, 2.67%, and 1.65% against AlexNet, VGG16, ResNet18, GoogleNet, and ResNet50 models, respectively. Similarly, the ML-MajVot algorithm also gave an improved accuracy of 1.91%, 13.36%, 3.25%, and 0.95% against K-nearest neighbor, naive Bayes, decision tree, and linear discrimination models, respectively, and 0.18% lesser accuracy than support vector machine. Also, the proposed method is validated on synthetic face data; here the accuracy improvements of 2.88%, 0.71%, 1.90%, 2.24%, and 0.35% is achieved against AlexNet, VGG16, ResNet18, GoogleNet, and ResNet50, respectively.

Yin et al. (2020) introduced a new metaheuristic-based methodology for the timely diagnosis of brain tumor. The proposed method consists of three main phases, namely, background removing, feature extraction, and classification based on multilayer perceptron neural network. Further, an improved model of the whale optimization algorithm based on the chaos theory and logistic mapping technique is also employed for the optimal feature selection and for the classification of stages. Furthermore, by analyzing the performance of the proposed method, 87% of CDR, 8% of FAR, and 5% of FRR were achieved. While comparing this proposed method with some existing methods like SVM, NNPSO, MLPPSOBBO, and CNN, the proposed method has better results. The three different database classes, FLAIR, T1, andT2, are employed for brain tumor diagnosis.

Jude Hemanth and Anitha (2019) introduced modified genetic algorithm (GA) approaches to overcome the drawback of the conventional approaches. In this, appropriate modifications are made in the existing GA to minimize the random nature of conventional GA. The main focus of this work is to develop the modified reproduction operators which form the core part of this algorithm. Various binary operations are employed in this work to generate offspring in the crossover and mutation processes. Abnormal brain images from four different classes are used here. The experimental result shows that the proposed method produces 98% of accuracy in comparison to other methods.

Shafi et al. (2021) presented an ensemble learning method to classify the brain tumors or neoplasms and autoimmune disease lesion using magnetic resonance imaging (MRI) of brain tumors and multiple sclerosis patients. The proposed method consists of preprocessing, feature extraction, feature selection, and classification stages. The preprocessing stage uses region of interest (ROI) of both tumor and lesion, Collewet normalization, and Lloyd Max quantization. The base learner is formed using a support vector machine (SVM) classifier and prediction model with greater voting. The proposed system yields weighted sensitivity, specificity, precision, and accuracy of 97.5%, 98.838%, 98.011%, and 98.719%, respectively. The overall training and testing accuracies of this model are 97.957% and 97.744%, respectively. This can be used for detecting the presence of lesions coexisting with the tumors in neuro-medicine diagnosis. The proposed method operates well in contrast to other state-of-the-art methods.

Navid Ghassemi et al. (2020) proposed a new deep learning method for tumor classification in MR images. In this proposed method, the deep neural network is first pretrained as a discriminator in a generative adversarial network (GAN) on different datasets of MR images to extract the features and to learn the structure of MR images in its convolutional layers. Further, the fully connected layers are replaced and the whole deep neural network is trained as a classifier to classify the tumor classes. The proposed deep neural network classifier has about six layers and 1.7 million weight parameters. Pretraining as a discriminator of a GAN, together with other techniques, such as data augmentation techniques, like image rotation and mirroring and dropout, prevent the network from overtraining on a relatively small dataset. The proposed method is tested on contrast-enhanced MRI dataset consisting of 3,064 T1-weighted CE MR images from 233 patients, 13 images from each patient on average, with three different brain tumor types: 708 images of meningioma, 1,426 images of glioma, and 930 images of pituitary tumor. The overall design performance of the proposed method is evaluated using fivefold cross-validation and achieved greater accuracy as compared to the existing methods.

Mehrotra et al. (2020) proposed deep learning methods and transfer learning techniques to classify the two types of brain tumors, called malignant and benign. CNN models are used for transfer learning techniques to extract features that are visually distinguishable and essential. The proposed method is tested on the Cancer Imaging Archive (TCIA) public access repository dataset consisting of 696 images, of which 224 images are labeled as benign and 472 are malignant images on T1-weighted images. The preprocessing of the images is done, followed by data division and augmentation. The data obtained is compared with the trained data, then the results are verified and validated. By using deep learning methods, the brain tumor is differentiated as malignant and benign, with the classification accuracy of 99.04%.

Ratna Raju et al. (2018) proposed the harmony crow search optimization algorithm to train the classifier, which is automatic in nature. In this proposed method, the brain segments of the individuals are collected and are classified into four modalities, namely, T1, T2, T1C, and flair. Each modalities comprises huge number of slices generated as segments using Bayesian fuzzy method. The level of the brain tumor is determined by the nature of the brain segments generated by the Bayasian fuzzy clustering method. The features of the brain segments are collected by using the information theoretic measures, scattering transform, and wavelet transform. The proposed method achieved accuracy of 0.93.

Bhattacharjee et al. (2019) presented the five versions of hybrid PSO-GA algorithms for training the MLP and for classifying different sets of problems, including six benchmark datasets (three benchmark function approximation datasets—sigmoid, sphere, Rastrigin—and three benchmark classification datasets—three-bit XOR, Iris, breast cancer) and molecular brain neoplasia data. The proposed hybrid variants are implemented on three test problems, viz sigmoid, sphere, and Rastrigin function, reveal that the classification accuracy of GA and the four proposed hybrid variants is 100% for sigmoid function. The main objective of this research is to distinguish human glioma from neoplasia. This is achieved by multilayer perception technique along with particle swarm optimization training method and genetic algorithm. If the complexity of the datasets is increased, none of the datasets are able to achieve 100% classification accuracy. By considering all the five versions, it was observed that the best hybrid algorithm in each experiment gives 15–60% better MSE values and 5–30% better classification accuracy than GA. The excellence of the algorithms is also claimed for high-dimensional datasets, like that of molecular brain neoplasia data.

Tandel et al. (2020) proposed a system that uses transfer learning–based artificial intelligence paradigm using a convolutional neural network (CCN) which leads to improved performance in brain tumor classification using MRI images. The transfer learning–based CNN model is benchmarked against six different machine learning (ML) classification methods, namely, decision tree, linear discrimination, naive Bayes, support vector machine, K-nearest neighbor, and ensemble. The proposed CNN-based deep learning (DL) model outperforms well on the six types of ML models when considering five types of multiclass tumor datasets, called two-, three-, four-, five-, and six-class. The CNN-based AlexNet transfer learning system yielded mean accuracies derived from three kinds of cross-validation protocols (K2, K5, and K10) of 100%, 95.97%, 96.65%, 87.14%, and 93.74%, respectively. The mean areas under the curve of DL and ML were found to be 0.99% and 0.87%, respectively, for

$p < 0.0001$, and DL showed a 12.12% improvement over ML. Multiclass datasets were benchmarked against the TT protocol (where training and testing samples are the same). The optimal model was validated using a statistical method of a tumor separation index and verified on synthetic data consisting of eight classes. The transfer learning–based AI system can be applied in multiclass brain tumor grading, and it offers best performance when compared to ML systems.

Mesut et al. (2020) proposed a convolutional neural network (CNN) model that is combined with the hypercolumn technique, pretrained AlexNet and VGG-16 networks, recursive feature elimination (RFE), and support vector machine (SVM). The advantage of this proposed model is that it can keep the local discriminative features, which are extracted from the layers located at the different levels of the deep architectures, with the help of the hypercolumn technique. Eventually, this model achieved the generalization abilities of both AlexNet and VGG-16 networks by combining the deep features achieved from the last fully connected layers of the networks. In addition to this, the differentiative capacity of this model is increased using RFE; hence, the most prominent features are revealed. The proposed model achieved an accuracy of 96.77% without using any handcrafted feature engine. This model can be used for realizing the unbiased object evaluation in the clinics. The proposed method reduces human diagnostic errors and helps experts for decision-making in clinical diagnosis.

Shresta et al. (2020) proposed an advanced cascaded anisotropic convolutional neural network (CA-CNN) architecture with an optimized feature selection method for effective tumor segmentation and detection process which comprises of data acquisition, data preprocessing, feature extraction, selection, and prediction methods. This architecture increased the accuracy of prediction, and it used genetic algorithm for effective features selection; by doing this, it prevents redundancy of data and decreases the delay in tumor detection. The usage of genetic algorithm puts redundancy within input voxels and facilitates in the optimal selection of features, which improves the classification accuracy of the solution. The proposed method improved brain tumor segmentation and detection process in terms of accuracy, specificity, and sensitivity using multiscale prediction and cross-validation. Also, it minimized the time consumption.

Kiruthika Lakshmi et al. (2018) proposed the adaptive threshold algorithm, which helps in improving the efficiency and accuracy of the tumor segmentation process. The MRI brain images are given as input image. The color image is converted into gray image, and then the preprocessing and segmentation of the image are carried out to get rid of noise. Genetic algorithm is used for feature extraction process. Classification is based on locust-based genetic algorithm. The data obtained from the test and trained data can be compared using deep learning CNN classifier. Based on the features obtained, the CNN classifier compared the trained and test datasets, and at the end, the output image is generated, confirming whether brain tumor is present or not. MATLAB software is used for the identification, feature extraction, and classification of MR brain images. The proposed method is assessed towards parameters such as accuracy, sensitivity, and specificity over 20, 60, and 100 iterations. This method achieved accuracy of 41.8%, 41.7%,41.9%, respectively, in 20,60, and 100 iterations.

Kesav et al. (2021) proposed the two-channel CNN technique to classify glioma images from MRI samples. Region-based convolutional neural network (RCNN) is used to detect the tumor-affected region. Feature extraction layer of RCNN is replaced with the two-channel CNN, which has better performance in our experimental studies. The same two-channel CNN is used to extract features for detecting tumor regions of the glioma MRI sample, and then the tumor region is differentiated by bounding it with the boxes. Extensively, this technique is used for the detection of meningioma and pituitary tumor also. Comparing with the previous techniques, RCNN consumes low execution time. The RCNN technique detects and classifies the brain tumor with an accuracy of 98.21% and consumes less execution time at 64.5 ns. This technique is used to decrease the total number of parameters and execution time so that it could use the devices, which has decreased computational arrangement. Comparing with the existing technologies, the proposed RCNN method's performance is good.

Khairandish et al. (2021) presented the scenario for the classification of benign and malignant tumors from brain MRI images, which are done by considering the performance of CNN on a public dataset. Along with several methods, convolution neural network (CNN) can be used for extracting the features without using handcrafted models, and accuracy of classification is also higher. The hybrid model comprised CNN and support vector machine (SVM) for classification and threshold-based segmentation for detection. Accuracy of different models achieved are listed here: rough extreme learning machine (RELM), 94.233%; deep CNN (DCNN), 95%; deep neural network (DNN) and discrete wavelet autoencoder (DWA), 96%; k-nearest neighbors (kNN), 96.6%; CNN, 97.5%. The accuracy of the hybrid CNN-SVM is 98.4959%.

From the literature, it is observed that deep learning methods give promising results for image classification problems, in particular, for brain tumor classification from MR brain images. To improve the performance of deep learning algorithms, one of the methods identified is tuning of hyperparameters. In the proposed work, genetic algorithm is used to tune the parameters of CNN—number of filters and window size. The contributions of the work include (i) implementation of CNN for classification of brain tumors from MR images, (ii) optimization of CNN with hyperparameter tuning using genetic algorithm, and (iii) evaluation of performance metrics and comparison with state-of-the art methods.

4.3 MATERIALS AND METHODS

4.3.1 Convolutional Neural Networks

In deep learning, convolutional neural network (CNN) is a class of artificial neural network. The term *convolution* in CNN represents the mathematical function of convolution, which is a linear mathematical operation; here, two functions are multiplied to produce a third function, which indicates how the shape of the one function is modified by the other, while considering images, which are represented as matrices, when the matrices of two images are multiplied to produce an output which is used to extract the features from the image. Neural networks are used for classification, forecasting, image and speech recognition, textual character recognition, and the domains of human expertise, like medical diagnosis and financial market indicator prediction.

The layered architecture of CNN is as follows: The first layer of the CNN architecture is called the input layer. The convolutional layer extracts features from input

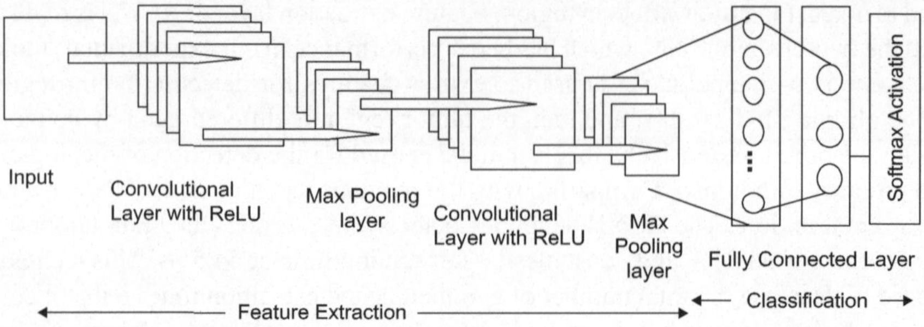

FIGURE 4.4 The architecture of convolutional neural network.

data. This layer contains a different number of filters for extracting features. The activation function used in this hidden layer is ReLU. The pooling layer is used to decrease the dimension of input for the next layer. This layer is usually placed after the convolutional layer. Dropout and fully connected layers are the middle layers to distinguish the patterns of features. The activation function used for the output layer is softmax. The classification layer is the final layer of the network. The CNN architecture is shown in Figure 4.4.

4.3.2 Genetic Algorithm

Genetic algorithm (GA) is the natural selection process in which the fittest individual is chosen for reproduction in order to produce offspring of the next generation. Mostly, GA is used to generate high-quality solutions for optimization and search problems. There are two main genetic operators, named crossover and mutation. The genetic operator crossover is also called recombination, which combines the genetic information of two parents and produces a new offspring. It is one way to generate new solutions from an existing population. Mutation has to be done after carrying out the crossover process. Mutation operator applies the changes randomly in one or more genes to produce a new offspring, by which it produces new adaptive solution. The flowchart of typical GA is shown in Figure 4.5. In this, first, the initial population is randomly created, and it can be of any size, from a few individuals to thousands. This evaluation measures how the individual is fulfilling the target problem. In the selection process, the main idea is to increase the chances for fitter individuals to be preserved over the next generation.

Genetic algorithm is based on searching the optimal solution through simulation of natural evolution method. The number of filters in the convolution layer and the filter size are selected as optimal values through genetic algorithm. A fitness function is used to select the best parameters, and its value is validation accuracy in percentage. The initial population is randomly generated, and the convolutional neural network is trained with the training images of brain images dataset. Based on the fitness values generated, the parents for the next step are selected. The entity of the population with the highest fitness value is selected as the parent. The process is repeated to obtain

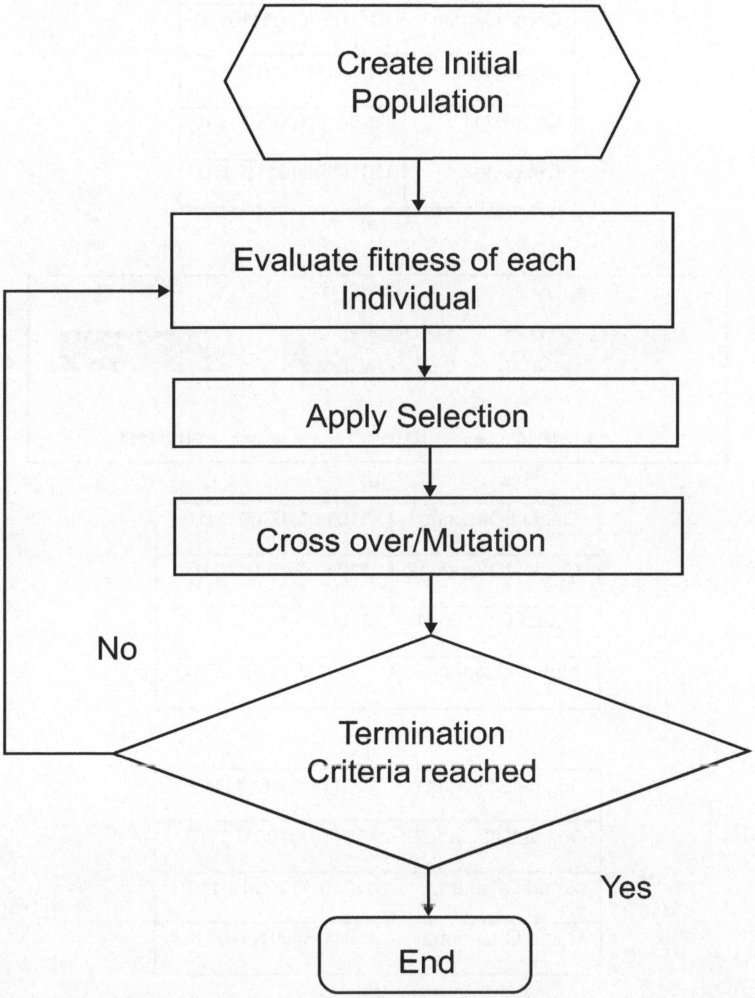

FIGURE 4.5 Flowchart of typical genetic algorithm.

the required number of parents. The subsequent steps are crossover and mutation, in which a new child is created by crossover from two parents. In the mutation step, new values are introduced by alteration with addition or subtraction of numbers. Thus, a new population is created, and the process is repeated for maximum fitness value. The hyperparameters of CNN—learning rate, type of activation function, and number of units in the dense layer—are also optimized.

4.3.3 Crossover and Mutation

In genetic algorithms and evolutionary computation, the two genetic operators, called crossover and mutation, are used to choose the fittest individual for reproduction in

Chromosome1	11001100101110110
Chromosome2	11010111000011010
Offspring1	11001 111000011010
Offspring2	11010100101110110

(a)

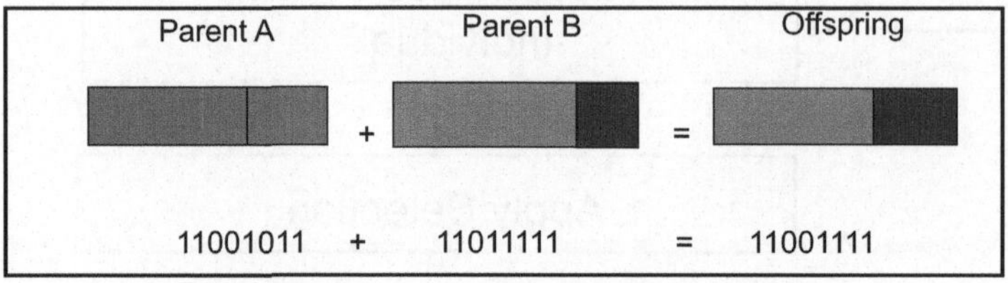

(b)

OriginalOffspring1	11011 11000011110
OriginalOffspring2	11011 00100110110
Muted Offspring1	11001 11000011110
Muted Offspring2	11011 01100110100

OriginalOffspring1	11011 11000011110
OriginalOffspring2	11011 00100110110
Muted Offspring1	11010 11000011110
Muted Offspring2	11011 00101110111

(c)

FIGURE 4.6 Examples of (a) crossover, (b) one-point crossover, and (c) mutation operator.

order to produce the offspring of the next generation. The genetic operator crossover or recombination is used to combine the genetic information of two parents to generate new offspring. The traditional genetic algorithms store the genetic information in a chromosome, which is represented by a bit array. Points on both the parents' chromosomes are picked randomly and designated a "crossover point." Bits to the right of that point are swapped between the two parent chromosomes. These produce two offspring, each carrying some genetic information from both the parents. Another genetic operator, referred to as mutation, is described as a small random twist within the chromosome to get a new solution. The mutation of bit strings is guaranteed through the bit flips at random positions. It is used to maintain and introduce diversity in the genetic population and is usually applied with a low probability. If the probability is very high, the genetic algorithm gets reduced to a random search. Figure 4.6 shows the examples of crossover and mutation processes.

4.4 PROPOSED METHOD

4.4.1 Convolutional Neural Network Architecture

The CNN architecture consists of convolutional, max pooling, flatten, dropout, and dense layers. In this proposed method, population is generated for a maximum of 128 number of filters with a window size of 21. The architecture of convolutional neural network is shown in Figure 4.7. The input layer of the architecture is fed with the input RGB images of size 224×224×3. The network consists of nine weight layers comprising of three convolutional layers, three max pooling layers, one flatten layer, one dropout layer, and one fully connected layer. The convolutional layers and pooling layers are used for feature extraction. The convolutional layer with rectified linear unit (ReLU) activation function is used, and the softmax activation function is used for fully connected layer for classification.

4.4.2 Flowchart of the Proposed Method

The proposed method is shown in Figure 4.8. The input MRI brain images from the dataset are normalized using min–max normalization process. Normalization provides value between 0 and 1 to each pixel in the image value. It provides a linear transformation of the original input data, and it conserves the relationships between

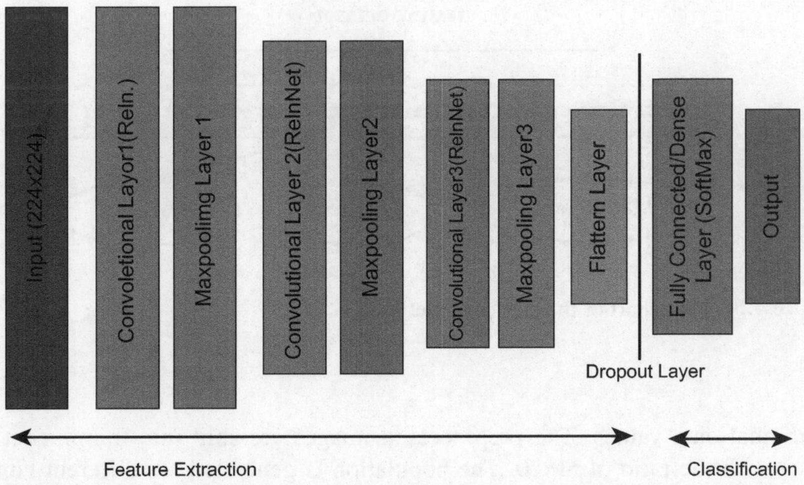

FIGURE 4.7 Convolutional neural network architecture.

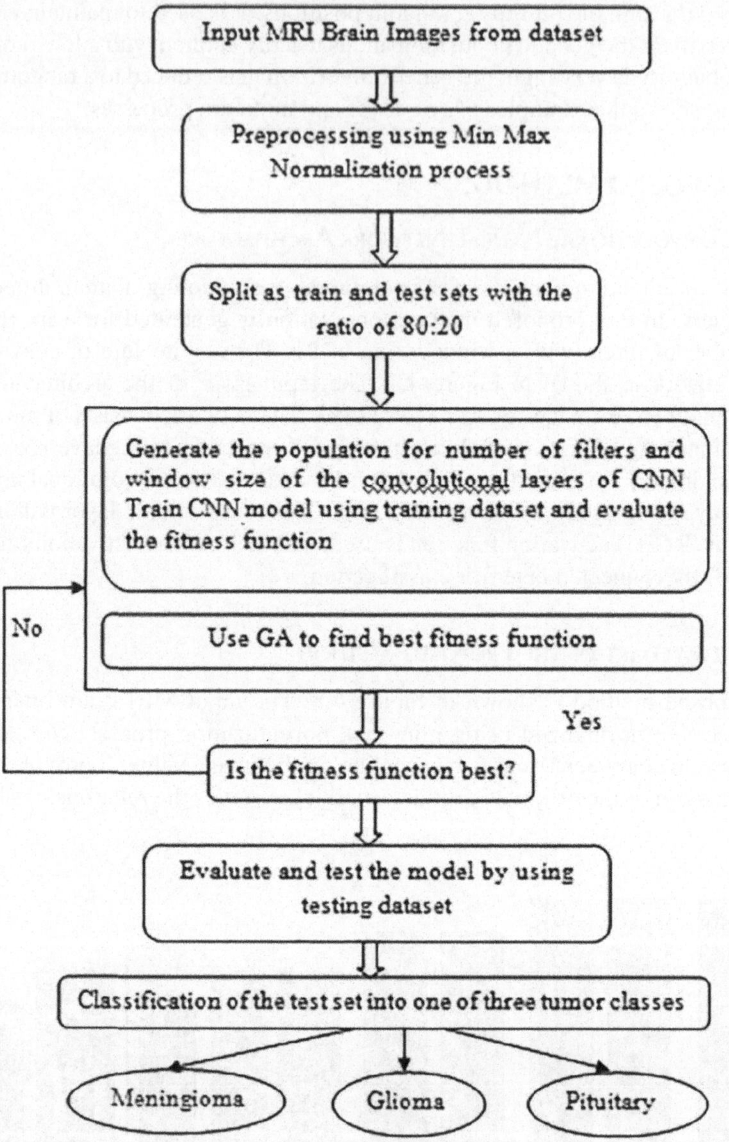

FIGURE 4.8 Flowchart of the proposed method.

the original data values. The preprocessed images are split into training set and test set with the ratio of 80:20. The population is generated for different number of filters and window size of the convolutional layers. CNN model is trained using training dataset, and it will evaluate the fitness function. Genetic algorithm is used to find the best fitness function. If the fitness function is best, then the model is tested and evaluated using testing dataset; otherwise, till the best fitness function is met, the training process is to be carried out. Among these fitness functions, the best five fitness functions are chosen as parents. Further, the crossover and mutation processes are carried out, in which the genetic information of two parents are combined and new offspring are produced to generate optimized solutions from an existing population. Finally, the test set is classified into one of three tumor classes, called meningioma, glioma, and pituitary. The model is compiled with Adam optimizer with the learning rate of 0.001. The batch size is 32, and the number of epochs is 20. The proposed method is evaluated on T1-weighted contrast-enhanced magnetic resonance brain tumor images (CE-MRI) benchmark dataset containing 3,064 T1-weighted contrast-enhanced images from 233 patients with three kinds of brain tumor: meningioma (708 slices), glioma (1,426 slices), and pituitary tumor (930 slices).

4.5 EXPERIMENTS AND RESULTS

4.5.1 Dataset

The brain tumor dataset called contrast-enhanced magnetic resonance images (CE-MRI) benchmark dataset (Swatia et al., 2019) is used to evaluate the proposed model. This methodology is tested with 3,064 T1-weighted contrast-enhanced MR images from 233 patients with three kind of brain tumor, consisting of 708 slices of meningioma, 1,426 slices of glioma, and 930 slices of pituitary brain tumors. The resolution of the image is 512×512 pixels; further, it can be reduced to 64 × 64 pixels to decrease the computational cost of the model.

4.5.2 Experiments and Results

The proposed model for the classification of brain tumor images has been developed in Python due to the availability of the most common machine learning and neural network libraries. For this work, Tensorflow and Keras libraries have been used, and it can be run on both the central processing unit (CPU) and graphical processing unit (GPU). The cloud service called Kaggle that offers Jupyter Notebook environment is used for implementing the proposed classification model. The method is implemented in Kaggle GPU environment.

4.5.2.1 Result

Genetic algorithm is based on searching the optimal solution through the simulation of natural selection method. The total number of filters in the convolution layer and the window size of the filter are selected as optimal values through genetic algorithm. A fitness function is used to select the best parameters, and it assesses the validation accuracy in percentage. The initial population is randomly generated for the 128 number of filters and the window size of 21. Convolutional neural network is trained with the training images of contrast-enhanced MRI brain images dataset. Based on the fitness values generated, the parents for the next step are selected. The unit of the population with the greatest fitness value is chosen as the parent. Table 4.1 shows the initial population generated

Here, the unit of the population with the greatest five fitness values is chosen as the parent. Table 4.2 gives the information about the five best fitness function chosen as parents.

TABLE 4.1
Initial population Generated

Population	Parameters	Accuracy
Population 1	[45, 48, 118, 12, 19, 3]	0.9575163125991821
Population 2	[65, 68, 124, 1, 1, 5]	0.906862735748291
Population 3	[68, 104, 10, 6, 7, 9]	0.9313725233078003
Population 4	[84, 22, 115, 18, 16, 5]	0.9395424723625183
Population 5	[37, 88, 71, 10, 11, 2]	0.9346405267715454
Population 6	[89, 89, 13, 2, 8, 10]	0.9281045794487
Population 7	[59, 66, 103, 4, 7, 12]	0.9607843160629272
Population 8	[40, 88, 47, 15, 19, 1]	0.936274528503418
Population 9	[89, 82, 38, 15, 4, 13]	0.9607843160629272
Population 10	[26, 78, 73, 11, 12, 5],	0.9379084706306458
Population 11	[10, 21, 116, 7, 5, 16]	0.9444444179534912
Population 12	[81, 116, 70, 4, 13, 5]	0.9297385811805725
Population 13	[127, 80, 48, 9, 15, 16]	0.9509803652763367
Population 14	[65, 83, 100, 4, 16, 14]	0.9460784196853638,
Population 15	[89, 50, 116, 17, 18, 6]	0.4656862616539015
Population 16	[30, 20, 20, 10, 4, 1]	0.9264705777168274
Population 17	[15, 40, 33, 6, 1, 18]	0.9133986830711365
Population 18	[66, 10, 58, 19, 5, 3]	0.9248365759849548
Population 19	[33, 32, 75, 17, 4, 3]	0.9297385811805725
Population 20	[117, 24, 36, 11, 14, 17]	0.9084967374801636

TABLE 4.2
Five Best Fitness Function Chosen as Parents

Parent	Parameters
Parent 1	[59.,66., 103.,4.,7.,12.]
Parent 2	[89.,82.,38.,15.,4.,13.]
Parent 3	[45.,48., 118.,12.,19.,3.]
Parent 4	[127.,80.,48.,9.,15.,16.]
Parent 5	[65.,83., 100.,4.,16.,14.]

TABLE 4.3
Generating Child by Crossover

Child	Parameters
Child 0	[59., 66., 38., 4., 7., 13.]
Child 1	[89., 82., 118., 15., 4., 3.]
Child 2	[45., 48., 48., 12., 19., 16.]
Child 3	[127., 80., 100., 9., 15., 14.]
Child 4	[65., 83., 103., 4.,16., 12.]

TABLE 4.4
Generating Child after Mutation

Child	Parameters	Validation Accuracy
Child 0	**[59 71 38 4 9 13]**	**0.9444**
Child 1	[89 87 118 15 4 5]	0.9346
Child 2	[45 49 48 12 17 16]	0.9395
Child 3	[125 80 100 9 18 14]	0.8284
Child 4	[65 87 103 4 16 14]	0.9281

Subsequently, new children are created by crossover from two parents. Table 4.3 gives the child information after the crossover.

In the mutation step, new values are introduced by alteration with addition or subtraction of numbers. Table 4.4 gives the child information after the mutation process.

The plot of validation accuracy and validation loss of child 0, child1, child 2, child 3, and child 4 is shown in Figure 4.9.

From the previous results, it is inferred that child 0 has the highest accuracy of 0.9444. The proposed method is evaluated using the performance metric called classification accuracy, which is given in the following:

$$\text{Classification Accuracy}=\frac{\text{Number of correct predictions}}{\text{Total Number of predictions}}$$

The proposed method is compared with the existing method and shows improved classification accuracy of 0.9444 over state-of-the-art method. Table 4.5 shows the comparison results of the proposed method with the existing (Cheng et al. 2015) method.

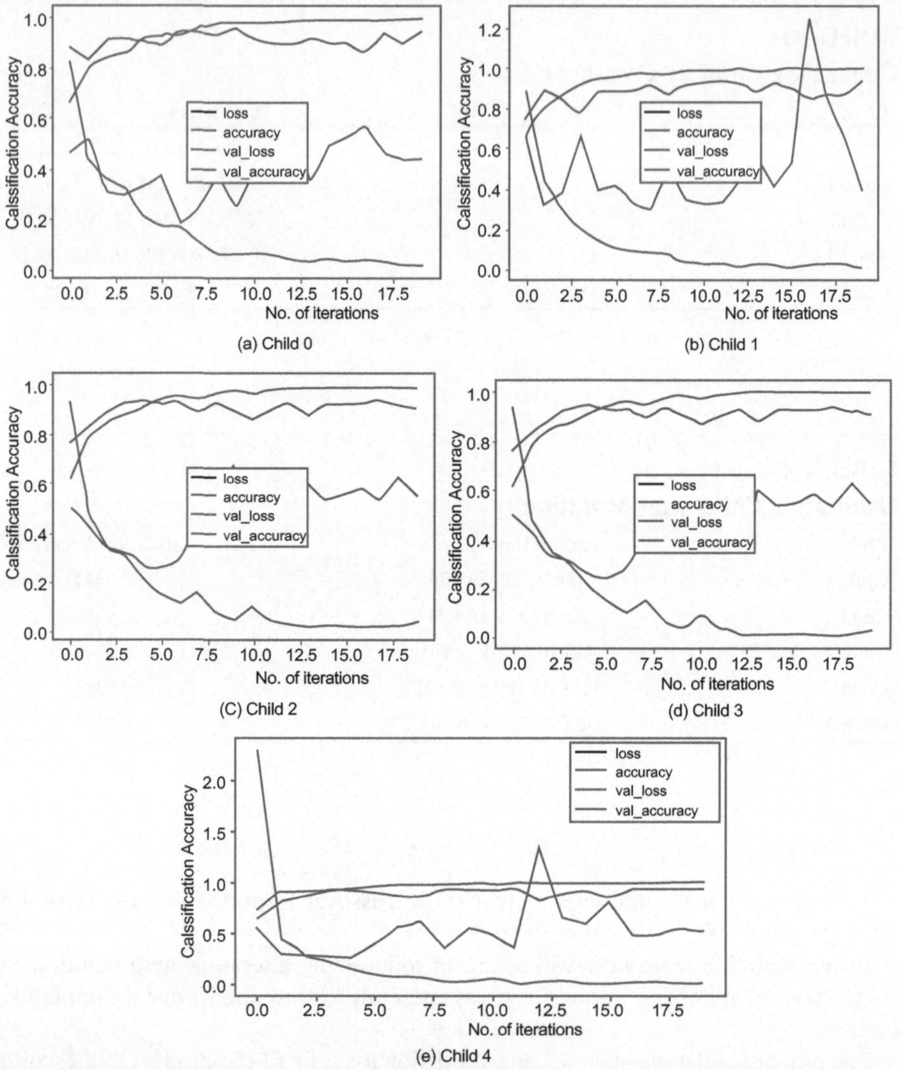

FIGURE 4.9 The plot of validation accuracy and validation loss of child 0, child 1, child 2, child 3, and child 4.

TABLE 4.5
Comparison of Proposed Method with the Existing Method

	Classification accuracy
Existing method (Cheng et al. 2015)	0.9128
Proposed method	**0.9444**

4.6 CONCLUSION

In this chapter, optimized convolutional neural network for classification of MR brain images is proposed. The CNN consists of convolutional layers, max pooling layers, flatten layers, dense layers, and dropout layer, and softmax activation ^s added for classification of meningiomas, gliomas, and pituitary tumors. The convolution layers and max pooling layers are used for feature extraction. The dense layer with softmax activation is used for classification. In the proposed method, the CNN architecture is optimized with genetic algorithm to improve performance. From the experimental results, it is inferred that the proposed method gives the highest classification accuracy of 0.9444 compared to the other models.

REFERENCES

Bhattacharjee, K., & Pant, M. (2019). Hybrid particle swarm optimization-genetic algorithm trained multi-layer perceptron for classification of human glioma from molecular brain neoplasia data. *Cognitive Systems Research*, *58*, 173–194. https://doi.org/10.1016/j.cogsys.2019.06.003

Cheng, J., Huang, W., Cao, S., Yang, R., Yang, W., Yun, Z., Wang, Z., & Feng, Q. (2015). Enhanced performance of brain tumor classification via tumor region augmentation and partition. *PLoS ONE*, *10*(10), e0140381. https://doi.org/10.1371/journal.pone.0140381

Ghassemi, N., Shoeibi, A., & Rouhani, M. (2020). Deep neural network with generative adversarial networks pre-training for brain tumor classification based on MR images. *Biomedical Signal Processing and Control*, *57*, 101678. https://doi.org/10.1016/j.bspc.2019.101678

Harish, P., & Baskar, S. (2020). MRI based detection and classification of brain tumor using enhanced faster R-CNN and Alex net model. *Materials Today: Proceedings*. https://doi.org/10.1016/j.matpr.2020.11.495

Jude Hemanth, D., & Anitha, J. (2019). Modified genetic algorithm approaches for classification of abnormal magnetic resonance brain tumor images. *Applied Soft Computing Journal*, *75*, 21–28. https://doi.org/10.1016/j.asoc.2018.10.054

KabirAnaraki, A., Ayati, M., & Kazemi, F. (2019). Magnetic resonance imaging-based brain tumor grades classification and grading via convolutional neural networks and genetic algorithms. *Biocybernetics and Biomedical Engineering*, *39*(1), 63–74. https://doi.org/10.1016/j.bbe.2018.10.004

Kesav, N., & Jibukumar, M. G. (2021). Efficient and low complex architecture for detection and classification of brain tumor using RCNN with two channel CNN. *Journal of King Saud University—Computer and Information Sciences*, *34*(8), 6229–6242. https://doi.org/10.1016/j.jksuci.2021.05.008

Khairandish, M. O., et al. (2021). A hybrid CNN-SVM threshold segmentation approach for tumor detection and classification of MRI brain images. *IRBM*, *43*(4), 290–299. https://doi.org/10.1016/j.irbm.2021.06.003

Kiruthika Lakshmi V., Amarsingh Feroz, C., & Asha Jenia Merlin, J. (2018). Automated detection and segmentation of brain tumor using genetic algorithm. *International Conference on Smart Systems and Inventive Technology (ICSSIT 2018) IEEE Xplore* (Part Number: CFP18P17-ART).

Mehrotra, R., Ansari, M. A., Agrawal, R., & Anand, R. S. (2020). A transfer learning approach for AI-based classification of brain tumors. *Machine Learning with Applications*, *2*, 100003. https://doi.org/10.1016/j.mlwa.2020.100003

Nazir, M., ShaKil, S., & Khurshid, K. (2021). Role of deep learning in brain tumor detection and classification (2015 to 2020): A review. *Computerized Medical Imaging and Graphics, 91*, 101940. https://doi.org/10.1016/j.compmedimag.2021.101940

Raju, A. R., Suresh, P., & Rao, R. R. (2018). Bayesian HCS-based multi-SVNN: A classification approach for Brain Tumor segmentation and classification using Bayesian fuzzy clustering. *Biocybernetics and Biomedical Engineering*, *38*(3), 646–660. https://doi.org/10.1016/j.bbe.2018.05.001

Sajjad, M., Khan, S., Muhammad, K., Wu, W., UllAh, A., & Baik, S. W. (2018). Multi-grade brain tumor classification using deep CNN with extensive data augmentation. *Journal of Computational Science*, *30*, 174–182. https://doi.org/10.1016/j.jocs.2018.12.003.

Shafi, A. S. M., Rahman, M. B., Anwar, T., Halder, R. S., & Kays, H. M. E. (2021). Classification of brain tumors and auto-immune disease using ensemble learning. *Informatics in Medicine Unlocked*, *24*, 100608. https://doi.org/10.1016/j.imu.2021.100608

Shresta, S., AroshaSenanayake, S. M. N., & Triloka, J. (2020). Advanced cascaded anisotropic convolutional neural network architecture based optimized feature selection brain tumour segmentation and classification. *2020 5thInternational Conference on Innovative Technologies in Intelligent Systems and Industrial Applications (CITISIA)*. https://doi.org/10.1109/CITISIA50690.2020.9371807

Swatia, Z. N. K., Zhao, Q., Kabir, M., Ali, F., Ali, Z., Ahmed, S., & Lu, J. (2019). Brain tumor classification for MR images using transfer learning and fine-tuning. *Computerized Medical Imaging and Graphics*, *75*, 34–46. https://doi.org/10.1016/j.compmedimag.2019.05.001

Tandel, G. S., Balestrieri, A., Jujaray, T., Khanna, N. N., Saba, L., & Suri, J. S. (2020). Multiclass magnetic resonance imaging brain tumor classification using artificial intelligence paradigm. *Computers in Biology and Medicine*, *122*, 103804. https://doi.org/10.1016/j.compbiomed.2020.103804.

Tandel, G. S., Tiwari, A., & Kakde, O. G. (2021). Performance optimisation of deep learning models using majority voting algorithm for brain tumour classification. *Computers in Biology and Medicine*, *135*, 104564. https://doi.org/10.1016/j.compbiomed.2021.104564

Toğaçar, M., Cömert, Z., & Ergen, B. (2020). Classification of brain MRI using hyper column technique with convolutional neural network and feature selection method. *Expert Systems with Applications*, *149*, 113274. https://doi.org/10.1016/j.eswa.2020.113274

Yin, B., Wang, C., & Abza, F. (2020). New brain tumor classification method based on an improved version of whale optimization algorithm. *Biomedical Signal Processing and Control*, *56*, 101728. https://doi.org/10.1016/j.bspc.2019.101728

5 Predictive Modeling of Epidemic Diseases Based on Vector-Borne Diseases Using Artificial Intelligence Techniques

Inderpreet Kaur, Yogesh Kumar, Amanpreet Kaur Sandhu, and Muhammad Fazal Ijaz

CONTENTS

5.1 INTRODUCTION

In the late nineteenth century, Theobald Smith proved that a protozoan parasite caused Texas cattle illness and spread it to animals via ticks. It was the first proof of arthropod

DOI: 10.1201/9781003309451-5

vector-borne transmission [1]. Scientists discovered that vectors might spread human disease agents like filariasis and malaria within a few years. Vector-borne illnesses continue to threaten human health. Vector-borne diseases (VBDs) have recently gained global and regional relevance [2]. The fluctuating vector, pathogen, and host population abundances throughout time and space influence the ages of VBD illness incidence. The vector's activity and life history are essential factors of VBD dynamics, affecting pathogen transmission rates between vector and host [3]. Controlling and preventing such illnesses needs complete knowledge of the complex interactions between pathogens, carriers, and their surroundings. Vector-borne diseases are a subgroup of hazardous arthropod-borne diseases. Vector-borne diseases such as parasitic, viral, and bacterial infections are rising globally [4]. The harmful bacteria interphase with mortal hosts (human hosts) and need continuing connection with an invertebrate vector. Several variables affect the kinetics of vector-borne illnesses. Vector competency is governed by fundamental factors, such as vector immunization, feeding habits, and the microbiome [3]. Meanwhile, intrinsic microbial eugenics influences infections' ability to accept and inhabit arthropod vectors. Finally, extrinsic and environmental factors may alter vectorial capacity, which evaluates vectorial efficacy in nature. These include heat, sickness reservoirs, vector ecology, life span, and biting frequency.

Vector infection is unpredictable and may lead to new illnesses. The variable count modifies the vector. An arthropod vector infecting a vertebrate is vector-borne [5–7]. However, a delicate balance exists between the vector, parasite, and host. While the disease's features and care strategies may change, the interactions between the three components may not. The condition tends to change with time. As demonstrated in Figure 5.1, travel and trade create environmental changes, such as climate change, habitat loss, and population movement.

Around 80% of the world's population lives in high-risk areas, with an annual death toll of 700,000. The WHO prioritizes entomology research in reducing vector-borne illness occurrence and death [8]. Malaria, the most common VBD, is a major cause of sickness and death. NTDs (neglected tropical illnesses) are many VBDs [9]. The impact of neglected tropical diseases is underestimated, and funds and priority have been given to other diseases. Based on the yearly DALY (disability-adjusted life year) burden, malaria cases were 4,64,37,811 in 2019 [10]. While vector-borne NTDs cause fewer deaths than malaria, they generate significant morbidity and are a major public health concern [11]. Moreover, several zoonotic NTDs strain veterinary health [12]. Climate change seems to be linked to the growth of vector-borne diseases. Clearly, climate change is affecting all of nature [13]. In the twenty-first century, climate change is crucial. According to previous study, global average temperatures would rise by 1.0–3.5°C by 2100 [14], increasing the risk of vector-borne diseases. Climate and weather changes directly affect vectors and disease transmission patterns. They are concerned about the long-term effects of human activities, like dam construction and irrigation systems, on food and energy security [15].

Increased water supply and movement of people may cause schistosomiasis to spread to nonendemic locations [16]. Thus, continual awareness of infectious diseases and improvements in control efforts are essential to ensure public safety [17]. Thus, this chapter will examine new and re-emerging essential vector-borne infectious diseases as well as management challenges.

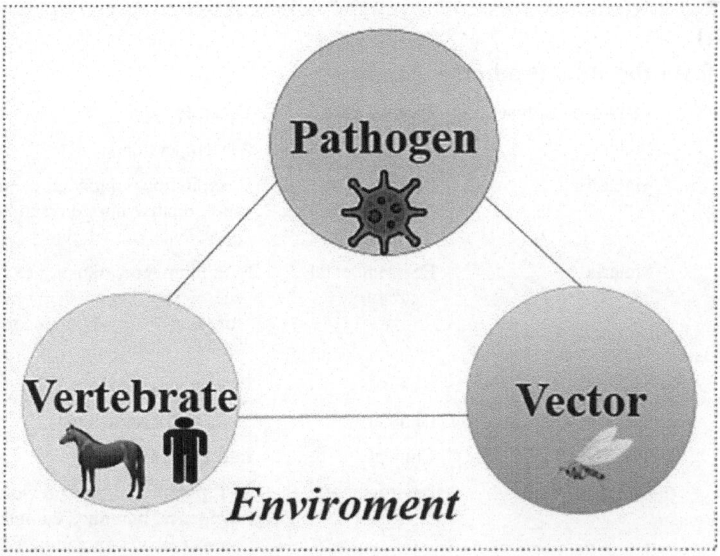

FIGURE 5.1 Vector–pathogen–vertebrate relationship.

5.2 DATA SOURCES

This section comprises a variety of data sources that may be used to extract VBD datasets. Table 5.1 displays the kinds of data utilized in some of the VBD datasets examined in this study.

5.3 FACTORS THAT AFFECT THE EMERGENCE AND RECURRENCE OF VECTOR-BORNE DISEASES

Infectious illnesses emerge and re-emerge throughout time. Before causing an epidemic, viruses undergo many adaption phases before contact with a new host [18]. Interactions between pathogens, carriers, and the environment significantly affect the development of infectious illnesses. Diverse environmental, ecological, and demographic factors may additionally perform a part in this adaptation plus the eventual emergence of illness. These factors have an impact by creating ideal conditions for contact with an unidentified pathogen or its natural host and encouraging spread. When these elements are joined with the constant evolution of pathogenic and microbial alterations and the problem of treatment endurance, an infection may occur frequently. In addition, various factors, both climate-related and non-climate-related, may facilitate the spread and settling of emerging or resurgent viruses [19].

5.3.1 Climate Change and Its Implications

Climatic change refers to long-term shifts in climate patterns and extremes. Another variable increases the likelihood of vector-borne infections [20]. It is possible that

TABLE 5.1
Dataset from the VBD Prediction Study

Reference	VBD type dataset	Type of data	Variable type
[48]	Malaria	Climate	Air temperatures
[49]	Malaria	Clinical data, Climate	Precipitation, temperature, air pressure, relative humidity, and the prevalence of malaria
[50]	Malaria	Environmental variables	Maximum temperature, precipitation seasonality, land cover, temperature seasonality, human population density, altitude
[81]	Malaria	Clinical	Parasitized, uninfected
[82]	Malaria	Clinical	Parasitized, uninfected
[83]	Malaria	Clinical	Infected, uninfected
[59]	Dengue	Environmental	SST, population size, previous dengue incidence, humidity, cumulative one-week precipitation, minimum air temperature, and date
[84]	Yellow fever	Clinical	Fever, body pain, headache, vomiting, jaundice, bleeding, and organ failure
[66]	West Nile fever	Historical weather	Precipitation, humidity, wind speed, max, min, average temperatures, and barometric pressure
[85]	West Nile fever	Climatic, environmental	Temperature, vegetation index, water index, precipitation, land use, year, area
[73]	Zika	Climate	Minimum and maximum yearly temperatures, as well as annual cumulative precipitation
[86]	Zika	Clinical	Infected, not infected
[72]	Zika	Demographic, clinical	Age, gender, fever, joint pain, conjunctivitis, rash
[87]	Lyme disease	Anatomical, biological,	Larvae, nymph, and adult, life history metrics, and biomes
[88]	Lyme disease	Clinical	"Better," "worse," or "unchanged"

changes in disease and vector interaction patterns directly or indirectly affect pathogen survival, reproduction, and life duration. Life and development need a global temperature range. Temperature impacts vector growth and survival [21]. The temperature ranges of 21–23°C is for mosquito growth and 24–26°C for JEV transmission [22]. Heat promotes the spread of insect-borne diseases [23]. Temperature changes directly affect habitats, and hence ecosystems, whereas human incursion into natural environments harms biodiversity and provides a risk to disease outbreak [24]. While vector-borne illnesses are innately sensitive to changes in climate and weather [25], the debate is on how much weather patterns and weather impact infections and associated vectors. Human control of some public health vector-borne diseases is ineffective [26]. Others cited land use changes as the main culprit, citing business as responsible for 11% and worldwide trade for 14% of vector-borne zoonotic infections. However, environmental and weather-related variables account for 10% of the etiology [27].

5.3.2 Nonclimatic Factors

Aside from the weather, several factors have been linked to the onset and re-emergence of vector-borne diseases. Nonclimatic influences include worldwide civilized populations, urbanization, global travel and commerce, rapid agricultural expansion and technology, population increase, and antimicrobial medicine intake [28, 29]. In addition, complex human and environmental factors have been related to the emergence of uncommon illnesses. For example, the host and environment may be harmed by chemical and antimicrobial exposures, causing gene damage [30], and the emergence of drug-resistant pathogen variants capable of causing new, dangerous agents may also proliferate and adapt to new environments via mechanisms including hereditary deviation [31].

5.3.3 Urbanization

Rural-to-urban migration is linked to vector breeding in peri-urban slums. As a consequence of inadequate housing and absence of essential services like water and waste disposal, vector populations thrive [32]. According to recent research, an emerging disease or cross-species dissemination may be facilitated by urbanization [33]. Urbanization demonstrates the ecological intricacy of wildlife–livestock–human relationships [34].

5.3.4 Globalization

Changing human behavior and activities, pathogen development, poverty, changing ecosystems, and changing human–animal connections have all been linked to the rise and spread of deadly diseases [31]. The emergence and comeback of infectious illnesses seem to be facilitated by social architecture and pathogenic mechanisms. Globalization is being blamed for changing VBD disease ecology. Many VBDs are getting increasingly hazardous globally [35] due to urbanization and globalization spreading vectors and microorganisms [36].

5.3.5 Agriculture and Development Programs

Agricultural operations and development initiatives have contributed to the formation of new illnesses or the re-emergence of old ones due to human-induced ecological shifts. Numerous studies have shown a substantial correlation between infectious illnesses and human-induced land use changes in agricultural practices [37–39]. For example, agricultural and land use activities that intrude on the natural environment and wildlife expose humans and their pets to a broader spectrum of vectors.

Furthermore, as the amount of irrigated farmland increased, so did the number of outbreaks of vector-borne diseases. Agricultural variables are linked to more than 25% of infectious diseases emerging in humans and more than 50% of zoonotic infectious infections starting in people, according to study [40]. In addition, dam development, irrigation, and other comparable improvement schemes have been shown to impact vector population densities, affecting the emergence of new illnesses and the revival of old ones [41].

5.4 PROMINENT VECTOR-BORNE DISEASES THAT ARE EMERGING OR RE-EMERGING

Not all disease vectors are the same [42]. Even more dangerous than grizzly bears and lions are hippos. Mosquitoes attack mice, rats, rabbits, and frogs. Cross-contamination has caused new ailments in recent decades. Plasmodium and Zika virus are two current and ancient mosquito-borne illnesses. Therefore, infectious infections have become a major worldwide public health issue [43]. This section will cover some of the most common vector-borne diseases. AI may predict vector-borne diseases. For example, computational modeling learning systems have shown promise in vector-borne diseases [89, 90]. Machine and deep learning can categorize VBDs (Table 5.3). The Plasmodium parasite causes malaria. Malaria parasites are spread by infected Anopheles mosquitoes, often known as "malaria vectors." Various parasites cause malaria (five in all). The 2018 Global Health Malaria Report estimates 219 million malaria infections and 435,000 deaths in 2017. In 2017, 17 African nations accounted for nearly 80% of worldwide malaria mortality. Terrain, precipitation, heat, and the population's socioeconomic condition have all been shown to influence malaria transmission [44]. Malaria is a highly contagious febrile sickness. In non-immune adults, symptoms generally appear 10–15 days after the infected mosquito bite. Early malaria symptoms, including headache, nausea, and chills, might be subtle. Adults are prone to multiorgan failure. Residents in malaria-endemic areas may acquire partial immunity, causing asymptomatic illness [45]. With images of Giemsa-stained thin blood film specimens [51], Tek et al. detected several malaria parasites. The authors evaluated the KNN and the back-propagation neural network for identifying genera and life stages (BPNN). The second and third models looked at binary identification, followed by one (16-class) or two (4-class) classifications for identification. On the stated attributes, KNN outperformed BPNN and FLD by 93.3 percentile (20-class KNN classification). From January 2009 to December 2015, Flaviviridae-infected mosquitoes spread dengue (genus *Flavivirus*). The disease gained popularity in the eighteenth and nineteenth centuries due to worldwide maritime trade and port construction [52].

It is possible to contract DENV four times in one's lifetime. Dengue fever occurs in over 100 countries, and 40% of the population lives in dengue-endemic regions [53]. From 2007 to 2017, the DALY rate for dengue fever rose by 26% [43]. Since the 1950s, urbanization, globalization, and poor vector management have increased viral infection and transmission [54]. Yellow fever is carried by *Aedes* and *Haemagogus* mosquitos and is caused by an arbovirus flavivirus. Mosquitoes may be found in various habitats, including dwellings, woods, and even both (semidomestic). *Yellow fever* is a disease that develops in the sylvatic (or jungle) environment. First, wild

Aedes mosquitoes feed on monkeys and transmit yellow fever to other monkeys in tropical jungles. Second, yellow fever is transmitted by mosquitoes that bite or fly through the forest. In Africa, an intermediary (savannah) loop includes mosquito-to-human viral transmission. During this time, mosquitoes may spread the virus from human to human or monkey to human. The urban cycle is characterized by virus transmission between people and *Aedes aegypti* mosquitoes [61]. The Zika virus was found in Uganda in 1947 and has since spread globally. The Zika arbovirus infects Aedes mosquitos, which spreads the virus. The Zika virus has been known to infect humans for over 50 years. In 2007, the Zika virus caused an epidemic outside Africa and Asia. Even when a human infection is present, symptoms and clinical features similar to dengue virus may vary from mild to self-limiting acute fever sickness [105]. People became sick after being bitten by infected black-legged ticks. Symptoms include fever, cough, vomiting, erythema migraines, and a characteristic skin rash. The infection may spread to the nerves, heart, and even brain if left untreated. Antibiotics may cure Lyme disease for a few weeks. Prevent Lyme fever by applying insect repellent, removing ticks quickly, spraying pesticides, and reducing tick habitat [104]. Research analysis of vector-borne diseases is listed in Table 5.2.

5.5 CONCEPT EXAMPLE OF AI APPLICATIONS IN DISEASE EPIDEMIOLOGY

The public's response to illness outbreaks may be unpredictable. Public health professionals employed BDA (big data analytics) and groups of AI (artificial intelligence)–based apps to correlate population behavior during an epidemic [91]. AI may be employed in real-world circumstances, like illness epidemics, as shown by the tweets below from Twitter. Many individuals have been capable of sending and reading each other's "tweets" on Twitter; Twitter is a free microblogging service (short, 140-character messages). To monitor and forecast illness epidemics, researchers think that Twitter users' reactions might be utilized as a source of information from which to conclude.

5.5.1 The Zika Virus Epidemic

The majority of people on Earth live in mosquito-infested environments. Zika was the most dangerous epidemic in 2015 and 2016, necessitating worldwide public health interventions. Population health approaches are critical in these instances [92]. They detected and characterized self-disclosures of behavior change during disease transmission using data from Twitter from 2015 and 2016. They utilized keywords screening and machine learning algorithms to discover first-person answers to Zika. Keywords: travel, trip arrangements, and cancellations. Participants' demographics, networking habits, and linguistic preferences were compared to controls' attributes, social network factors, and language patterns were found in 1,567 Twitter users. They altered or considered altering their trip plans due to Zika. There were significant disparities across geographic locations in the United States, with women discussing Zika more than males and some variances in exposure to Zika-related material. This study suggests that using AI ideas might help researchers better understand how the general population perceives and responds to the hazards of infectious disease outbreaks.

TABLE 5.2
Research Analysis of VBD Diseases

References	Illness type	Datasets	Techniques used	Challenges	Findings
[46]	Malaria	CFSR, CHIRPS	Logistic regression, chi-square test	Failure to consider the ramifications of malaria resistance circumstances of pesticide resistance.	No one measure reduced the likelihood of a malaria resurgence in the same location.
[47]		MATLAB 2015 tool	Optimized genetic algorithm, K-nearest neighbor, principal component analysis, independent component analysis	The applications of RNA-seq necessitated effective dimension reduction and classification techniques.	Genetic algorithm with independent component analysis and KNN outperformed with 90% accuracy than genetic algorithm with principal 4 component analysis and KNN.
[48]		Open European Union Data Portal	Artificial neural networks (ANNs), RNN (recurrent neural network)	Limited Dataset, no standardized way for building network architecture.	NARX model predicted slightly better (R = 0.623) than the FTD model (R = 0.534).
[49]		WHO, NCAR	Extreme gradient boosting (XGBoost), K-means clustering	Scarcity of daily updates of malaria incidence; only the year data was taken into consideration.	XGBoost outperformed as compared with other machine learning models.
[50]		WorldClim, SRTM, ORNL, GBIF	K-means clustering	Inadequate prevalence data from communities.	Unsupervised land cover classification outperformed.
[55]	Dengue	Hospitals (5,000 instances)	SVM (support vector machine)	High computational time.	Quality measure of SVM (support vector machine) sensitivity, 0.4723; specificity, 0.9759; accuracy, 0.9042.
[56]		Hospital data	ANN (artificial neural network)	Limited dataset.	ANN (artificial neural network) has achieved accuracy of 96.27%.

[57]		Real-time dataset	ALCD—Aedes larvae classification and detection system ANN	The predictive accuracy of the ALCD method is not as good as the laboratory test performance.	ALCD model outperformed highest accuracy match based on the larvae sort.
[58]		Real-time dataset	Computer simulation, ANN	Experimental tests and simulation evaluations are expensive.	Performance measured in terms of ROC curve (receiver operating characteristic curve).
[59]		CDC (USA)	ANN models, multiple linear regression models (MLR)	ANN models are not accurately predicting dengue fever outbreaks.	ANN models had a higher statistical power compared to other MLR models.
[60]		Hospitals (654 instances)	NB (naive Bayes), DT (decision tree), K-nearest neighbor, multilayered perception algorithm, support vector machines	Lack of benchmark dataset.,	Bioimpedance measurements are only Applicable to extracellular cells.
[62]	**Yellow fever**	UCI repository	Neural network, multilayer perceptron	Limited dataset.	Model achieved accuracy of 88%.
[63]		ChEMBL database	Naive Bayes	Unbalanced distribution of inhibitors and noninhibitors in the datasets.	Model achieved accuracy of more than 90%.
[64]		Google Trends and Google Street view	Propagation model	Lack of computational power.	The computer vision model based on Google Streetview requires more study
[65]		Generated RGB color images camera	KNN and back-propagation neural network (BPNN)	Limited dataset.	KNN (K-nearest neighbor) outperformed by achieving more than 93% accuracy.
[66]		(GPW) dataset	Logistic regression	Limited data.	AUC, 0.819 (2012); AUC, 0.853 (2013)

(Continued)

TABLE 5.2 Continued
Research Analysis of VBD Diseases

References	Illness type	Datasets	Techniques used	Challenges	Findings
[67]	West Nile fever	CDC	Decision trees	Spatial resolution data was not available.	The annual results were extremely inconsistent: R = 0–0.84; tested over a six-year period: R=0.86.
[68]		Health website	Partial Least Square Regression	Absence of climate data in ND Department of Health.	Statistical accuracy of MAE = 3.3 achieved.
[69]		Review article	Agent-based model, individual-based model	Insufficient human element and field data.	NA.
[70]		Website (3186)	RF ensemble learning	Few sporadic links between human WNND instances and biotic and abiotic predictors.	RF (random forest) obtained 94% accuracy, 91.1 percent sensitivity, and 96.7 percent specificity.
[71]	Zika	http://dados.recife.pe.gov.br/	Ensemble methods (DTs, bootstrap aggregation algorithm)	Insufficient data and has failed to collect data (demographic, environmental, and behavioral data).	Model achieved 62.5% accuracy and performed excellently in terms of AUC = 0.91.
[72]		Hospital data	RF, SVM (support vector machine), SMO (sequential minimal optimization), ISDA (iterative single data algorithm)	The only criterion is that the information found by the mass spectrometer be present in the serum of disease patients.	RF (random forest) model outperformed as compared to other models.
[73]		WorldClim	GBM (gradient boosting machine), BPNN (backward-propagation neural network), RF (random forest)	Single machine learning method is not accurately forecasting possible transmission risk distributions.	BPNN model outperformed predictive accuracy (AUC) of 0.966.

[74]		Data captured by sensor	k-nearest neighbors, Gaussian mixture models	Constraint of space.	KNN obtained the highest results (greater than 90% accuracy).
[75]		Laboratory data	SVM (support vector machine), KNN (K-nearest neighbors), decision trees, and Gaussian mixture model (GMM), autoencoder	Low detection time is a major concern (online classification).	SAE + SVM (50) earns the top result of 92.1%.
[76]		Data captured by sensor	SVM, RBF, RF, KNN	Using several classifiers with the same features proved ineffective.	Support vector machines achieved an accuracy of 87.33%.
[77]		Website (Nigeria Ministry of Health), WorldClim, CGIAR-CSI	Generalized boosted model (GBM), random forest (RF)	Lack of independent dataset.	Random forest and generalized boosted regression models outperformed than others, with ROC and TSS values greater than 0.95 and 0.75, respectively.
[78]	**Lyme disease**	Human serous samples (106)	ANN	Two-tier test has low sensitivity in early-stage LD.	SVM achieved highest accuracy of 0.747.
[79]		MyLymeData set (private survey dataset)	MLR (multivariate linear regression), SVM (support vector machine), decision tree, and K-nearest neighbors	It is very difficult or challenging to identify the features which contributed most to global rating of change (GROC).	SVM achieved highest accuracy of 0.747.
[80]		Online images, clinical images	DCNN, ResNet50, ResNet152, Inception v3, Inception-ResNetV2, and DenseNet121	Bias study was unable to establish whether our dataset included diversity like gender and age.	ResNet152 achieved the highest accuracy of 82.88%.

5.5.2 Pandemic Influenza

Signorini and collaborators utilized Twitter-embedded data in 2011 to track both the quick surge in public worry about H1N1 and actual illness activity. The researchers gathered tweets on H1N1 and added phrases about illness transmission, prevention, and treatment and consumption of food in the United States. Then, they employed supervised machine learning to forecast an estimated model utilizing intelligence information on influenza-like symptoms. Twitter may be used to gauge public concern about H1N1-related health issues. These include periodic spikes in user Twitter activity connected with preventive measures (hand hygiene, mask usage), travel and eating habits, and specific antiviral and vaccine adoption. They also discovered Twitter's AI applications might anticipate influenza epidemics [93].

5.6 TRACKING HEALTH BEHAVIORS USING INFODEMIOLOGY

The objective of the internet is to enlighten individuals regarding public health and public policies [94]. Google Trends, for instance, is a widely used open-source program that provides traffic data on patterns, trends, and changes in online interests over time. As a method for tracking human health-seeking behavior during epidemics, it has been praised [95]. The 2017 Chikungunya outbreak in Italy prompted an increase in internet inquiries and social media engagement. We forecasted Italian public responses to Chikungunya outbreaks using Twitter activity, Google Trends, Google Headlines, Wikipedia views and modifications, and PubMed publications. On the other hand, epidemiological data did not significantly mediate tweet production; instead, search terms posed prominent mediating tweets. Similar findings were found after revising the internet penetration index with the modified model [96].

5.7 EXPERT SYSTEMS AIDED BY COMPUTER

Expert systems are used to identify infectious disease epidemics faster by measuring the intensity of vector agents within infectious disease dynamics triads. Nigerian researchers created an intelligent system for plasmodium environmental detection to help researchers and policymakers [97]. They developed a prototype comprising of "knowledge," "applications," "system database," "user visual interface," and "user components," since present malaria preventive tactics are inadequate. The user component was built in Java, and the application component in JESS, using the Netbeans Java IDE. The instrument might be used to test the strength of plasmodium (malaria), in particular African locales. The proposed device successfully stopped malaria spread at a low cost [28].

5.8 THE OBSTACLES OF AI APPLICATIONS DURING EPIDEMICS

Artificial intelligence is being more commonly employed in pandemic planning, although it has limits. Even in resource-rich environments, updating expert system knowledge bases, providing high-quality datasets for machine learning techniques, and adhering to ethical data ownership is problematic [98]. The limitations of poorly organized and integrated health systems, inadequate information communications infrastructure, and social, economic, and cultural settings [99] contribute to AI systems' failure to be adopted successfully in resource-constrained scenarios. Additional variables might influence the accuracy of epidemic disease modeling frameworks and the behavior of individuals seeking medical care during outbreaks [100]. Public health specialists and computer specialists must address health coverage scalability, privacy, anonymity, data aggregation and portability, and the morality of collected electronic data. For example, the "black box" nature of predictive intelligent systems has been questioned, leading in biases in environments with a high level of unfairness [101].

5.9 THE FUTURE DIRECTION OF ARTIFICIAL INTELLIGENCE APPLICATIONS IN DISEASE EPIDEMICS

Despite present challenges, including budget restrictions, the application of artificial intelligence to track people's health, anticipate public health threats, and improve pandemic preparation is expected to skyrocket shortly. Deployment of mobile phones is quickly rising, as is investment in cloud-based solutions and m-health (mobile health) applications in commodity configurations. [102]. Infodemiology will aid in the analysis of public health policy, practice, and studies in the future [94, 103]. It's hardly surprising that several studies have produced very accurate prediction models using artificial intelligence techniques. Diverse future directions are necessary to expand this field. Most of our research initiatives examined selected models (ML and DL). Their use in future machine and deep learning systems for vector-borne illness prediction should be fostered. Using deep learning and AI innovations: deep learning's multimodality performance may be beneficial for VBD prediction. Using feature engineering to establish a balance may save time and money.

5.10 CONCLUSION

Artificial intelligence technologies are increasingly being ingrained in our daily lives, transforming the health-care industry to benefit global populations. Technological and digital advancements have permitted meteorological data into surveillance systems, allowing for a more accurate prediction of vector-borne disease outbreaks than was previously possible using standard monitoring models. Early detection of an epidemic allows communities and decision-makers to ramp up prevention and preparedness measures and reallocate response and recovery resources to the most at-risk places. With the long-term viability of the innovation, there is a need to ensure that technology is advancing instead of dividing surveillance systems, and the fundamental obligation to ensure that populations infected by the disease participate in the technology's design and benefit directly from its implementation are all critical considerations in the advancement of these new technologies. Apart from temperature and meteorological data, additional consequences of climate change, such as world population mobility and local biodiversity, should be included into new surveillance models to identify and predict vector-borne disease outbreaks more precisely than classic surveillance models did. The use of AI and machine learning is auspicious, as shown by the examples mentioned in this section. Using both established and innovative machine learning methodologies, the use of AI may be critical in combating infectious illnesses. Along with advancements in biomedical studies, AI applications enable a speedier analysis of large volumes of infectious illness data, enabling policymakers, medical experts, and health-care institutions to react more quickly to emerging diseases.

REFERENCES

1. Claborn, David. "Introductory chapter: Vector-Borne diseases." *Vector-Borne Diseases: Recent Developments in Epidemiology and Control* (2020): 3
2. Valenzuela, Jesus G., and Serap Aksoy. "Impact of vector biology research on old and emerging neglected tropical diseases." *PLOS Neglected Tropical Diseases* 12, no. 5 (2018): e0006365.
3. Gibert, Jean P., Anthony I. Dell, John P. DeLong, and Samraat Pawar. "Scaling-up trait variation from individuals to ecosystems." *Advances in Ecological Research* 52 (2015): 1–17.
4. Kaur, Inderpreet, Amanpreet Kaur Sandhu, and Yogesh Kumar. "Analyzing and minimizing the effects of vector-borne diseases using machine and deep learning techniques: A systematic review." In *2021 Sixth International Conference on Image Information Processing (ICIIP)*, vol. 6, pp. 69–74. IEEE, 2021.
5. Eder, Marcus, Fanny Cortes, Noêmia Teixeira de Siqueira Filha, Giovanny Vinícius Araújo de França, Stéphanie Degroote, Cynthia Braga, Valéry Ridde, et al. "Scoping review on vector-borne diseases in urban areas: Transmission dynamics, vectorial capacity and co-infection." *Infectious Diseases of Poverty* 7, no. 1 (2018): 1–24.
6. La Deau, Shannon L., Brian F. Allan, Paul T. Leisnham, and Michael Z. Levy. "The ecological foundations of transmission potential and vector-borne disease in urban landscapes." *Functional Ecology* 29, no. 7 (2015): 889–901.
7. Magori, Krisztian, and John M. Drake. "The population dynamics of vector-borne diseases." *Nature Education Knowledge* 4, no. 4 (2013): 14.
8. Powell, Jeffrey R. "An evolutionary perspective on vector-borne diseases." *Frontiers in Genetics* 10 (2019): 1266.
9. Hotez, Peter J. *Forgotten people, forgotten diseases: The neglected tropical diseases and their impact on global health and development*. John Wiley & Sons, 2021.
10. Athni, Tejas S., Marta S. Shocket, Lisa I. Couper, Nicole Nova, Iain R. Caldwell, Jamie M. Caldwell, Jasmine N. Childress, et al. "The influence of vector-borne disease on human history: Socio-ecological mechanisms." *Ecology Letters* 24, no. 4 (2021): 829–846.
11. Ung, Lawson, Nisha R. Acharya, Tushar Agarwal, Eduardo C. Alfonso, Bhupesh Bagga, Paulo J. M. Bispo, Matthew J. Burton, et al. "Infectious corneal ulceration: A proposal for neglected tropical disease status." *Bulletin of the World Health Organization* 97, no. 12 (2019): 854.
12. Fene, Fato, María Jesús Ríos-Blancas, James Lachaud, Christian Razo, Hector Lamadrid-Figueroa, Michael Liu, Jacob Michel, et al. "Life expectancy, death, and disability in Haiti, 1990–2017: A systematic analysis from the Global Burden of Disease Study 2017." *Revistapanamericana de salud publica* 44 (2020).
13. Vonesch, Nicoletta, Maria Concetta D'Ovidio, Paola Melis, Maria Elena Remoli, Maria grazia Ciufolini, and Paola Tomao. "Climate change, vector-borne diseases and working population." *Annalidell' Istitutosuperiore di sanita* 52, no. 3 (2016): 397–405.
14. Faburay, Bonto. "The case for a 'one health' approach to combating vector-borne diseases." *Infection ecology & Epidemiology* 5, no. 1 (2015): 28132.
15. Kibret, Solomon, Jonathan Lautze, Matthew McCartney, Luxon Nhamo, and G. Glenn Wilson. "Malaria and large dams in sub-Saharan Africa: Future impacts in a changing climate." *Malaria Journal* 15, no. 1 (2016): 1–14.

16. Chala, Bayissa, and Workineh Torben. "An epidemiological trend of urogenital schistosomiasis in Ethiopia." *Frontiers in Public Health* 6 (2018): 60.
17. Balogun, Emmanuel O., Andrew J. Nok, and Kiyoshi Kita. "Global warming and the possible globalization of vector-borne diseases: A call for increased awareness and action." *Tropical Medicine and Health* 44, no. 1 (2016): 1–3.
18. Savić, Sara, Branka Vidić, ZivoslavGrgić, Aleksandar Potkonjak, and Ljubica Spasojevic. "Emerging vector-borne diseases—incidence through vectors." *Frontiers in Public Health* 2 (2014): 267.
19. Kulkarni, Manisha A., Lea Berrang-Ford, Peter A. Buck, Michael A. Drebot, L. Robbin Lindsay, and Nicholas H. Ogden. "Major emerging vector-borne zoonotic diseases of public health importance in Canada." *Emerging Microbes & Infections* 4, no. 1 (2015): 1–7.
20. Field, Christopher B., and Vicente R. Barros, eds. *Climate change 2014—Impacts, adaptation and vulnerability: Regional aspects*. Cambridge University Press, 2014.
21. Brady, Oliver J., Michael A. Johansson, Carlos A. Guerra, Samir Bhatt, Nick Golding, David M. Pigott, Hélène Delatte, et al. "Modelling adult Aedes aegypti and Aedes albopictus survival at different temperatures in laboratory and field settings." *Parasites & vectors* 6, no. 1 (2013): 1–12.
22. Tian, Huaiyu, Sen Zhou, Lu Dong, Thomas P. Van Boeckel, Yujun Cui, Scott H. Newman, John Y. Takekawa, et al. "Avian influenza H5N1 viral and bird migration networks in Asia." *Proceedings of the National Academy of Sciences* 112, no. 1 (2015): 172–177.
23. Carpenter, Simon, Anthony Wilson, James Barber, Eva Veronesi, Philip Mellor, Gert Venter, and Simon Gubbins. "Temperature dependence of the extrinsic incubation period of orbiviruses in Culicoides biting midges." *PLoS One* 6, no. 11 (2011): e27987.
24. Jones, Bryony A., Delia Grace, Richard Kock, Silvia Alonso, Jonathan Rushton, Mohammed Y. Said, Declan McKeever, et al. "Zoonosis emergence linked to agricultural intensification and environmental change." *Proceedings of the National Academy of Sciences* 110, no. 21 (2013): 8399–8404.
25. Medlock, Jolyon M., and Steve A. Leach. "Effect of climate change on vector-borne disease risk in the UK." *The Lancet Infectious Diseases* 15, no. 6 (2015): 721–730.
26. Ogden, Nicholas H., Milka Radojevic, Xiaotian Wu, Venkata R. Duvvuri, Patrick A. Leighton, and Jianhong Wu. "Estimated effects of projected climate change on the basic reproductive number of the Lyme disease vector Ixodes scapularis." *Environmental Health Perspectives* 122, no. 6 (2014): 631–638.
27. Swei, Andrea, Lisa I. Couper, Lark L. Coffey, Durrell Kapan, and Shannon Bennett. "Patterns, drivers, and challenges of vector-borne disease emergence." *Vector-Borne and Zoonotic Diseases* 20, no. 3 (2020): 159–170.
28. Jones, Bryony A., Delia Grace, Richard Kock, Silvia Alonso, Jonathan Rushton, Mohammed Y. Said, Declan McKeever, et al. "Zoonosis emergence linked to agricultural intensification and environmental change." *Proceedings of the National Academy of Sciences* 110, no. 21 (2013): 8399–8404.
29. Tong, Michael Xiaoliang, Alana Hansen, Scott Hanson-Easey, Scott Cameron, Jianjun Xiang, Qiyong Liu, Yehuan Sun, et al. "Infectious diseases, urbanization and climate change: Challenges in future China." *International Journal of Environmental Research and Public Health* 12, no. 9 (2015): 11025–11036.
30. Rivero, Ana, Julien Vezilier, Mylene Weill, Andrew F. Read, and Sylvain Gandon. "Insecticide control of vector-borne diseases: When is insecticide resistance a problem?" *PLoS Pathogens* 6, no. 8 (2010): e1001000.
31. Nii-Trebi, Nicholas Israel. "Emerging and neglected infectious diseases: Insights, advances, and challenges." *BioMed Research International* 2017 (2017).
32. Hassell, James M., Michael Begon, Melissa J. Ward, and Eric M. Fèvre. "Urbanization and disease emergence: Dynamics at the wildlife—livestock—human interface." *Trends in Ecology & Evolution* 32, no. 1 (2017): 55–67.

33. Mishra, Charudutt, Gustaf Samelius, Munib Khanyari, Prashanth Nuggehalli Srinivas, Matthew Low, Carol Esson, Suri Venkatachalam, et al. "Increasing risks for emerging infectious diseases within a rapidly changing High Asia." *Ambio* (2021): 1–14.
34. Devaux, Christian A., Oleg Mediannikov, Hacene Medkour, and Didier Raoult. "Infectious disease risk across the growing human-non human primate interface: A review of the evidence." *Frontiers in Public Health* 7 (2019): 305.
35. Gubler, Duane J. "Dengue, urbanization and globalization: The unholy trinity of the 21st century." *Tropical Medicine and Health* 39, no. 4 (2011): S3–S11.
36. Lowe, Rachel, Christovam Barcellos, Patrícia Brasil, Oswaldo G. Cruz, Nildimar Alves Honório, Hannah Kuper, and Marilia Sá Carvalho. "The Zika virus epidemic in Brazil: From discovery to future implications." *International Journal of Environmental Research and Public Health* 15, no. 1 (2018): 96.
37. N'Dri, Bédjou P., Kathrin Heitz-Tokpa, Mouhamadou Chouaïbou, Giovanna Raso, Amoin J. Koffi, Jean T. Coulibaly, Richard B. Yapi, et al. "Use of insecticides in agriculture and the prevention of vector-borne diseases: Population knowledge, attitudes, practices and beliefs in Elibou, South Côte d'Ivoire." *Tropical Medicine and Infectious Disease* 5, no. 1 (2020): 36.
38. McFarlane, Rosemary A., Adrian C. Sleigh, and Anthony J. McMichael. "Land-use change and emerging infectious disease on an island continent." *International Journal of Environmental Research and Public Health* 10, no. 7 (2013): 2699–2719.
39. Gottdenker, Nicole L., Daniel G. Streicker, Christina L. Faust, and C. R. Carroll. "Anthropogenic land use change and infectious diseases: A review of the evidence." *EcoHealth* 11, no. 4 (2014): 619–632.
40. Shah, Hiral A., Paul Huxley, Jocelyn Elmes, and Kris A. Murray. "Agricultural land-uses consistently exacerbate infectious disease risks in Southeast Asia." *Nature communications* 10, no. 1 (2019): 1–13.
41. Sarkar, Atanu, Kristan J. Aronson, Shantagouda Patil, and Lingappa B. Hugar. "Emerging health risks associated with modern agriculture practices: A comprehensive study in India." *Environmental Research* 115 (2012): 37–50.
42. Kim, Kyukwang, Jieum Hyun, Hyeongkeun Kim, Hwijoon Lim, and Hyun Myung. "A deep learning-based automatic mosquito sensing and control system for urban mosquito habitats." *Sensors* 19, no. 12 (2019): 2785.
43. Roth, Gregory A., Degu Abate, Kalkidan Hassen Abate, Solomon M. Abay, Cristiana Abbafati, Nooshin Abbasi, HedayatAbbastabar, et al. "Global, regional, and national age-sex-specific mortality for 282 causes of death in 195 countries and territories, 1980–2017: A systematic analysis for the Global Burden of Disease Study 2017." *The Lancet* 392, no. 10159 (2018): 1736–1788.
44. Wilson, Anne L., Orin Courtenay, Louise A. Kelly-Hope, Thomas W. Scott, Willem Takken, Steve J. Torr, and Steve W. Lindsay. "The importance of vector control for the control and elimination of vector-borne diseases." *PLoS Neglected Tropical Diseases* 14, no. 1 (2020): e0007831.
45. Nambunga, Ismail H., Halfan S. Ngowo, Salum A. Mapua, Emmanuel E. Hape, Betwel J. Msugupakulya, Dickson S. Msaky, Nicolaus T. Mhumbira, et al. "Aquatic habitats of the malaria vector Anopheles funestus in rural south-eastern Tanzania." *Malaria Journal* 19, no. 1 (2020): 1–11.
46. Baghbanzadeh, Mahdi, Dewesh Kumar, Sare I. Yavasoglu, Sydney Manning, Ahmad Ali Hanafi-Bojd, Hassan Ghasemzadeh, Ifthekar Sikder, et al. "Malaria epidemics in India: Role of climatic condition and control measures." *Science of The Total Environment* 712 (2020): 136368.
47. Arowolo, Micheal Olaolu, Marion Olubunmi Adebiyi, Ayodele Ariyo Adebiyi, and Oludayo Olugbara. "Optimized hybrid investigative based dimensionality reduction methods

for malaria vector using KNN classifier." *Journal of Big Data* 8, no. 1 (2021): 1–14.

48. Damos, Petros, José Tuells, and Pablo Caballero. "Soft computing of a medically important arthropod vector with autoregressive recurrent and focused time delay artificial neural networks." *Insects* 12, no. 6 (2021): 503.
49. Nkiruka, Odu, Rajesh Prasad, and Onime Clement. "Prediction of malaria incidence using climate variability and machine learning." *Informatics in Medicine Unlocked* 22 (2021): 100508.
50. Kulkarni, Manisha A., Rachelle E. Desrochers, and Jeremy T. Kerr. "High resolution niche models of malaria vectors in northern Tanzania: A new capacity to predict malaria risk?." *PLoS One* 5, no. 2 (2010): e9396.
51. Tek, F. Boray, Andrew G. Dempster, and Izzet Kale. "Parasite detection and identification for automated thin blood film malaria diagnosis." *Computer Vision and Image Understanding* 114, no. 1 (2010): 21–32.
52. Laughlin, Catherine A., David M. Morens, M. Cristina Cassetti, Adriana Costero-Saint Denis, Jose-Luis San Martin, Stephen S. Whitehead, and Anthony S. Fauci. "Dengue research opportunities in the Americas." *The Journal of Infectious Diseases* 206, no. 7 (2012): 1121–1127.
53. Licciardi, Séverine, Etienne Loire, Eric Cardinale, Marie Gislard, Emeric Dubois, and Catherine Cêtre-Sossah. "In vitro shared transcriptomic responses of Aedes aegypti to arboviral infections: Example of dengue and Rift Valley fever viruses." *Parasites & Vectors* 13, no. 1 (2020): 1–10.
54. Lowe, Rachel, Christovam Barcellos, Patrícia Brasil, Oswaldo G. Cruz, Nildimar Alves Honório, Hannah Kuper, and Marilia Sá Carvalho. "The Zika virus epidemic in Brazil: From discovery to future implications." *International Journal of Environmental Research and Public Health* 15, no. 1 (2018): 96.
55. Fathima, A., and D. Manimegalai. "Predictive analysis for the arbovirus-dengue using SVM classification." *International Journal of Engineering and Technology* 2, no. 3 (2012): 521–7.
56. Ibrahim, Fatimah, Tarig Faisal, M. I. Mohamad Salim, and Mohd Nasir Taib. "Non-invasive diagnosis of risk in dengue patients using bioelectrical impedance analysis and artificial neural network." *Medical & Biological Engineering & Computing* 48, no. 11 (2010): 1141–1148.
57. Azman, Muhammad Izzul, Azri Bin Zainol, and Aliza Binti Sarlan. "Aedes Larvae Classification and Detection (ALCD) system by using deep learning." In *2020 International Conference on Computational Intelligence (ICCI)*. IEEE, 2020. doi:10.1109/icci51257.2020.9247647.
58. Sohail, Ayesha, Mehwish Iftikhar, RobiaArif, Hijaz Ahmad, Khaled A. Gepreel, and Sahrish Iftikhar. "Dengue control measures via cytoplasmic incompatibility and modern programming tools." *Results in Physics* 21 (2021): 103819.
59. Laureano-Rosario, Abdiel E., Andrew P. Duncan, Pablo A. Mendez-Lazaro, Julian E. Garcia-Rejon, Salvador Gomez-Carro, Jose Farfan-Ale, Dragan A. Savic, et al. "Application of artificial neural networks for dengue fever outbreak predictions in the northwest coast of Yucatan, Mexico and San Juan, Puerto Rico." *Tropical Medicine and Infectious Disease* 3, no. 1 (2018): 5.
60. Farooqi, Wajeeha, and Sadaf Ali. "A critical study of selected classification algorithms for dengue fever and dengue hemorrhagic fever." In *2013 11th International Conference on Frontiers of Information Technology (FIT)*. IEEE, 2013. doi:10.1109/fit.2013.33.
61. Abílio, Ana Paula, Ayubo Kampango, Eliseu J. Armando, Eduardo S. Gudo, Luís C. B. Das Neves, Ricardo Parreira, Mohsin Sidat, et al. "First confirmed occurrence of the yellow fever virus and dengue virus vector Aedes (Stegomyia) luteocephalus (Newstead, 1907) in Mozambique." *Parasites & Vectors* 13, no. 1 (2020): 1–8.

62. Amadin, F. I., and M. E. Bello. "Prediction of yellow fever using multilayer perceptron neural network classifier." *Journal of Emerging Trends in Engineering and Applied Sciences* 9, no. 6 (2018): 282–286.
63. Moorthy, N. H. N., and V. Poongavanam. "The KNIME based classification models for yellow fever virus inhibition." *RSC Advances* 5, no. 19 (2015): 14663–14669.
64. De Silva, Shalen, Ramya Pinnamaneni, Kavya Ravichandran, Alaa Fadaq, Yun Mei, and Vincent Sin. "Yellow fever in Brazil: Using novel data sources to produce localized policy recommendations." In *Leveraging data science for global health*, pp. 417–428. Springer, 2020.
65. Anderson, Michelle E., Jessica Mavica, Lewis Shackleford, Ilona Flis, Sophia Fochler, Sanjay Basu, and Luke Alphey. "CRISPR/Cas9 gene editing in the West Nile Virus vector, Culex quinquefasciatus Say." *PLoS One* 14, no. 11 (2019): e0224857.
66. Tran, Annelise, Bertrand Sudre, Shlomit Paz, Massimiliano Rossi, Annie Desbrosse, Véronique Chevalier, and Jan C. Semenza. "Environmental predictors of West Nile fever risk in Europe." *International Journal of Health Geographics* 13, no. 1 (2014): 1–11.
67. Young, Sean G., Jason A. Tullis, and Jackson Cothren. "A remote sensing and GIS-assisted landscape epidemiology approach to West Nile virus." *Applied Geography* 45 (2013): 241–249.
68. Campion, Mitch, Calvin Bina, Martin Pozniak, Todd Hanson, Jeff Vaughan, Joseph Mehus, Scott Hanson, et al. "Predicting West Nile Virus (WNV) occurrences in North Dakota using data mining techniques." In *2016 Future Technologies Conference (FTC)*. IEEE, 2016. doi:10.1109/ftc.2016.7821628.
69. Nasrinpour, Hamid Reza, Marcia R. Friesen, and Robert D. McLeod. "Agent based modelling and West Nile Virus: A survey." *Journal of Medical and Biological Engineering* 39, no. 2 (2019): 178–183.
70. Coroian, Mircea, Mina Petrić, Adriana Pistol, Anca Sirbu, Cristian Domşa, and Andrei Daniel Mihalca. "Human West Nile meningo-encephalitis in a highly endemic country: A complex epidemiological analysis on biotic and abiotic risk factors." *International Journal of Environmental Research and Public Health* 17, no. 21 (2020): 8250.
71. Moreira, Mario W. L., Joel J. P. C. Rodrigues, Francisco H. C. Carvalho, Jalal Al-Muhtadi, Sergey Kozlov, and Ricardo AL Rabelo. "Classification of risk areas using a bootstrap-aggregated ensemble approach for reducing Zika virus infection in pregnant women." *Pattern Recognition Letters* 125 (2019): 289–294.
72. Melo, Carlos Fernando Odir Rodrigues, Luiz Claudio Navarro, Diogo Noin De Oliveira, Tatiane Melina Guerreiro, Estela de Oliveira Lima, Jeany Delafiori, Mohamed Ziad Dabaja, et al. "A machine learning application based in random forest for integrating mass spectrometry-based metabolomic data: A simple screening method for patients with zika virus." *Frontiers in Bioengineering and Biotechnology* 6 (2018): 31.
73. Jiang, Dong, Mengmeng Hao, Fangyu Ding, Jingying Fu, and Meng Li. "Mapping the transmission risk of Zika virus using machine learning models." *Acta Tropica* 185 (2018): 391–399.
74. de Souza, Vinicius M. A., Diego F. Silva, and Gustavo E. A. P. A. Batista. "Classification of data streams applied to insect recognition: Initial results." In *2013 Brazilian Conference on Intelligent Systems (BRACIS)*. IEEE. 2013. doi:10.1109/bracis.2013.21.
75. Qi, Yu, Goktug T. Cinar, Vinicius M. A. Souza, Gustavo E. A. P. A. Batista, Yueming Wang, and Jose C. Principe. "Effective insect recognition using a stacked autoencoder with maximum correntropy criterion." In *2015 International Joint Conference on Neural Networks (IJCNN)*. IEEE, 2015. doi:10.1109/ijcnn.2015.7280418.
76. Silva, Diego F., Vinicius M. A. De Souza, Gustavo E. A. P. A. Batista, Eamonn Keogh, and Daniel P. W. Ellis. "Applying machine learning and audio analysis techniques to insect recognition in intelligent traps." In *2013 12th International Conference on Machine Learning and Applications (ICMLA)*. IEEE. 2013. doi:10.1109/icmla.2013.24.

77. Eneanya, Obiora A., Jorge Cano, Ilaria Dorigatti, Ifeoma Anagbogu, Chukwu Okoronkwo, Tini Garske, and Christl A. Donnelly. "Environmental suitability for lymphatic filariasis in Nigeria." *Parasites & Vectors* 11, no. 1 (2018): 1–13.
78. Joung, Hyou-Arm, Zachary S. Ballard, Jing Wu, Derek K. Tseng, Hailemariam Teshome, Linghao Zhang, Elizabeth J. Horn, et al. "Point-of-care serodiagnostic test for early-stage lyme disease using a multiplexed paper-based immunoassay and machine learning." *ACS Nano* 14, no. 1 (2019): 229–240.
79. Vendrow, Joshua, Jamie Haddock, Deanna Needell, and Lorraine Johnson. "Feature selection from lyme disease patient survey using machine learning." *Algorithms* 13, no. 12 (2020): 334.
80. Burlina, P. M., N. J. Joshi, P. A. Mathew, W. Paul, A. W. Rebman, and J. N. Aucott. AI-based detection of erythema migrans and disambiguation against other skin lesions. *Computers in Biology and Medicine* 125 (2020): 103977.
81. Maqsood, Asma, Muhammad Shahid Farid, Muhammad Hassan Khan, and Marcin Grzegorzek. "Deep malaria parasite detection in thin blood smear microscopic images." *Applied Sciences* 11, no. 5 (2021): 2284.
82. Telang, Hrishikesh, and Kavita Sonawane. "Effective performance of bins approach for classification of malaria Parasite using Machine Learning." In *2020 IEEE 5th International Conference on Computing Communication and Automation (ICCCA)*. IEEE, 2020. doi:10.1109/iccca49541.2020.9250789.
83. Sriporn, Krit, Cheng-Fa Tsai, Chia-En Tsai, and Paohsi Wang. "Analyzing Malaria disease using effective deep learning approach." *Diagnostics* 10, no. 10 (2020): 744.
84. Childs, Marissa L., Nicole Nova, Justine Colvin, and Erin A. Mordecai. "Mosquito and primate ecology predict human risk of yellow fever virus spillover in Brazil." *Philosophical Transactions of the Royal Society B* 374, no. 1782 (2019): 20180335.
85. Marcantonio, Matteo, Annapaola Rizzoli, Markus Metz, Roberto Rosà, Giovanni Marini, Elizabeth Chadwick, and Markus Neteler. "Identifying the environmental conditions favouring West Nile virus outbreaks in Europe." *PLoS One* 10, no. 3 (2015): e0121158.
86. Soliman, Marwah, Vyacheslav Lyubchich, and Yulia R. Gel. "Ensemble forecasting of the Zika space-time spread with topological data analysis." *Environmetrics* 31, no. 7 (2020): e2629.
87. Han, Barbara A., and Laura Yang. "Predicting novel tick vectors of zoonotic disease." *arXiv preprint arXiv:1606.06323* (2016).
88. Walter, Melanie, Janna R. Vogelgesang, Franz Rubel, and Katharina Brugger. "Tick-borne encephalitis virus and its European distribution in ticks and endothermic mammals." *Microorganisms* 8, no. 7 (2020): 1065.
89. Bhatia, Gresha, Shravan Bhat, Vivek Choudhary, Aditya Deopurkar, and Sahil Talreja. "Disease prediction using deep learning." In *2021 2nd International Conference for Emerging Technology (INCET)*. IEEE. 2021. doi:10.1109/incet51464.2021.9456172.
90. Jain, Kamal. "Artificial intelligence applications in handling the infectious diseases." *Primary Health Care: Open Access* (2020): 1–3.
91. Wong, Zoie S. Y., Jiaqi Zhou, and Qingpeng Zhang. "Artificial intelligence for infectious disease big data analytics." *Infection, Disease & Health* 24, no. 1 (2019): 44–48.
92. Kaur, Inderpreet, Amanpreet Kaur Sandhu, and Yogesh Kumar. "Artificial intelligence techniques for predictive modeling of vector-borne diseases and its pathogens: A systematic review." *Archives of Computational Methods in Engineering* (2022): 1–31.
93. Signorini, Alessio, Alberto Maria Segre, and Philip M. Polgreen. "The use of Twitter to track levels of disease activity and public concern in the US during the influenza A H1N1 pandemic." *PLoS One* 6, no. 5 (2011): e19467.
94. Eysenbach, Gunther. "Infodemiology and infoveillance: Framework for an emerging set of public health informatics methods to analyze search, communication and publication behavior on the Internet." *Journal of Medical Internet Research* 11, no. 1 (2009): e1157.

95. Mavragani, Amaryllis, and Gabriela Ochoa. “Google Trends in infodemiology and infoveillance: Methodology framework.” JMIR Public Health and Surveillance 5, no. 2 (2019): e13439.
96. Mahroum, Naim, Mohammad Adawi, Kassem Sharif, Roy Waknin, Hussein Mahagna, Bishara Bisharat, Mahmud Mahamid, et al. “Public reaction to Chikungunya outbreaks in Italy—Insights from an extensive novel data streams-based structural equation modeling analysis.” PLoS One 13, no. 5 (2018): e0197337.
97. Sheikhtaheri, Abbas, Farahnaz Sadoughi, and Zahra Hashemi Dehaghi. “Developing and using expert systems and neural networks in medicine: A review on benefits and challenges.” Journal of Medical Systems 38, no. 9 (2014): 1–6.
98. Oluwagbemi, O. O., Esther Adeoye, and Segun Fatumo. “Building a computer-based expert system for malaria environmental diagnosis: An alternative malaria control strategy.” Egyptian Computer Science Journal 33, no. 1 (2009): 55–69.
99. Caliskan, Aylin, Joanna J. Bryson, and Arvind Narayanan. “Semantics derived automatically from language corpora contain human-like biases.” *Science* 356, no. 6334 (2017): 183–186.
100. Moss, Robert, Alexander E. Zarebski, Sandra J. Carlson, and James M. McCaw. “Accounting for healthcare-seeking behaviours and testing practices in real-time influenza forecasts.” *Tropical Medicine and Infectious Disease* 4, no. 1 (2019): 12.
101 Datta, Anupam, Shayak Sen, and Yair Zick. “Algorithmic transparency via quantitative input influence: Theory and experiments with learning systems.” In *2016 IEEE Symposium on Security and Privacy (SP)*. IEEE, 2016. doi:10.1109/sp.2016.42.
102. Shaban-Nejad, A., M. Michalowski, and D. L. Buckeridge. “Health intelligence: How artificial intelligence transforms population and personalized health.” *NPJ Digital Medicine*, no. 1 (2018): 53.
103. Eysenbach, Gunther. “Infodemiology and infoveillance: Tracking online health information and cyberbehavior for public health.” *American Journal of Preventive Medicine* 40, no. 5 (2011): S154–S158.
104. Obregón Alvarez, Dasiel, Belkis Corona-González, Alina Rodríguez-Mallón, Islay Rodríguez Gonzalez, Pastor Alfonso, Angel A. Noda Ramos, Adrian A. Díaz-Sánchez, et al. “Ticks and tick-borne diseases in Cuba, half a century of scientific research.” *Pathogens* 9, no. 8 (2020): 616.
105. Sakkas, Hercules, Vangelis Economou, and Chrissanthy Papadopoulou. “Zika virus infection: Past and present of another emerging vector-borne disease.” *Journal of Vector Borne Diseases* 53, no. 4 (2016): 305.

6 Hybrid Neural Network-Based Fuzzy Inference System Combined with Machine Learning to Detect and Segment Kidney Tumor

P Srinivasa Rao, Pradeep Kumar Bheemavarapu, D Swapna, Subba Rao Polamuri, and M. Madhusudhana Subramanyam

CONTENTS

6.1 INTRODUCTION

Kidney cancer [1] arises when healthy cells in any one of the kidneys [2] start growing disorderly, forming a lump, called tumor. During the initial phases, the symptoms cannot be identified. In diagnosing other complaints by abdominal imaging, kidney cancer is identified. As the lump grows, symptoms like lower back pain, blood in urine, weight loss, fever, etc. are seen. With the increase in tumor size, the criticality in diagnosing the patient increases. Hence, an automated process [3, 27] to identify kidney cancer at the preliminary stage is to be developed to increase the chances of diagnosing the disease in its preliminary state.

Deep learning is a class of ML [9] imitating the learning process of the human brain. The deep learning algorithms are hierarchically stacked with increasing

DOI: 10.1201/9781003309451-6

complexity and abstraction, while state-of-the-art ML algorithms are linear. CNNs are a category of ANN designed for image processing, image recognition, and most commonly used to analyze visual images. The CNNs add values to objects in the image as learnable weights and biases and can distinguish them from others. CNNs use smaller preprocessing techniques than traditional image processing algorithms.

In the traditional method of feature extraction [5, 7], handmade filters [4, 6] are trained independently, which is now automated by neural networks. This is a major advantage of deep learning over traditional methods [8, 21, 28]. This automated feature engineering is achieved with the convolution, pooling, dropout, activation function, and the fully connected layer as a whole. The automation of manual processes in the classification and tumor segmentation in kidneys by nephrologists is a challenging task. To address this challenge, deep learning and machine learning [26] techniques are to be investigated for efficient segmentation of kidney tumors. Contributions of this chapter are:

- End-to-end system for significant classification and segmentation of kidney tumor using CT scans.
- A novel classifier associating CNN with DT for classification of CT scans into tumor and kidney classes is designed and implemented.

The paper is further ordered as: Section 2 describes the literature overview, Section 3 elucidates the methodology of the proposed system, Section 4 details the experimental setup, Section 5 explains the performance analysis and experimentation results, and Section 6 gives out the conclusion and future work.

6.2 LITERATURE OVERVIEW

Segmentation of kidney tumors focuses on accurate identification of kidney and tumor contours, which has extreme importance in diagnosing through a computer-aided methodology. Recently, Siriapisith et al. [10] explored a variable neighborhood search concept for eliminating local minima in segmentation by interchanging gradient and intensity search spaces. Also, they worked on graph-cut probability density function and graph-cut-based active contour methods for kidney tumor segmentation. The results showed their proposed method outperformed other competing methodologies for segmentation. Koyuncu and Ceylan [11] described a tool for enhancing CT images for the segmentation process. The processes involved are denoising, removal of internal fat issues, and removal of redundant parts in an image. Results stated that their methodology showed better outcome when compared to fast linking spiking cortical model (FL-SCM) and block matching (BM3D) and FL-SCM methods.

Pan et al. [12] experimented with a single network model called segmentation and classification convolutional neural network for classification and segmentation of tumor in the kidneys. The experimented single neural network model performed better than the competing algorithms with a dice coefficient of 0.882. Also, this method performed well when compared to individual classification and segmentation models. Hou et al. [13] elucidated a self-guided neural network model using the three-stage process for segmentation of tumor in kidneys. Low-resolution net, full-resolution net, and refine net are examined to extract clear boundaries of tumor and kidney. Also, they suggested a dilated convolutional block as a replacement for pooling operation. They achieved 0.8454 and 0.9674 average dice for tumor and kidney.

Yan et al. [14] described a two-step process: First, the kidneys with tumors are segmented using a foreground segmentation network. Secondly, a sparse point cloud segmentation is examined to fine-grain the tumor in the kidney. They achieved a model which consumes reduced GPU resources when compared to state-of-the-art methodologies. Causey et al. [15] demonstrated an ensemble of the U-Net model to segment tumor in kidneys with CT images. The analyzed ensemble model has obtained a 0.949 dice score for kidney and tumor segmentation. Zhou and Chen [16] addressed the challenges in automatic segmentation of kidney tumor images by using single atlas and super voxel segmentation techniques. First, the single atlas technique is used for segmenting kidneys from the image, and super voxel segmentation is used for estimating each voxel probability. The results indicate that they have obtained a flexible and reliable system when compared to other automatic methods.

Qin et al. [17] combined the data augmentation model with a deep reinforcement learning to segment tumors in kidney. After multiple experimentations, they obtained a model giving promising results to segment tumors in kidney on CT images. Yang et al. [18] proposed a fully convolutional neural network model of 3D by combining a pyramid pooling module for segmentation of kidney and renal tumors. They improved kidney segmentation by considering 3D spatial contextual information and achieved a 0.931 dice score for kidneys. Hussain et al. [19] described a selection convolutional neural network model for localization of kidney in CT scans. To detect organ bounding boxes, they used cross-sectional 2D slices combined with Mask-RCNN. Their method resulted in ~2.4mm of localization error.

The proposed method centers on a new hybrid [20, 29] classification technique called neural fuzzy inference classification system to classify kidney and tumor images and segmentation of tumors using modified U-Net. This system is developed by integrating convolutional neural networks with decision tree classifiers for the efficient classification of kidney tumors. Then U-Net-based segmentation technique is adopted for accurate segmentation of identified kidney tumor images.

6.3 METHODOLOGY

In order to detect the kidney tumor, blood tests and CT scan analysis are opted by nephrologists, where the tumor can only be detected when its size is noticeable by the human eye. To detect the tumor in the kidney at an early stage, when it cannot be recognized by the human eye, an automated end-to-end tumor classification and segmentation model using a neural fuzzy inference system is investigated in this chapter.

The proposed end-to-end classification and segmentation of kidney tumors by a hybrid neural network–based fuzzy inference system is highlighted in Figure 6.1. In the simulation, the end-to-end process of the proposed system works as follows: The preprocessed images are fed into a hybrid neural fuzzy classifier to detect whether an image belongs to a tumor or kidney class. If the image belongs to the kidney class, then no further action is taken. If the image is categorized as tumor class, then it is

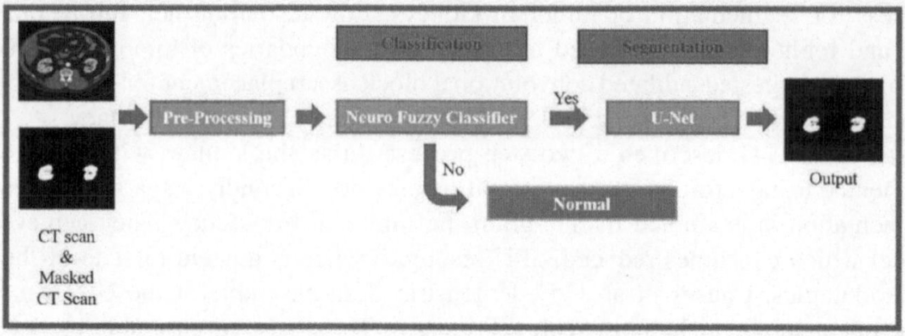

FIGURE 6.1 Architecture of proposed kidney tumor classification and detection system.

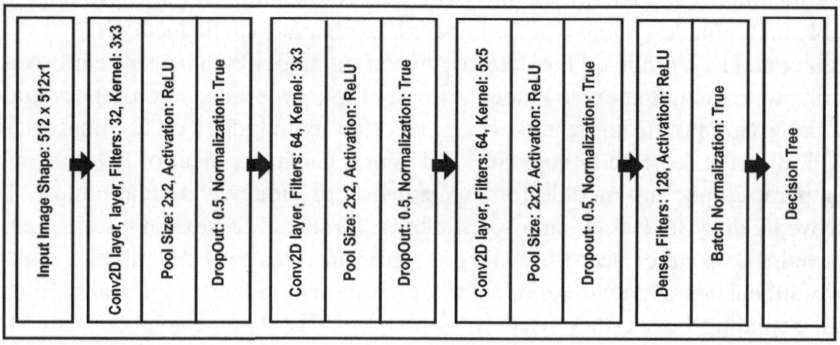

FIGURE 6.2 Architecture of neural fuzzy classifier.

passed onto the modified U-Net segmentation model for segmenting the tumor area. At first, the dataset is factionalized into 70% training and 30% testing data. Resizing the shape of images into 512 × 512 × 1 is reckoned as a preprocessing step.

The neural fuzzy classifier shown in Figure 6.2 comprises an input layer of 512 × 512 × 1 size, three hidden Conv2D layers built up with 32, 64, 64 filter having 3 × 3, 3 × 3, 5 × 5 kernel sizes with 2 × 2 pooling size, and one full connection hidden layer with 128 filters, and a decision tree classifier as an output layer. The activation function used in all the layers is ReLU, which is shown in equation (1), except for the decision tree layer.

$$ReLU : f\left(h_\theta\left(x\right)\right) = h_\theta\left(x\right)^+ = Max\left(0, h_\theta\left(x\right)\right) \quad (1)$$

Where, $h_\theta\left(x\right)$ is obtained using the equation (2).

$$\sum_{i=1}^{n} w_i x_i + bias = w_1 x_1 + w_2 x_2 + bias \quad (2)$$

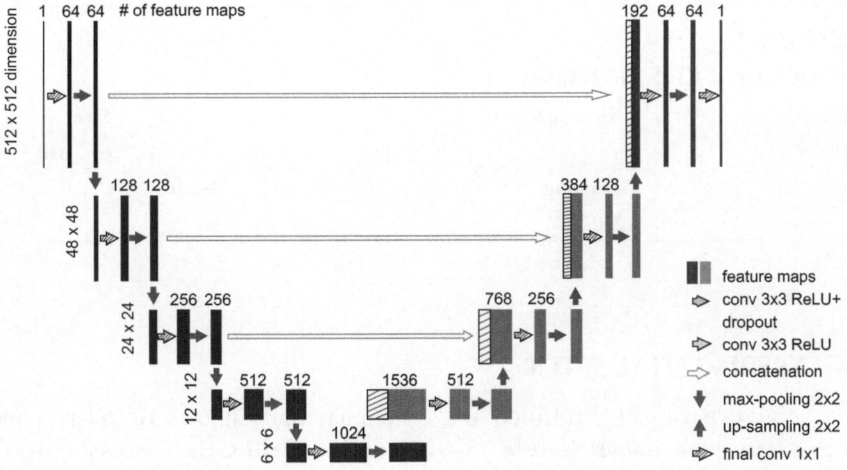

FIGURE 6.3 Architecture of modified U-Net model with different input patch size for segmentation.

Where w_i is the weight assigned to the neuron i, x_i represents the input at i^{th} neuron, and $bias$ ensures that the parameters don't pass through the origin. The neuron will be fired when the output of an activation function is above the threshold; otherwise, the neuron will not be activated. Example of triggering a neuron is given in equation (3).

$$\text{Output} = f(x) = \begin{cases} \sum_{i=1}^{n} w_i x_i + bias \geq threshold \\ \sum_{i=1}^{n} w_i x_i + bias < threshold \end{cases} \tag{3}$$

The modified convolutional neural network–based U-Net architecture represented in Figure 6.3 is adopted for the segmentation of tumors in the kidney. There are two parts in the architecture. The left-side part is called the encoding part or contracting path, and the right-side part is called the decoding part or expansive path. During the encoding process, at every down-sampling step, the feature channels are doubled and the dimensions of the image are reduced to half. Two Conv2D layers of 3 × 3 with stride 1, 2 × 2 max pooling operation with stride 2, and in all layers ReLU activation function are used on four levels. The last level comprises two 3 × 3 Conv2D layers. In the decoding process, the actual dimensions of the input images are recovered by up-sampling of feature maps. One Conv2D layer with ReLU activation function and a dropout layer and another Conv2D layer with ReLU activation function concatenated with their respective feature channels are used in the architecture.

TABLE 6.1
Description of KiTS19 Dataset

Dataset	No. of patients		Size
KiTS19	300		63.2 MB
	Training	Testing	
	210	90	

6.4 EXPERIMENTAL SETUP

The overall experiment is fulfilled in a system having Windows 10 (64-bit) operating system with Intel® Core™ i5–8250 CPU @2.30 GHz Processor, 16 GB Ram, 1TB HDD, and 256 GB SSD as its configuration. The system is installed with an Anaconda environment having Python as a programming language and supported machine learning and deep learning packages with Tensorflow as its back end.

The KiTS19 Kidney Tumor Segmentation Challenge dataset [22] is used for experimentation. It contains data of 300 patients who subjected themselves to nephrectomy for kidney tumors. The dataset was collected between 2010 and 2018. The dataset comprises segmented and actual CT scan images in nii format; 70% of the dataset is used to train the mode, and the remaining 30% of the dataset is used to test the model. The description of the KiTS19 dataset is given in Table 6.1.

6.5 PERFORMANCE ANALYSIS AND EXPERIMENTATION RESULTS

The proposed neural fuzzy classifier is tested for its classification accuracy by examining different classification analysis metrics, like MSE, RMSE, precision, recall, F-score, and accuracy, defined as follows:

$$MSE = \frac{1}{n}\sum_{i=1}^{n}\left(Y_i - Y_i\right)^2 \tag{4}$$

$$RMSE = \sqrt{\frac{1}{n}\sum_{i=1}^{n}\left(Y_i - Y_i\right)^2} \tag{5}$$

$$Precision = \frac{T_{positive}}{F_{positive} + T_{positive}} \tag{6}$$

$$Recall = \frac{T_{positive}}{F_{negative} + T_{positive}} \tag{7}$$

$$F1 - Score = \frac{2\,x\,Precision\,x\,Recall}{Precision + Recall} \tag{8}$$

$$Accuracy = \frac{T_{positive} + T_{negative}}{T_{positive} + F_{positive} + T_{negative} + F_{negative}} \quad (9)$$

Where $T_{positive}$ represents the samples classified correctly as tumor, $T_{negative}$ indicates the samples incorrectly classified as tumor, $F_{negative}$ defines the samples classified correctly as kidney and Y_i represents the samples classified incorrectly as kidney, Y_i is the actual output, n is the predicted output of classifier, and [Insert Equation Here] represents total samples. The evaluation metrics of the proposed neural fuzzy classifier are shown in Figure 6.4.

The related work comparison derived from accuracies of kidney tumor classification is presented in Table 6.2. The model accuracy during training and testing phase is shown in Figure 6.5. The contrast between similar work and the proposed work developed on dice coefficient is represented in Table 6.3.

Figure 6.6 represents an example of input CT scan image, actual masked image of input CT scan, and the model generated output.

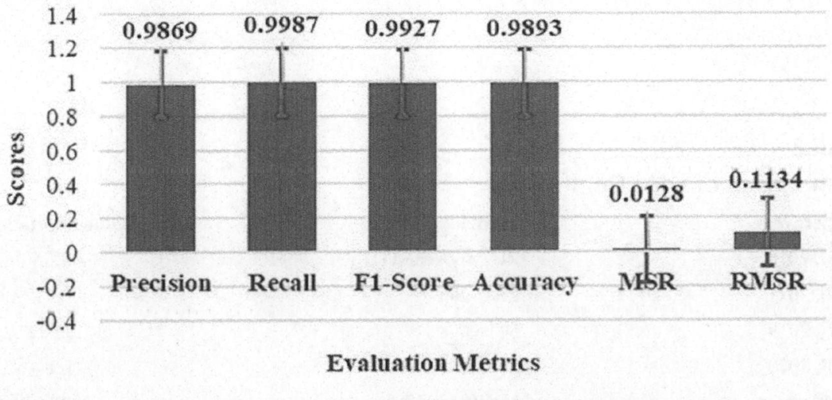

FIGURE 6.4 Evaluation metrics the proposed neuro fuzzy classifier.

TABLE 6.2
Accuracy Comparison of Similar Works

Paper, year	Classifier	Accuracy
[23], 2020	ResNet (ensemble)	0.70
[24], 2019	Custom CNN model	0.85
[25], 2020	Custom CNN model	0.90
[26], 2021	Pyramidal deep learning	0.92
[27], 2020	Text analysis with deep learning	0.93
This paper	**Neuro fuzzy classifier**	**0.98**

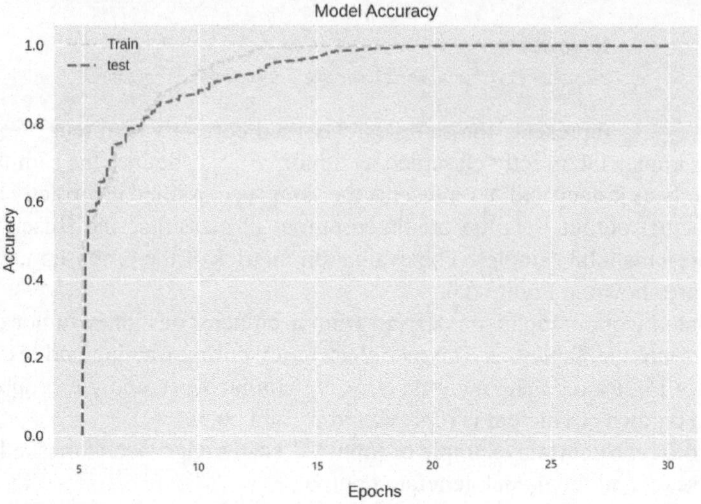

FIGURE 6.5 U-Net model accuracy.

TABLE 6.3
Comparison of Dice Coefficient

Paper, year	Model	Dice coefficient
[28], 2016	RUSBoost and DT	0.85
[29], 2021	RAU-Net	0.77
[12], 2019	SCNet	0.882
[13], 2020	Modified U-Net	0.84
This paper	**Modified U-Net**	**0.8923**

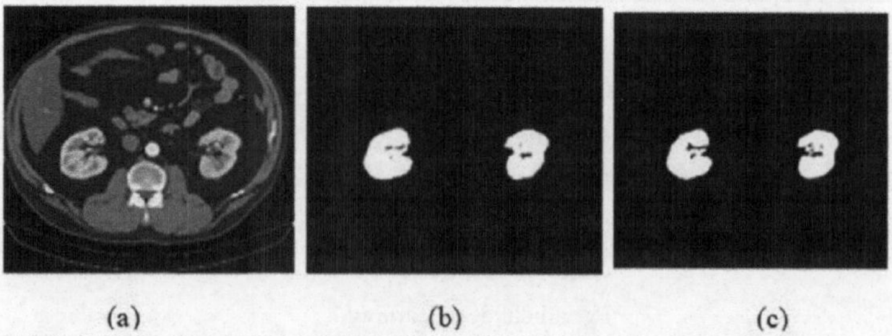

FIGURE 6.6 (a) CT scan image, (b) masked CT scan image, (c) output of the model.

6.6 CONCLUSION AND FUTURE WORK

The ninth most common cancer in men and fourteenth in women is recognized as kidney cancer. The detection of tumors at an early stage, before the expansion of tumor dimensionality, is a challenging task for nephrologists. An end-to-end model for the classification and segmentation of tumors in kidneys is proposed in this chapter. The publicly available KiTS19 Kidney Tumor Segmentation Challenge dataset is used for experimentation. A hybrid neural fuzzy inference classifier is proposed for the classification of CT scans into the kidney and tumor class. The neural fuzzy inference classifier is the combination of convolutional neural network having three Conv2D layers and one full connection layer with decision tree classifier as output layer. This hybrid neural fuzzy inference classifier obtained an accuracy of 98.69% in classifying kidney and tumor classes when compared to other state-of-the-art methodologies. The modified U-Net architecture with different input patching sizes and different artificial neurons in different layers resulted in a dice coefficient of 0.8923. The experimental results showcased that the proposed end-to-end kidney tumor detection system outperformed the competing algorithms. The design and implementation of different hybrid architectures with advanced deep learning techniques and exploration of high-volume datasets for enhanced classification and segmentation of tumors in kidneys are considered as future work.

REFERENCES

[1] Q. Yu, Y. Shi, J. Sun, Y. Gao, J. Zhu and Y. Dai, "Crossbar-net: A novel convolutional neural network for kidney tumor segmentation in CT images," in *IEEE Transactions on Image Processing*, vol. 28, no. 8, pp. 4060–4074, August 2019, doi: 10.1109/TIP.2019.2905537.

[2] Xi Huang, Xiangrui Jin, Jianyong Guo, Guangxin Zhao, Yanrong Tan, Feiming Chen and Geli Zhu, "Kidney damage associated with tumor," *2009 3rd International Conference on Bioinformatics and Biomedical Engineering*, IEEE, Beijing, China, 2009, pp. 1–4, doi: 10.1109/ICBBE.2009.5163002.

[3] G. Tulum, U. Teomete, T. Ergin, Ö. Dandin, F. Cüce, and O. Osman, "Computed aided detection of traumatized kidneys in CT images," *2016 Medical Technologies National Congress (TIPTEKNO)*, 2016, pp. 1–4, doi: 10.1109/TIPTEKNO.2016.7863116.

[4] P. Srinivasa Rao, M. H. M. Krishna Prasad, and K. Thammi Reddy, "An efficient semantic ranked keyword search of big data using map reduce," *IJDTA*, vol.8, no. 6, pp. 47–56, 2015.

[5] S. Aadhirai and D. N. Jamal, "Feature extraction and analysis of renal abnormalities using fuzzy clustering segmentation and SIFT method," *2017 Third International Conference on Biosignals, Images and Instrumentation (ICBSII)*, IEEE, Chennai, India, 2017, pp. 1–5, doi: 10.1109/ICBSII.2017.8082279.

[6] Y. Hu, M. Grossberg, and G. Mageras, "Tumor segmentation with multi-modality image in Conditional Random Field framework with logistic regression models," *2014 36th Annual International Conference of the IEEE Engineering in Medicine and Biology Society*, 2014, pp. 6450–6454, doi: 10.1109/EMBC.2014.6945105.

[7] S. Nedevschi, A. Ciurte, and G. Mile, "Kidney CT image segmentation using multi-feature EM algorithm, based on Gabor filters," *2008 4th International Conference on*

Intelligent Computer Communication and Processing, 2008, pp. 283–286, doi: 10.1109/ICCP.2008.4648387.

[8] P. Srinivasa Rao and S. Satyanarayana, "Privacy preserving data publishing based on sensitivity in context of big data using hive", *Journal of Bigdata*, vol. 5, no. 20, July 2018.

[9] T. V. Madhusudhana Rao, P Srinivasa Rao, and P.S. Latha Kalyampudi, "Iridology based Vital Organs Malfunctioning identification using Machine learning Techniques", International Journal of Advanced Science and Technology, Volume: 29, No. 5, PP: 5544–5554,2020

[10] T. Siriapisith, W. Kusakunniran, and P. Haddawy, "A general approach to segmentation in CT grayscale images using variable neighborhood search," *2018 Digital Image Computing: Techniques and Applications (DICTA)*, 2018, pp. 1–7, doi: 10.1109/DICTA.2018.8615823.

[11] H. Koyuncu and R. Ceylan, "A hybrid tool on denoising and enhancement of abdominal CT images before organ & tumour segmentation," *2017 IEEE 37th International Conference on Electronics and Nanotechnology (ELNANO)*, IEEE, Kyiv, UKraine, 2017, pp. 249–254, doi: 10.1109/ELNANO.2017.7939757.

[12] T. Pan et al., "A multi-task convolutional neural network for renal tumor segmentation and classification using multi-phasic CT images," *2019 IEEE International Conference on Image Processing (ICIP)*, IEEE, Taipei, Taiwan, 2019, pp. 809–813, doi: 10.1109/ICIP.2019.8802924.

[13] X. Hou et al., "A triple-stage self-guided network for kidney tumor segmentation," *2020 IEEE 17th International Symposium on Biomedical Imaging (ISBI)*, IEEE, Iowa City, IA, USA, 2020, pp. 341–344, doi: 10.1109/ISBI45749.2020.9098609.

[14] X. Yan, K. Yuan, W. Zhao, S. Wang, Z. Li, and S. Cui, "An efficient hybrid model for kidney tumor segmentation in CT images," *2020 IEEE 17th International Symposium on Biomedical Imaging (ISBI)*, IEEE, Iowa City, IA, USA, 2020, pp. 333–336, doi: 10.1109/ISBI45749.2020.9098325.

[15] J. Causey et al., "An ensemble of U-net models for kidney tumor segmentation with CT images," in *IEEE/ACM Transactions on Computational Biology and Bioinformatics*, vol. 19, no. 3, pp. 1387–1392, doi: 10.1109/TCBB.2021.3085608.

[16] B. Zhou and L. Chen, "Atlas-based semi-automatic kidney tumor detection and segmentation in CT images," *2016 9th International Congress on Image and Signal Processing, BioMedical Engineering and Informatics (CISP-BMEI)*, IEEE, Datong, China, 2016, pp. 1397–1401, doi: 10.1109/CISP-BMEI.2016.7852935.

[17] T. Qin, Z. Wang, K. He, Y. Shi, Y. Gao, and D. Shen, "Automatic data augmentation via deep reinforcement learning for effective kidney tumor segmentation," *ICASSP 2020–2020 IEEE International Conference on Acoustics, Speech and Signal Processing (ICASSP)*, IEEE, Barcelona, Spain, 2020, pp. 1419–1423, doi: 10.1109/ICASSP40776.2020.9053403.

[18] G. Yang et al., "Automatic segmentation of kidney and renal tumor in CT images based on 3D fully convolutional neural network with pyramid pooling module," *2018 24th International Conference on Pattern Recognition (ICPR)*, IEEE, Beijing, China, 2018, pp. 3790–3795, doi: 10.1109/ICPR.2018.8545143.

[19] M. A. Hussain, G. Hamarneh, and R. Garbi, "Cascaded regression neural nets for kidney localization and segmentation-free volume estimation," in *IEEE Transactions on Medical Imaging*, vol. 40, no. 6, pp. 1555–1567, June 2021, doi: 10.1109/TMI.2021.3060465.

[20] A. Soni and A. Rai, "Kidney stone recognition and extraction using directional emboss & SVM from computed tomography images," *2020 Third International Conference on Multimedia Processing, Communication & Information Technology (MPCIT)*, IEEE, Shivamogga, India, 2020, pp. 57–62, doi: 10.1109/MPCIT51588.2020.9350388.

[21] P. Srinivasa Rao, M. H. M. Krishna Prasad, and K. Thammi Reddy, "A novel and efficient method for protecting internet usage from unauthorized access using map reduce," *IJITCS (MECS)*, vol. 5, no. 3, pp. 49–55, February 2013.

[22] https://kits19.grand-challenge.org/

[23] I. L. Xi, Y. Zhao, et al., "Deep learning to distinguish benign from malignant renal lesions based on routine MR imaging," in *Clinical Cancer Research*, vol. 26, no. 8, pp. 1944–1952, 2020. doi: 10.1158/1078-0432.ccr-19-0374.

[24] Han, S., S. I. Hwang, and H. J. Lee, "The classification of renal cancer in 3-phase CT images using a deep learning method," in *Journal of Digital Imaging*, vol. 32, no. 4, pp. 638–643, 2019. doi: 10.1007/s10278-019-00230-2.

[25] Pedersen, M., M. B. Andersen, H. Christiansen, and N. H. Azawi, "Classification of renal tumour using convolutional neural networks to detect oncocytoma", *European Journal of Radiology*, vol. 133, p. 109343, 2020, doi: 10.1016/j.ejrad.2020.109343

[26] Abdeltawab, H., F. Khalifa, M. Ghazal, L. Cheng, D. Gondim, and A. El-Baz, "A pyramidal deep learning pipeline for kidney whole-slide histology images classification", *Scientific Reports*, vol. 11, p. 20189, 2021, doi: 10.1038/s41598-021-99735-6

[27] Osowska-Kurczab, A. M., T. Markiewicz, M. Dziekiewicz, and M. Lorent, "Combining texture analysis and deep learning in renal tumour classification task", *2020 IEEE 21st International Conference on Computational Problems of Electrical Engineering (CPEE)*, IEEE, (Online Conference), Poland, 2020, doi:10.1109/cpee50798.2020.9238757

[28] Skalski, A., J. Jakubowski, and T. Drewniak, "Kidney tumor segmentation and detection on Computed Tomography data," *2016 IEEE International Conference on Imaging Systems and Techniques (IST)*, IEEE, Chania, Greece, 2016, pp. 238–242, doi: 10.1109/IST.2016.7738230.

[29] Guo, J., W. Zeng, S. Yu, and J. Xiao, "RAU-net: U-net model based on residual and attention for kidney and kidney tumor segmentation," *2021 IEEE International Conference on Consumer Electronics and Computer Engineering (ICCECE)*, IEEE, Guangzhou, China, 2021, pp. 353–356, doi: 10.1109/ICCECE51280.2021.9342530.

7 Classification of Breast Tumor from Histopathological Images with Transfer Learning

R. Meena Prakash, K. Ramalakshmi, S. Thayammal, R. Shantha Selva Kumari, and Henry Selvaraj

CONTENTS

7.1 INTRODUCTION

Breast cancer is one of the most common cancers occurring in women, and the death rate is high. Breast cancer arises from the uncontrolled proliferation of cells in the ducts and milk glands of breast portion. Early detection and diagnosis of breast cancer leads to successful treatment of the disease and increases survival rate. The tissues taken in biopsy are analyzed using microscope by pathologists, and these histopathological images are inspected to perform the diagnosis. Recently, deep learning techniques are successfully being employed for image classification tasks.

DOI: 10.1201/9781003309451-7

Transfer learning is one of the techniques of deep learning, where knowledge gained in solving one problem is saved and used to solve another related problem. The different classification techniques of histopathological images especially using deep learning available in literature are analyzed.

Alirezazadeh et al. (2018) proposed representation learning–based unsupervised domain adaption for classification of breast cancer histopathology images. The feature vectors were extracted from labeled histopathology training set using handcraft descriptors, such as local binary pattern (LBP). A projection matrix was formed through learning of the extracted feature vectors. A learning model was obtained using the basic classifiers and mapped labeled samples. The unlabeled histopathology test image feature vectors were extracted in the same way as the labeled training set. Then they were mapped to an invariant space through the projection matrix and finally classified. The method was tested on BreakHis dataset and obtained an average classification rate of 88.5%.

Boumaraf et al. (2021) proposed a transfer learning–based method for automated classification of breast cancer from histopathological images. The classification included both magnification-dependent and magnification-independent binary and eight class classifications. The deep neural network ResNet-18, which is pretrained on ImageNet, which is the large dataset of images, was used for classification. The method was based on block-wise fine-tuning strategy, in which the last two residual blocks of the deep network model were made more domain-specific to the target data. This helped avoid overfitting and speed up the training process. The method was tested on publicly available BreakHis dataset and achieved an accuracy of 92.03% for eight class classification and 98.42% for binary classification.

Hekler et al. (2019) performed a study in which it was proven that deep learning–based classification method outperformed 11 histopathologists in the classification of histopathological melanoma images.

Gandomkar et al. (2018) proposed a framework called multicategory classification of breast histopathological image using deep residual networks (MuDeRN) for classification of hematoxylin-eosin (H&E)–stained breast digital slides into benign and malignant tumors. Classification rates of 98.52%, 97.90%, 98.33%, and 97.66% were achieved for ×40, ×100, ×200, and ×400 magnification factors, respectively, on BreakHis dataset.

Öztürk and Akdemir (2019) proposed a convolutional neural network (CNN) model to automatically identify cancerous areas on whole-slide histopathological images (WSI). In the proposed method, an effective preprocessing step through the use of Gaussian filter and median filter was included to remove the background noise. The CNN consisted of six convolutional layers, six ReLU layers, six pooling layers, a fully connected layer, an L2 regularization layer, and a dropout layer. The method was tested on dataset comprising of 30,656 images and achieved an accuracy of 97.7%.

Carvalho et al. (2020) proposed a method in which phylogenetic diversity index was used to characterize images, and a model was created. The model was used to classify histopathological breast images into four categories—in situ carcinoma, invasive carcinoma, benign lesion, and normal tissue. Also, content-based image retrieval was performed to confirm the classification results, and a ranking was suggested for sets of images that were not labeled. The method was tested on the Breast Cancer Histology Challenge (BACH) 2018 dataset.

Kumar et al. (2020) proposed a framework based on transfer learning with VGGNet-16 and fine-tuning for classification of canine mammary tumors (CMTs) and human breast cancer. They have introduced a dataset of CMT histopathological images (CMTHis). The proposed framework was tested on CMTHis and BreakHis datasets and achieved an accuracy of 97% and 93%, respectively.

Li et al. (2020) proposed a framework for classification of histopathological images in which the discriminative feature learning and mutual information–based multi-channel joint sparse representation were included. Linear support vector machine (SVM) was used for classification. The method was tested on BreakHis and ADL histopathological datasets.

Sudharshan et al. (2019) proposed a method based on multiple instance learning (MIL) for computer-aided diagnosis of histopathological breast cancer images. The method was tested on BreakHis dataset comprising of 8,000 microscopic biopsy images of benign and malignant tumors which originated from 82 patients. The experimental results proved that nonparametric approach gives the best result among the different MIL methods, including axis-parallel rectangle (APR), diverse density, MI-SVM, citationK-nearestneighbor (CKNN), and MIL-CNN.

Hekler et al. (2019) implemented a deep learning–based classification of histopathological melanoma images, where 695 lesions were classified by an expert histopathologist as 350 nevi and 345 melanomas. Out of the 695 images, 595 H&E images were used for training and 100 images were used for testing the deep neural networks. The findings proved that the total discordance with the histopathologist was 18% for melanoma, and it was concluded that CNN could be a valuable tool to assist melanoma diagnosis in histopathological images.

Toğaçar et al. (2020) proposed a convolutional neural network model called BreastNet for the diagnosis of breast cancer through histopathological images. The general structure of the model was a residual architecture built on attention modules. Before applying as input to the model, data augmentation techniques such as flip, rotation, and shift were employed.

Öztürk and Akdemir (2018) employed different feature extraction algorithms, including GLCM, LBP, LBGLCM, GLRLM, and SFTA, to extract various features from histopathological images. The extracted features were classified with classifiers of SVM, KNN, LDA, and boosted tree classifiers. Best classification results were achieved for SVM and boosted tree.

Wang et al. (2021) proposed a classification of histopathological images based on deep feature fusion and enhanced routing, which took the advantages of both convolutional neural network and capsule network. A structure with two channels was designed to extract convolution features and capsule features simultaneously and then integrate the semantic features and spatial features into new capsules to obtain more discriminative information. By modifying the loss function, the routing coefficients were optimized indirectly and adaptively. The routing process was embedded into the entire optimization process. The proposed method was tested on the public BreakHis dataset.

Yan et al. (2018) proposed a hybrid deep neural network for classification of breast cancer histopathological images. The method integrated the advantages of convolutional and recurrent neural networks based on the multilevel feature representation of the histopathological image patches. The short-term and long-term spatial correlations between patches were preserved. The method was tested on the dataset of 3,771 breast cancer histopathological images by the authors.

Deniz et al. (2018) proposed transfer learning and deep feature extraction methods for breast cancer detection from histopathological images. AlexNet and VGG16 were used as pretrained CNN models for feature extraction. Classification was done with support vector machine (SVM). The method was tested on publicly available hithopathological breast cancer dataset, and the experiments showed that transfer learning produced better results compared to deep feature extraction and SVM classification.

Ahmed et al. (2021) proposed a transfer learning–based framework with two popular deep learning models, Inception-V3 and VGG-16. The pretrained weights of these models were used and concatenated with an image vector, which was used as input to train the network architecture. The Kimia Path24 dataset, comprising 23,916 histopathological patches with 24 tissue texture classes, was used to evaluate the performance of the model.

Xie et al. (2019) utilized transfer learning techniques for binary and multiclass classification of breast cancer histopathological images with Inception_V3 and Inception_ResNet_V2 architectures. Data augmentation techniques were employed to help the model generalize better, thus preventing overfitting. A new autoencoder network was also constructed to transform the features extracted by Inception_ResNet_V2 to a low dimensional space and to do clustering analysis of the images. The method was tested on publicly available BreakHis dataset.

Ahmad et al. (2019) proposed a transfer learning–based classification of histopathological images based on AlexNet, GoogleNet, and ResNet which classifies images at multiple cellular and nuclei configurations. The dataset comprises of 240 training and 20 test images for classification of histology images into four classes—normal, benign, insitu carcinoma, and invasive carcinoma. An accuracy of 85% was achieved with the case of ResNet.

Talo (2019) proposed automatic classification of histopathological images using transfer learning. ResNet-50 and DenseNet-161 pretrained models were employed for classification. The method was tested on publicly available Kimia Path24 histopathology dataset.

Murtaza et al. (2019) proposed a transfer learning–based approach for classification of digital biopsy histopathology images. The pretrained model of AlexNet was retained, with the last layer fine-tuned for binary classification into benign and malignant tumors. The preprocessed images were fed to the model for training. The model training was repeated several times by changing the values of hyperparameters randomly until minimum validation loss was achieved. The trained model was used for feature extraction, followed by classification with six machine learning classifiers, softmax, decision tree, naive Bayes, linear discriminant analysis, support vector machine, and K-nearest neighbor. The method was tested on the BreakHis dataset, which is publicly available.

Based on the literature review, it is proposed to implement a transfer learning–based classification of histopathological images into benign and malignant tumors. The main contributions of the chapter include (i) implementation of transfer learning–based method for classification of histopathological images, (ii) four popular pretrained models—AlexNet, VGG16, Inception v3, and DenseNet121—employed for classification, and (iii) comparison of the results obtained for the four different architectures and with other state-of-the-the art methods. The rest of the paper is organized as follows: The concepts of deep learning, CNN, transfer learning, and architecture of pretrained models and the proposed method are discussed in Section 2. The experiments and results are discussed in Section 3, with a conclusion in Section 4.

7.2 MATERIALS AND METHODS

7.2.1 Deep Learning

Deep learning is a subfield of machine learning which emphasizes learning representations from data with successive layers of increasingly meaningful representations. In deep learning, these layered representations are learned by means of models, called neural networks. The number of layers in the network is called the depth of the model. Figure 7.1 shows the concept of deep learning.

The weights of the layers of the network are initialized with random values. The loss function of the network is calculated as the distance between the predictions of the network and the true target. Thus, the loss score will be initially high. With every training data, the weights are adjusted a little in the right direction and the loss score decreases. The training loop is repeated a sufficient number of times (thousands of data and tens of iterations), and the weight values are obtained, which minimize the loss function.

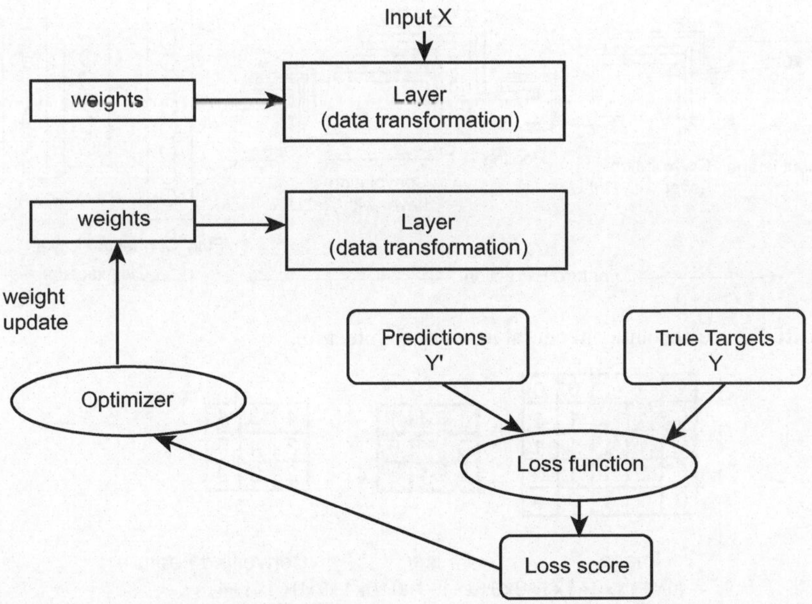

FIGURE 7.1 Concept of deep learning (Chollet, 2021).

The mechanism through which the network will update itself based on the training data and loss function is called the optimizer. A trained network will be obtained once the outputs are as close to the targets. Once the network is trained, it is used to evaluate the test data.

7.2.2 Convolutional Neural Networks

The convolutional neural network (CNN) comprises of the convolutional layer and pooling layer for feature extraction, and the fully connected layer for classification, which is shown in Figure 7.2.

The spatial and temporal dependencies in an image are successfully captured through the application of relevant filters in a CNN. The convolution operation is depicted in Figure 7.3.

The input image block of size 5 × 5 is convolved with a filter of size 3 × 3. The stride is 1, and hence, the filter shifts nine times. Every time the filter shifts, point-by-point matrix multiplication is taken between portion of the image and the filter, as shown in Figure 7.3. The process is repeated until the entire image is transversed. The high-level features, such as edges and curves, are extracted from the image through the convolution operation. The low-level features are extracted in the first convolutional layer, while the higher-level features are extracted in the successive convolutional layers. The objective of the pooling layer is to reduce the size of the feature map so that computational power is decreased. In max pooling operation, the maximum value of the portion of the image is taken, while in average pooling operation, the average of all the values in the portion of the image is taken. The illustration of pooling operation is shown in Figure 7.4.

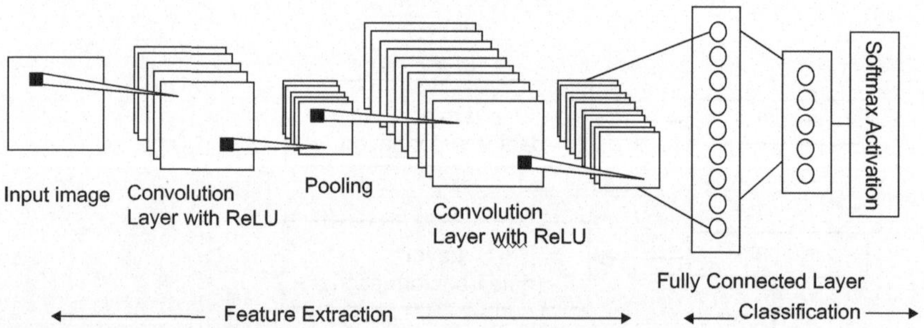

FIGURE 7.2 Convolutional neural network architecture.

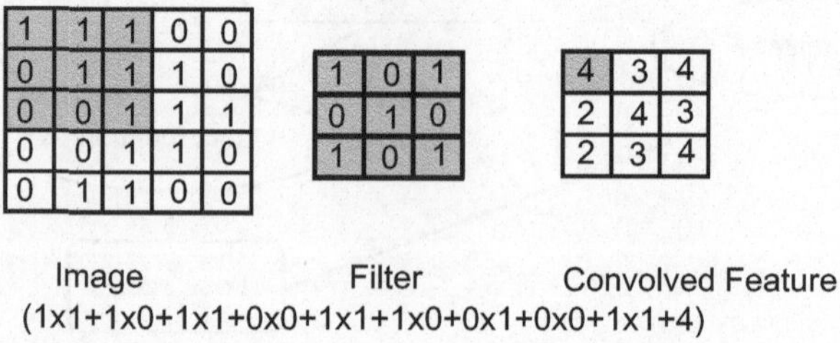

FIGURE 7.3 Illustration of convolution operation.

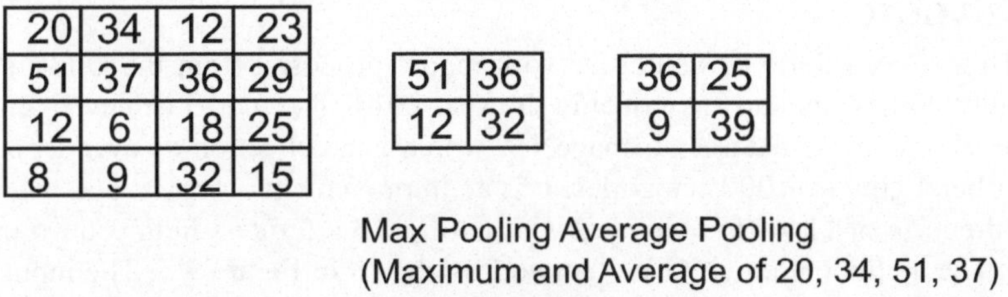

FIGURE 7.4 Illustration of pooling operation.

7.2.3 Transfer Learning

Transfer learning is the process of utilizing knowledge learned from one task to solve related tasks. It is referred to as a situation where what has been learned in one setting is exploited to improve generalization in another setting.

7.2.3.1 AlexNet

The AlexNet architecture consists of eight layers with learnable parameters. There are five convolutional layers and three max pooling layers. There are four fully connected layers and three dropout layers and softmax activation. The architecture of AlexNet is shown in Table 7.1

AlexNet was used to classify the images from the subset of ImageNet dataset with almost 1.2 million training images, 150,000 test images, and 50,000 validation images. The activation function used in the convolutional layers is rectified linear unit, and for classification in the final dense layer, softmax activation function is used.

TABLE 7.1
Architecture of AlexNet

Layer		Size	Feature map	Kernel size	Stride	Activation
Input	**Image**	**227 × 227 × 3**	**1**	**-**	**-**	**-**
	Convolution	55 × 55 × 96	96	11×11	4	ReLU
	Max pooling	27 × 27 × 96	96	3×3	2	ReLU
	Convolution	27 × 27 × 256	256	5×5	1	ReLU
	Max pooling	13 × 13 × 256	256	3×3	2	ReLU
	Convolution	13 × 13 × 384	384	3×3	1	ReLU
	Convolution	13 × 13 × 384	384	3×3	1	ReLU
	Convolution	13 × 13 × 256	256	3×3	1	ReLU
	Max pooling	6 × 6 × 256	256	3×3	2	ReLU
	FCL	-	9,216	-	-	ReLU
	FCL	-	4,096	-	-	ReLU
	FCL	-	4,096	-	-	ReLU
Output	FCL	-	1,000	-	-	Softmax

7.2.3.2 VGG16

VGG16 is a convolutional neural network model proposed from the University of Oxford by Simonyan and Zissermanin the year 2014. The model achieved an accuracy of 92.7% in the dataset of ImageNet, which is a collection of over 14 million images belonging to 1,000 categories. It is an improvement over Alexnet, where the large filter size of 11 × 11 is replaced with multiple 3 × 3 filters in the convolutional layers. The architecture of VGG16 network is shown in Figure 7.5. The input layer of VGG16 is fed with input RGB images of size 224 × 2243. The network consists of 16 weight layers comprising of 13 convolutional layers and 3 fully connected layers. The number of filters employed is 64 in the first layer, and then increasing by a factor of 2 after each max pooling layer. The last convolutional layer has 512 filters. The fully connected layer with softmax activation function is used for classification.

7.2.3.3 DenseNet121

In the dense convolutional network (DenseNet), each layer is connected to every other layer in a feed-forward fashion (Huang et al., 2017). Traditional convolutional networks have their connections between each layer and its succeeding layer. Hence, convolutional network with L layers have L connections. In DenseNet, there are $L(L+1)/2$ direct connections. The feature maps of all previous layers are used as input to the current layer, and the feature maps of the current layer are used as inputs to all the succeeding layers. The major advantages of DenseNet are to improve the problem of vanishing gradient, build up propagation of features, support feature reuse, and greatly reduce the number of parameters.

The network is called dense convolutional network due to its dense connectivity model. The network is divided into multiple densely connected dense blocks. The layers between the dense blocks are called transition layers, which perform convolution and pooling. The transition layer consists of a batch normalization layer, a 1 × 1 convolutional layer, and a 2 × 2 average pooling layer. Every dense layer receives outputs of all the previous layers. The input depth for the kth layer is ((k-1)×growth rate) + input depth of the first layer. A relatively small growth rate is enough to obtain

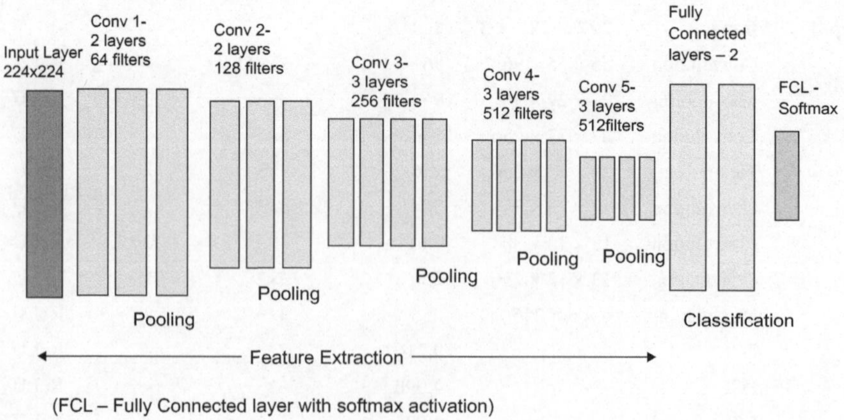

FIGURE 7.5 Architecture of VGG16.

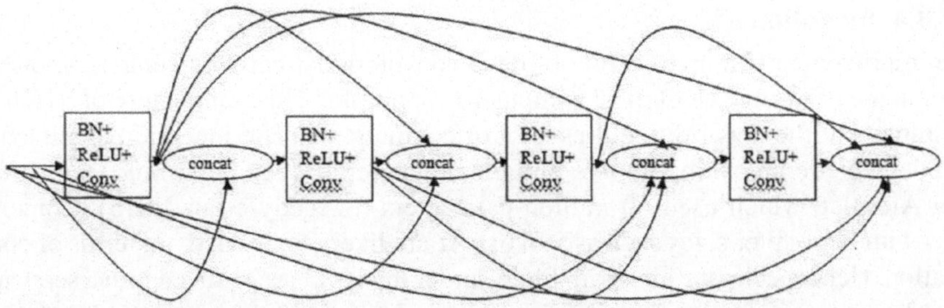

FIGURE 7.6 Four-layer dense block in DenseNet. The input to all layers is the concatenation of feature maps from all previous layers.

good results. The growth rate controls the amount of new information contributed to the global state by each layer. Figure 7.6 shows four-layer dense block in DenseNet.

DenseNet121 has four dense blocks, and the input size is 224×224. There are three transition layers and one fully connected layer. The main advantage of DenseNet is that it requires fewer parameters and less computation to achieve excellent performance. The architecture of DenseNet121 for ImageNet is shown in Figure 7.7. Each Conv layer corresponds to batch normalization, rectified linear unit, and convolutional layer.

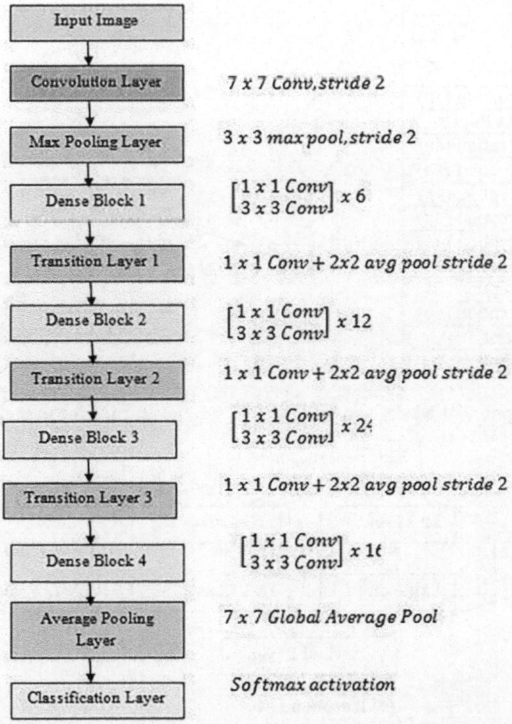

FIGURE 7.7 Architecture of DenseNet121.

7.2.3.4 Inception v3

The improvements in architecture of deep convolutional networks enable improvement in the performance of most computer vision tasks. The architecture of VGGNet is simple, but the network requires a lot of computation. The Inception architecture of GoogleNet employed only five million parameters, which is 12 times reductions over AlexNet, which used 60 million parameters (Szegedy et al., 2015). Convolutions with large filter sizes such as 5×5 or 7×7 are likely to be costly in terms of computation. Hence, convolutions with filters larger than 3×3 are reduced into a sequence of 3×3 convolutional layers. The architecture of Inception V1 is shown in Figure 7.8.

The inception layer is the center concept of lightly connected architecture. The Inception layer is a combination of 1×1, 3×3, and 5×5 convolutional layers. The output filter banks are concatenated into a single output vector which forms the input of the next stage, as shown in Figure 7.8 (b). Two additions have been made in the idea of inception module. A 1×1 convolutional layer is applied before another layer for dimensionality reduction. Also, a parallel max pooling layer is added, as shown in Figure 7.8 (c). The inception idea is based on the fact that when creating a succeeding layer in a network, attention should be paid to learn the previous layer. The Inception V1 architecture is basically a convolutional neural network with a depth of 27 layers. The inception layer is a combination of

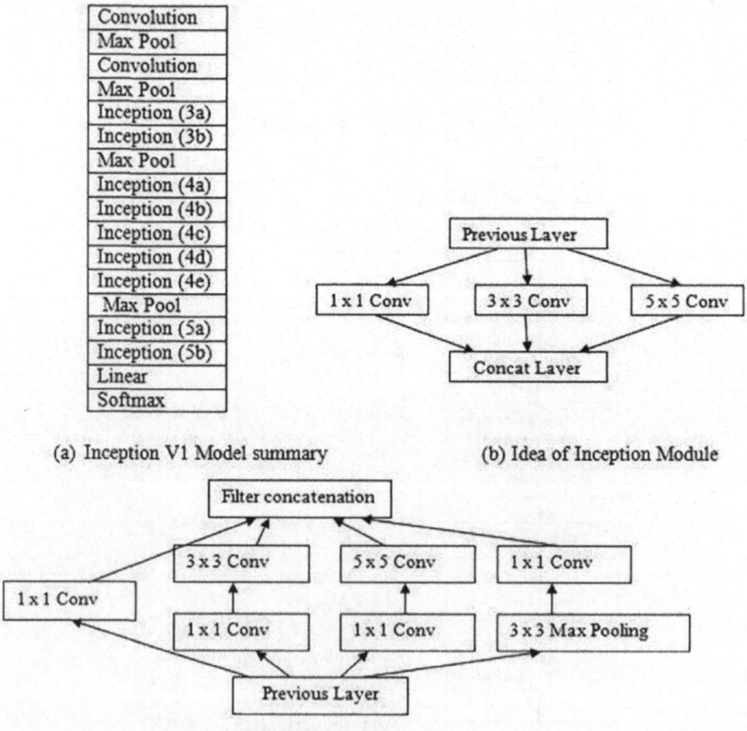

(a) Inception V1 Model summary (b) Idea of Inception Module

(c) Inception Layer

FIGURE 7.8 Inception V1 or GoogleNet architecture.

1 × 1, 3×3, and 5×5 convolutional layers, and the output filter banks are concatenated into a single output vector to form the input of the next stage (Szegedy et al., 2016).

7.2.4 Proposed Method

The proposed method is shown in Figure 7.9.

The input images are split into training set and test set. First, the input images are normalized so that each pixel in the image has the value between 0 and 1. The input images are fed to the pretrained models with the top dense layers removed. The pretrained models used in our method are AlexNet, VGG16, Inception v3, and DenseNet121. These models are already trained on ImageNet database, consisting of more than ten million images belonging to 1,000 classes. The weights of the layers are properly set through training, and these pretrained models are fine-tuned on the input images from BreakHis. The BreakHis dataset comprises 5,429 benign tumor images and 2,480 malignant images, with a total of 7,909 images. Since the dataset is imbalanced, up-sampling is done, and the number of malignant images also is made to 5,429; hence, the total number of images in the dataset has become 10,858. After, the weight layers of the pretrained models, flatten layer, dense layers (fully connected layer), and dropout layers are added. The final fully connected layer has the number of units as two for binary classification into benign and malignant tumors. The softmax activation function is used in this dense layer. The model is compiled using Adam optimizer with learning rate of 10e-6. The batch size is 256, and the number of epochs is 120, except for VGG16, where the number of epochs is set at 40 since the computation is heavy for the VGG16 pretrained model.

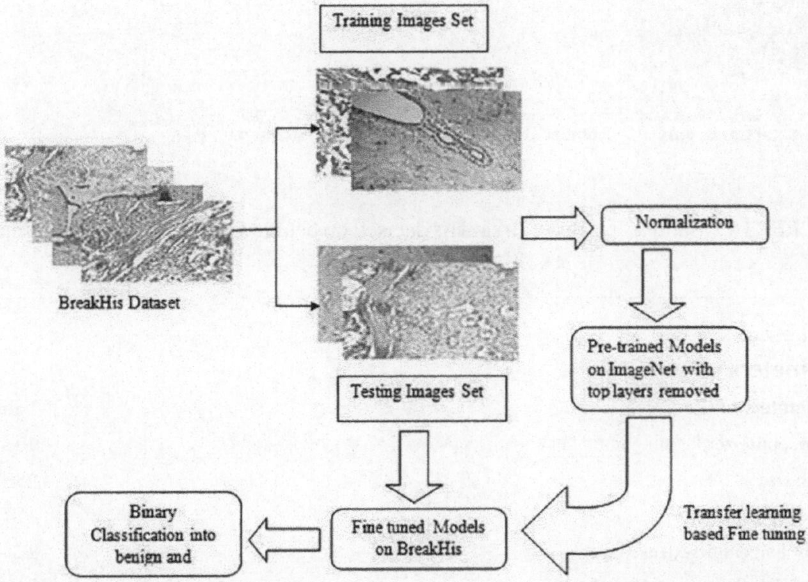

FIGURE 7.9 Proposed method.

7.3 EXPERIMENTS AND RESULTS

7.3.1 Dataset

The BreakHis dataset is used for the evaluation of the proposed method, which is provided by the P&D Laboratory, Pathological Anatomy and Cytopathology, Brazil. The samples of benign and malignant images of the dataset are shown in Figure 7.1. The dataset comprises of 7,909 breast histopathological images (H&E stained), with 40×, 100×, 200×, and 400× magnifying factors. The dataset includes two types of tumor images, benign and malignant, belonging to eight classes, as shown in Figure 7.10. The dataset is split in the ratio of 80:20 for training images and testing images.

7.3.2 Selection of Parameters and Performance Metrics

The software tool used is Python with Keras. The pretrained models used are AlexNet, VGG16, Inception v3, and DenseNet121. The input images are down-sampled to a size of 120×120 and fed to the network. The parameters of the CNN added after the pretrained models are shown in Table 7.2.

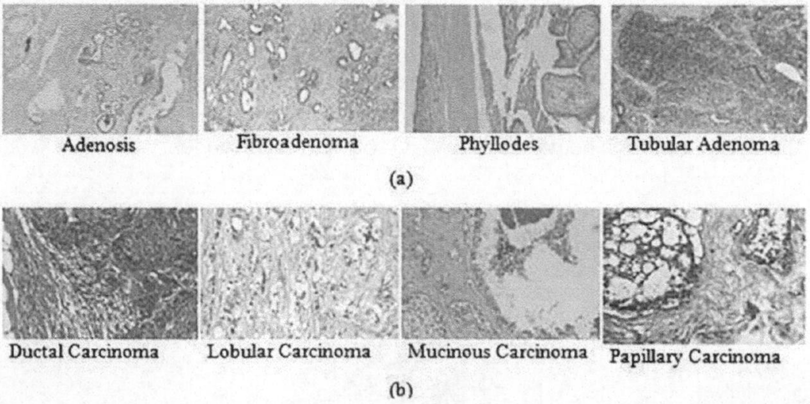

FIGURE 7.10 Sample images of BreakHis dataset: (a) benign images, (b) malignant images.

TABLE 7.2
Parameters of the CNN

Parameters of the CNN	Values
Adam optimizer learning rate	10e-6
Batch size	256
No. of epochs	120
No. of units in first dense layer	128
No. of units in second dense layer	2
Dropout rate	0.5

Comparing the total number of parameters of the three networks—VGG16, Inception v3, and DenseNet121—the pretrained model of DenseNet121 has fewer parameters, and hence, the computation requirement is less compared to other models and faster.

The performance metrics used to evaluate the proposed method is classification accuracy, which is given by:

$$Classification\ Accuracy = \frac{Number\ of\ correct\ predictions}{Total\ Number\ of\ predictions}$$

7.3.3 Experimental Results

The experimental results of the pretrained models AlexNet, VGG16, Inception v3, and DenseNet121 are shown in Table 7.3.

TABLE 7.3
Classification Results of the Proposed Method

Pretrained model	Classification accuracy
VGG16	0.7750
Inception v3	0.8700
AlexNet	0.8780
DenseNet121	**0.9520**

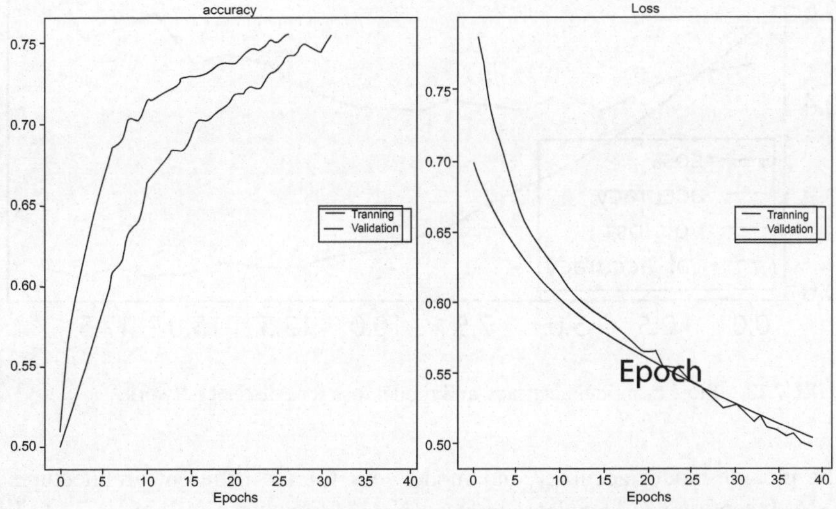

FIGURE 7.11 Plots of model accuracy and model loss for VGG16 network.

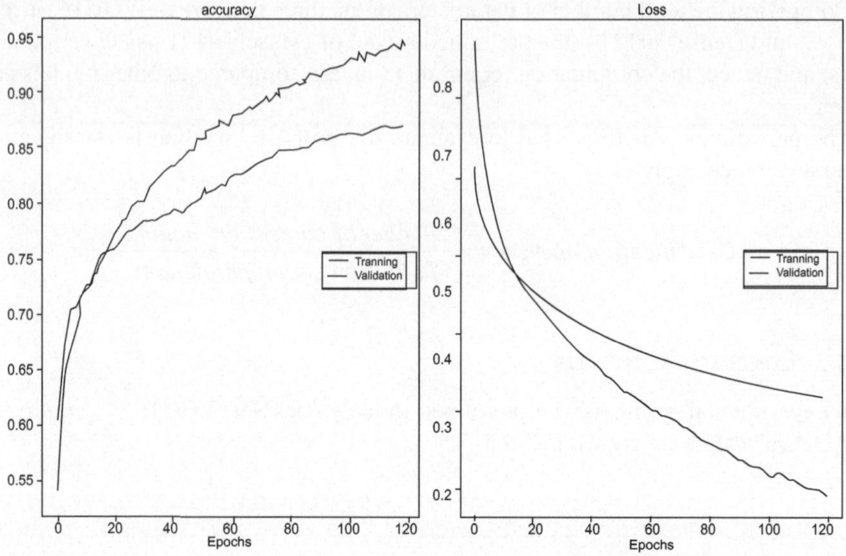

FIGURE 7.12 Plots of model accuracy and model loss for Inception v3 network.

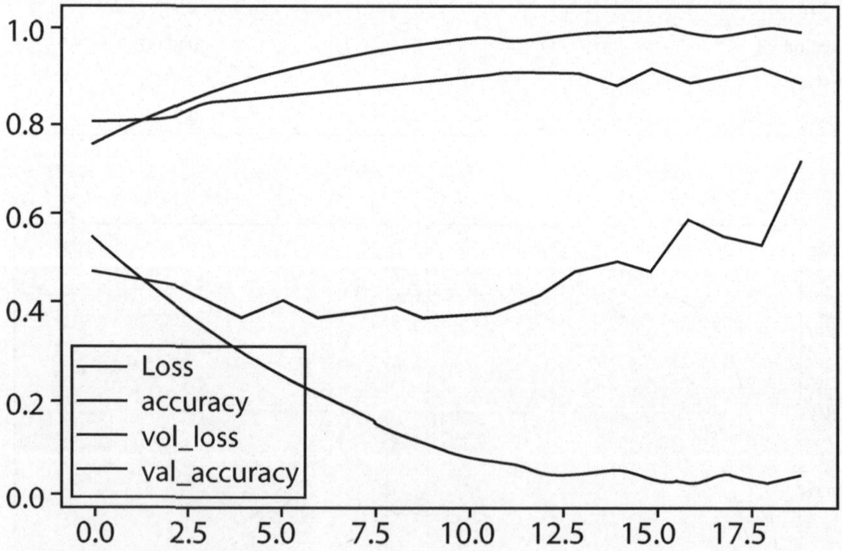

FIGURE 7.13 Plots of model accuracy and model loss for AlexNet network.

The plot of model accuracy and model loss for the different architectures of VGG16, Inception v3, AlexNet, and DenseNet121 are shown in Figures 7.11, 7.12, 7.13, and 7.14, respectively.

From the results, it is inferred that the DenseNet121 architecture gives the highest classification accuracy of 0.9520 compared to the other models. The number of

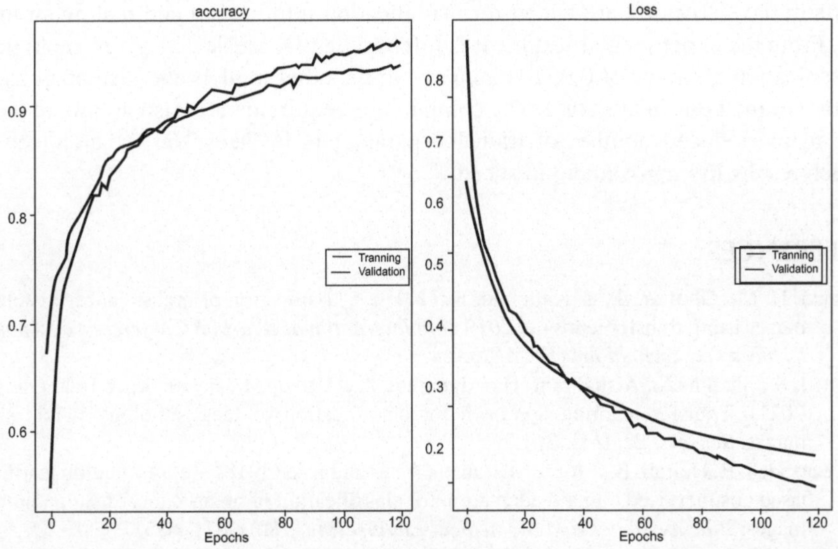

FIGURE 7.14 Plots of model accuracy and model loss for DenseNet121 network.

TABLE 7.4
Comparison of Results of Proposed Method with Other Method

Method	Model	Classification Accuracy
Deniz et al. (2018)	Fine-tuned AlexNet	0.9130
Proposed method	**DenseNet121**	**0.9520**

trainable parameters is also less in the case of DenseNet121 compared to other models, and hence faster. The comparison of the proposed method with the other method in literature is shown in Table 7.4.

It is inferred from Table 7.3 that DenseNet121 architecture gives a better classification accuracy of 0.9520 on BreakHis dataset compared to other state-of-the-art methods in literature.

7.4 CONCLUSION

In this chapter, a classification of breast tumor from histopathological images with transfer learning is proposed. Transfer learning requires less computation, as the model need not learn from scratch. The knowledge gained in one task is used to solve another related task. Four popular pretrained models—AlexNet, VGG16, Inception v3, and DenseNet121—are used for classification of the histopathological images form BreakHis dataset into benign and malignant tumors. The pretrained models are already trained on images from huge dataset of ImagNet with more than one million images belonging to 1,000 classes. The models are then fine-tuned on the BreakHis dataset, and features are extracted. Then, flatten layers, dense layers, dropout layer, and softmax activation are added for classification into benign and malignant tumors. From the experimental results, it is inferred that DenseNet121 gives the highest classification accuracy of 0.9520 compared to the other models and also other state-of-the-art methods in literature. The computation requirement is also greatly reduced due to the reduced number of trainable parameters in DenseNet121 architecture, which results in improvement in speed.

REFERENCES

Ahmad, H. M., Ghuffar, S., & Khurshid, K. (2019). Classification of breast cancer histology images using transfer learning. *2019 16th International Bhurban Conference on Applied Sciences and Technology (IBCAST).*

Ahmed, S., Shaikh, A., Alshahrani, H., Alghamdi, A., Alrizq, M., Baber, J., & Bakhtyar, M. (2021). Transfer learning approach for classification of histopathology whole slide images. *Sensors*, *21*(16), 5361.

Alirezazadeh, P., Hejrati, B., Monsef-Esfahani, A., & Fathi, A. (2018). Representation learning-based unsupervised domain adaptation for classification of breast cancer histopathology images. *Biocybernetics and Biomedical Engineering*, *38*(3), 671–683.

Boumaraf, S., Liu, X., Zheng, Z., Ma, X., & Ferkous, C. (2021). A new transfer learning based approach to magnification dependent and independent classification of breast cancer in histopathological images. *Biomedical Signal Processing and Control*, *63*, 102192.

Carvalho, E. D., Filho, A. O., Silva, R. R., Araújo, F. H., Diniz, J. O., Silva, A. C., . . . Gattass, M. (2020). Breast cancer diagnosis from histopathological images using textural features and CBIR. *Artificial Intelligence in Medicine*, *105*, 101845.

Chollet, F. (2021). *Deep learning with python*. Oreilly Media.

Deniz, E., Şengür, A., Kadiroğlu, Z., Guo, Y., Bajaj, V., & Budak, Ü. (2018). Transfer learning based histopathologic image classification for breast cancer detection. *Health Information Science and Systems*, *6*(1).

Gandomkar, Z., Brennan, P. C., & Mello-Thoms, C. (2018). MuDeRN: Multi-category classification of breast histopathological image using deep residual networks. *Artificial Intelligence in Medicine*, *88*, 14–24.

Hekler, A., Utikal, J. S., Enk, A. H., Berking, C., Klode, J., Schadendorf, D., . . . Brinker, T. J. (2019). Pathologist-level classification of histopathological melanoma images with deep neural networks. *European Journal of Cancer*, *115*, 79–83.

Hekler, A., Utikal, J. S., Enk, A. H., Solass, W., Schmitt, M., Klode, J., . . . Brinker, T. J. (2019). Deep learning outperformed 11 pathologists in the classification of histopathological melanoma images. *European Journal of Cancer*, *118*, 91–96.

Huang, G., Liu, Z., Maaten, L. V., & Weinberger, K. Q. (2017). Densely connected convolutional networks. *2017 IEEE Conference on Computer Vision and Pattern Recognition (CVPR).*

Kumar, A., Singh, S. K., Saxena, S., Lakshmanan, K., Sangaiah, A. K., Chauhan, H., . . . Singh, R. K. (2020). Deep feature learning for histopathological image classification of canine mammary tumors and human breast cancer. *Information Sciences*, *508*, 405–421.

Li, X., Tang, H., Zhang, D., Liu, T., Mao, L., & Chen, T. (2020). Histopathological image classification through discriminative feature learning and mutual information-based multi-channel joint sparse representation. *Journal of Visual Communication and Image Representation*, *70*, 102799.

Murtaza, G., Shuib, L., Wahab, A. W., Mujtaba, G., Mujtaba, G., Raza, G., & Azmi, N. A. (2019). Breast cancer classification using digital biopsy histopathology images through transfer learning. *Journal of Physics: Conference Series*, *1339*, 012035.

Öztürk, Ş, & Akdemir, B. (2018). Application of feature extraction and classification methods for histopathological image using GLCM, LBP, LBGLCM, GLRLM and SFTA. *Procedia Computer Science*, *132*, 40–46.

Öztürk, Ş, & Akdemir, B. (2019). HIC-net: A deep convolutional neural network model for classification of histopathological breast images. *Computers & Electrical Engineering*, *76*, 299–310.

Sudharshan, P., Petitjean, C., Spanhol, F., Oliveira, L. E., Heutte, L., & Honeine, P. (2019). Multiple instance learning for histopathological breast cancer image classification. *Expert Systems with Applications*, *117*, 103–111.

Szegedy, C., Liu, W., Jia, Y., Sermanet, P., Reed, S., Anguelov, D., . . . Rabinovich, A. (2015). Going deeper with convolutions. *2015 IEEE Conference on Computer Vision and Pattern Recognition (CVPR)*, IEEE, Boston, MA.

Szegedy, C., Vanhoucke, V., Ioffe, S., Shlens, J., & Wojna, Z. (2016). Rethinking the inception architecture for computer vision. *2016 IEEE Conference on Computer Vision and Pattern Recognition (CVPR)*, IEEE, Las Vegas, NV, USA.

Talo, M. (2019). Automated classification of histopathology images using transfer learning. *Artificial Intelligence in Medicine*, *101*, 101743.

Toğaçar, M., Özkurt, K. B., Ergen, B., & Cömert, Z. (2020). BreastNet: A novel convolutional neural network model through histopathological images for the diagnosis of breast cancer. *Physica A: Statistical Mechanics and Its Applications*, *545*, 123592.

Wang, P., Wang, J., Li, Y., Li, P., Li, L., & Jiang, M. (2021). Automatic classification of breast cancer histopathological images based on deep feature fusion and enhanced routing. *Biomedical Signal Processing and Control*, *65*, 102341.

Xie, J., Liu, R., Luttrell, J., & Zhang, C. (2019). Deep learning based analysis of histopathological images of breast cancer. *Frontiers in Genetics*, *10*.

Yan, R., Ren, F., Wang, Z., Wang, L., Ren, Y., Liu, Y., . . . Zhang, F. (2018). A hybrid convolutional and recurrent deep neural network for breast cancer pathological image classification. *2018 IEEE International Conference on Bioinformatics and Biomedicine (BIBM)*, IEEE, Spain.

8 Performance of IoT-Enabled Devices in Remote Health Monitoring Applications

D. Narendar Singh, Pavitra B, Ashish Singh, and Jayasimha Reddy A

CONTENTS

8.1 INTRODUCTION

Remote patient monitoring is the most common IoT use in health care. IoT devices can assess a patient's heart rate, blood pressure, and temperature remotely, saving them time and money. As demonstrated in Figure 1.1, health-care professionals and

DOI: 10.1201/9781003309451-8

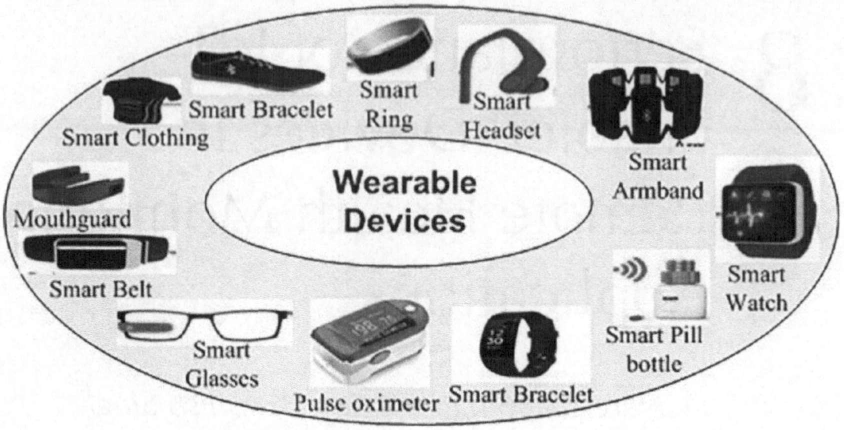

FIGURE 8.1 IOT smart wearable devices.

patients may access patient data via IoT devices. Based on data analysis, algorithms might suggest or warn about treatments. A patient's heart rate may be monitored by medical staff. It's difficult to keep very sensitive data collected by remote patient monitoring devices secure and private (Iranpak et al., 2021)

Diabetes affects almost 30 million Americans, making glucose monitoring a difficult task. Determining a patient's glucose levels at the moment of testing is difficult (Rowley et al., 2017). Even if levels don't fluctuate much, periodic testing may not show a problem.

Monitoring glucose levels in patients with IoT devices may help alleviate some of these concerns. Diabetics may be alerted to dangerously low blood sugar levels with glucose monitoring devices. It is critical for IoT glucose monitoring devices to be small enough to monitor constantly without disturbing patients while also not using excessive energy and requiring frequent recharging. Despite the obstacles, it is still conceivable to revolutionize the way diabetics check their blood glucose levels.

Even in the friendliest environments, it might be challenging to measure heart rate and blood sugar levels correctly. When a patient is continually connected to a monitoring equipment, their mobility is considerably restricted. Heart rate checks cannot safeguard against abrupt variations in heart rate. Patients may now move about while their heart rate is continually monitored, thanks to the development of affordable Internet of Things (IoT) sensors, as seen in Figure 1.2. While ultra-accurate measurements are still uncommon, the majority of existing technologies reach accuracy rates of 90% or better (Farahani et al., 2018).

When it comes to limiting the transmission of illness, doctors and patients in a health-care facility lack reliable ways of determining if they washed their hands sufficiently. In various hospitals and other health-care institutions, IoT devices remind patients to wash their hands before entering hospital rooms. Additionally, the gadgets may provide recommendations on cleaning in order to minimize a particular threat to a patient. Due to the fact that these devices might serve as a reminder to clean one's

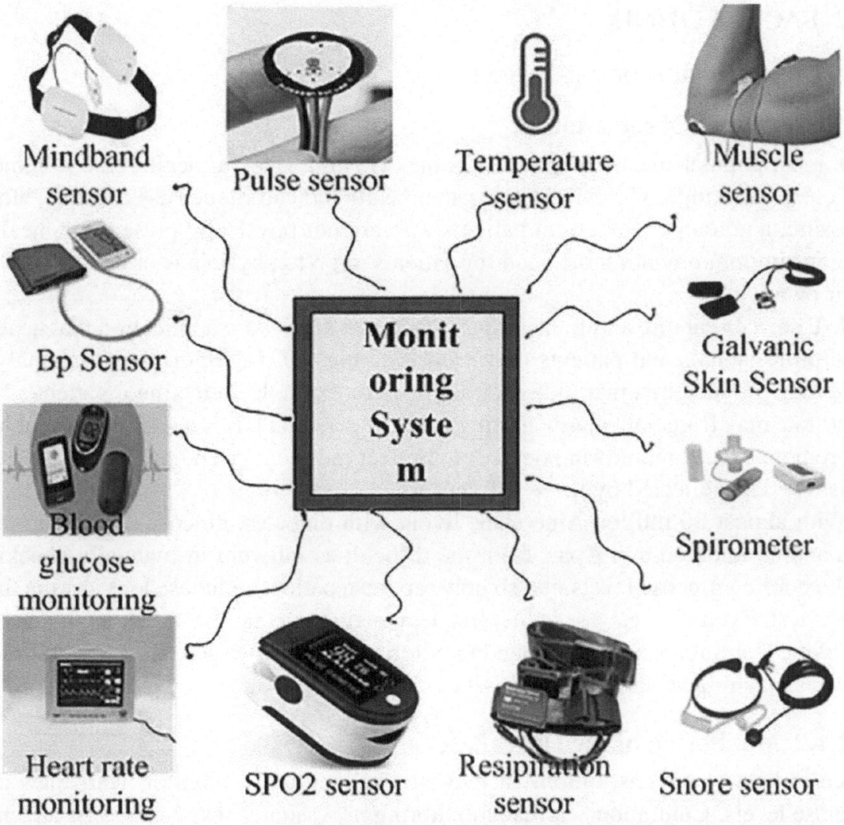

FIGURE 8.2 IOT biomedical sensor monitoring system.

hands rather than really cleaning them, they have a number of major downsides that must be addressed. According to study, these devices have the potential to reduce hospital infection rates by more than 60%.

Other sorts of data have proven more challenging to gather on a consistent basis, such as patients' depressed symptoms and general mood. Patients could speak with their physicians about their feelings on a frequent basis, but they could not foresee mood swings. Frequently, patient expressiveness is insufficient.

These issues might be solved by "mood-aware" Internet of Things devices. The mental state of a patient may be deduced from medical equipment–collected and analyzed information, such as heart rate and blood pressure. Advanced IoT mood-monitoring devices can track a patient's eye movement.

A host of new opportunities for health-care practitioners and patients to keep tabs on each other are opened up by IoT devices. As a consequence, both health-care professionals and patients may benefit from and have issues with different wearable Internet of Things (IoT) devices.

8.2 BACKGROUND

8.2.1 Observation of the Patient

8.2.1.1 Patient Observation

The most popular use of Internet of Things (IoT) devices in health care is remote patient monitoring. IoT devices may gather health indicators such as heart rate, blood pressure, and temperature from patients who are not physically present at a health-care institution, obviating the need for patients to go to physicians or collect data on their own.

IoT devices capture and transmit patient data to a software application that health-care professionals and patients may view (Narang, 2021). Algorithms that analyze data may produce treatment ideas or alarms. For example, detecting a patient's low heart rate may trigger an alarm alerting medical personnel. It is a significant problem for remote patient monitoring devices to protect the security and privacy of the very sensitive data collected by these IoT sensors.

With almost 30 million Americans living with diabetes, glucose monitoring is a demanding undertaking. Apart from the difficulties inherent in manually checking and recording glucose levels, it also only reports a patient's glucose levels at the time of the test. Even if levels are stable, frequent testing may not be sufficient to detect a problem. Continuous and automated monitoring of patients' glucose levels with IoT devices may relieve some of these issues.

8.2.1.2 Monitoring of the Heart Rate

Even in health-care institutions, it may be challenging to monitor heart rates and glucose levels. Continuous cardiac monitoring necessitates that patients be attached to connected devices at all times, greatly limiting their movement. Periodic heart rate checks will not protect you against rapid heart rate fluctuations. Patients may now roam freely while their heart rate is continually monitored, courtesy of a number of tiny Internet of Things devices. While ultra-accurate readings remain a challenge, the majority of current technologies can achieve accuracy rates of about 90% or greater.

8.2.1.3 Maintaining an Awareness of Your Mood

Depressive symptoms and patients' overall mood are two additional forms of data that have been challenging to obtain on a consistent basis in the past. Patients may be often asked how they are feeling, but health-care providers have been unable to predict mood fluctuations. Patients often lack the ability to effectively communicate their emotions.

8.3 THE IMPORTANCE OF IOT IN THE HEALTH-CARE INDUSTRY

- Faster diagnosis. Diseases and health disorders may be discovered earlier in the course of therapy with the aid of IoT-enabled monitoring equipment.
- Because data obtained by networked medical equipment is error-free, the possibility of human error is reduced. This increases the field's capacity to provide precise diagnoses.

- Medical equipment driven by the Internet of Things enables physicians to monitor patients continuously, enabling more specialized and customized therapies.
- The Internet of Things enables a decrease in medical waste caused by wasteful use of equipment and pharmaceuticals (IoT).
- Early detection of adverse effects. The cutting-edge technology utilized in IoT enables the early detection of adverse effects associated with pharmaceuticals.
- Consumers who adhere to treatment suggestions throughout the recovery period may receive awards from some insurers.
- Developing a proactive health-care system, as opposed to one that only offers reactive therapy in the case of disease discovery.
- The adoption of smart monitoring devices may assist in the early detection and treatment of symptoms.

8.4 ADVANTAGES

- Provision of feedback and monitoring in real time.
- Remote health monitoring through linked devices may aid patients with heart failure, diabetes, or asthma attacks.
- Connected medical devices may collect medical and other necessary health data and transmit it to a doctor or to a cloud platform using a smartphone app and a smart medical device connected to a smartphone's internet connection.
- Connectivity from end to end at a reasonable price.
- Through the deployment of health-care mobility solutions and other new IoT technologies, as well as next-generation health-care facilities, the Internet of Things may be able to automate patient care processes.
- Interoperability, artificial intelligence machine-to-machine connection, information sharing, and data transfer are all benefits of the Internet of Things in health care.
- Information gathering and analysis.
- A lack of cloud connectivity complicates the preservation and management of massive volumes of data transmitted by a real-time application on a health-care device.
- Medical practitioners face an uphill battle when physically collecting and analyzing data from a variety of sources and devices.
- Internet of Things (IoT) devices eliminate the requirement for raw data storage by collecting, reporting, and analyzing data in real time. All this may be accomplished in the cloud, with providers just seeing the end reports with graphs of the data.
- Additionally, health-care operations enable businesses to get critical health-care analytics and data-driven insights, facilitating faster decision-making and minimizing mistakes. This service includes alerts and tracking.
- Chronic illnesses need constant vigilance on the part of the patient. When combined with mobile applications and smart sensors, IoT medical equipment may enable doctors to monitor patients' vital signs in near real time while also notifying those in need.

- Reports and notifications provide reliable assessment of patient status independent of location or time.
- Additionally, it enables medical practitioners to make educated judgments and provide therapy on time.

As a consequence, the Internet of Things enables real-time alerting, tracking, and monitoring, allowing hands-on treatments, better precision, and appropriate medical intervention, as well as improved overall patient care outcomes (Bollipelly et al., 2021). Numerous IoT-based health-care delivery chains are also proposing to develop machines that may dispense drugs based on the patient's prescription and data about the patient's disease accessible through connected devices. Patients will benefit from improved therapy as a result of the Internet of Things (IoT). As a result, medical expenditures are reduced (Narendar Singh & Pavitra, 2021)

8.5 CASE STUDY OF AN IOT-ENABLED REMOTE MONITORING SYSTEM FOR A HEALTH-CARE FACILITY

It is essential that Alzheimer's disease monitoring be handled with the greatest care since it has caused several complications in the past. In the absence of a physician, patients with Alzheimer's disease cannot be diagnosed on their own. Furthermore, even for members of the family, it is difficult to keep track of the illness. They can only discover that the patient has the condition when the patient exhibits unusual behavior and loses consciousness while doing routine chores. The authors think that the Internet of Things (IoT) may play a significant role in reducing or improving the quality of life of those suffering from Alzheimer's disease (Pavitra et al., 2020). There are three key signs and symptoms of Alzheimer's disease that are extremely hazardous:

1. Extensive forgetfulness
2. Wandering
3. Dementia

In order to combat this, the usage of wearables that are linked to the BAN through the use of communication protocols such as MQTT, Zigbee, and others is being promoted. When a patient wanders out in the city, the sensors may send out an alert to caregivers to let them know where they are, making it much easier for them to locate them. In fact, patients may be taught to utilize the Internet of Information framework to receive notes or recall things such as their name, the name of the doctor who is caring for them, or their location. Furthermore, as technology continues to advance, the collection of vital signs from patients via the use of multiple servers and databases might make it easier for physicians to monitor their patients even when they are not there.

8.6 EMG REMOTE DATA ACQUISITION SYSTEM

When it comes to muscle health and the nerve cells that govern them, electromyography (EMG) is a diagnostic tool that may be used (motor neurons). When an EMG is performed, it may indicate nerve dysfunction, muscle dysfunction, and issues with nerve-to-muscle signal transmission. Muscle contraction is caused by

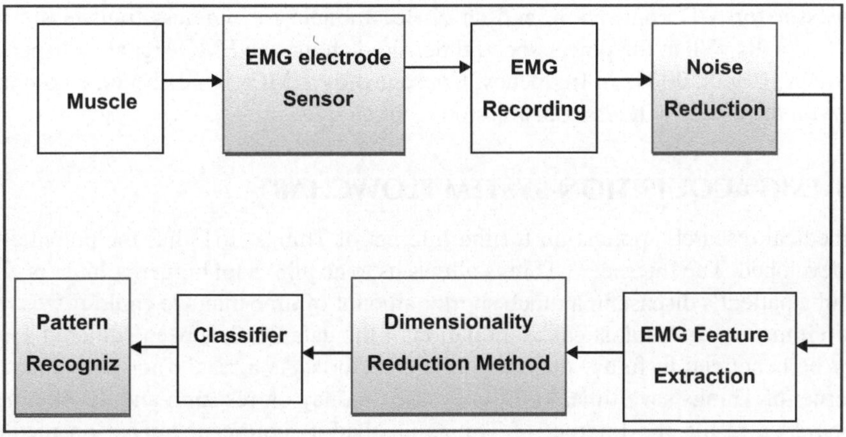

FIGURE 8.3 EEG block diagram.

electrical impulses sent by motor neurons, as seen in Figure 8.3. An EMG makes use of small devices known as electrodes to transform these impulses into graphs, noises, or numerical values, which are then analyzed by a trained professional. A needle EMG is a procedure in which a needle electrode is placed directly into a muscle and the electrical activity in that muscle is recorded. Surface electrodes are put to the skin (surface electrodes) in order to evaluate the speed and intensity of signals passing between two or more sites during a nerve conduction study, which is another component of an EMG test procedure.

The neurologist will analyze and create a report based on the results of the tests. Alternatively, the doctor who ordered the EMG will review the results with you during a follow-up consultation that will be conducted through the internet.

8.7 EMG ACQUISITION BLOCK DIAGRAM

Currently, wireless technology plays an important part in today's society. Many applications, such as health monitoring, make use of wireless sensor networks. The application to assess muscle response or electrical activity as a result of nerve stimulation of muscle is one such example. There are several critical components in the development of wireless electromyography (EMG) for the remote monitoring of muscle actions, and this is only one of them in the field of biomedical technology (Singh et al., 2020). This wireless EMG is capable of transmitting signals to a computer for monitoring purposes using wireless transmission technology. Doctors treating patients with muscle complication may monitor their patients from the comfort of their own offices, while the patients are free to walk about in their own rooms or treatment rooms, thanks to this technology. This technology is critical for patients who need home-based monitoring as well as cost-effective medical services.

The design of wireless EMG system includes a preamplifier and an electrode for the measurement of EMG signal, a main amplifier for signal processing, a DSP processor for A/D conversion, and an X-Bee module for wireless transmission of EMG signals. When the power spectral density is high, the EMG signal is dispersed between 10 and 500 Hz in frequency. Consequently, EMG signals become compatible with the use of wireless transmission technologies.

8.8 EMG ACQUISITION SYSTEM FLOWCHART

A medical research application for the Internet of Things (IoT) has the potential to be developed. The Internet of Things allows us to acquire a lot bigger volume of data about a patient's disease in a much shorter amount of time than we could if we were to do it manually. A statistical examination of the data that has been gathered so far may be beneficial in future medical research efforts. As a result, not only does the Internet of Things save time, but it also saves money on research and development costs. As a result, the Internet of Things has had a significant impact on medical research in recent years. As a consequence of this development, it is possible to produce medical treatments that are larger and more effective.

The Internet of Things, as seen in Figure 8.4, assists health-care practitioners in providing better treatment to their patients in a number of different ways. The Internet

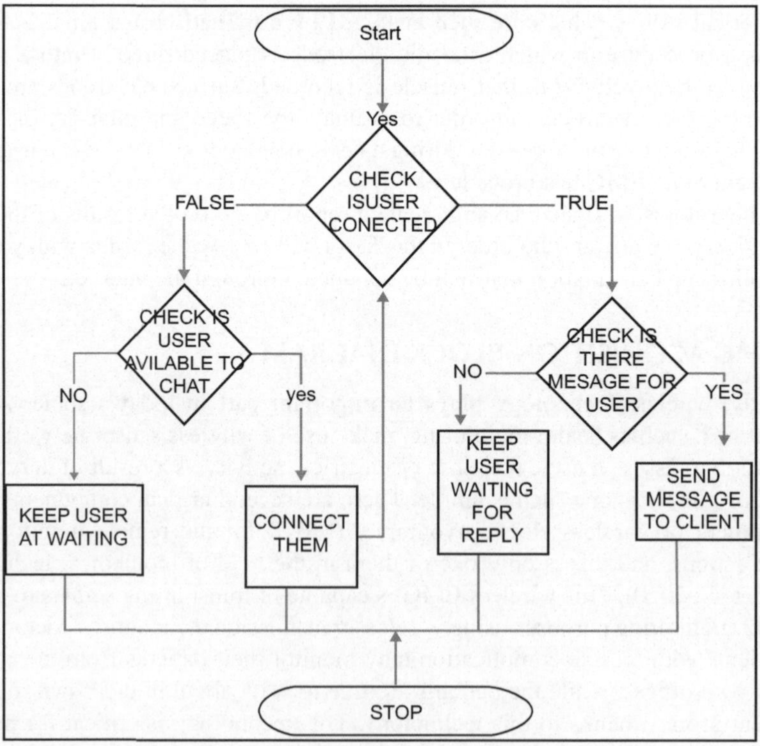

FIGURE 8.4 EMG acquisition system flowchart.

of Things (IoT) is increasingly being used to improve current equipment by simply introducing chips from smart health-care goods. The chip makes it simpler for anyone suffering from Alzheimer's disease to get medical attention and treatment.

8.9 CHALLENGES

8.9.1 Privacy and Security of Personal Information

IoT in health care raises severe issues about data security and privacy. An IoT security system collects and transmits real-time data from sensors. Data standards and security criteria are absent from the majority of IoT devices. In addition, there is a great deal of ambiguity about electronic device data ownership legislation. The electronic health information of a patient may be used by cybercriminals to create fake identifications and then used to acquire drugs and medical equipment that they can resell. Using a patient's information, a hacker might submit a fake insurance claim on their behalf.

8.9.2 Integration of External Device

Devices and protocols from a variety of sources are combined. Integrating a variety of devices into a health-care system might be a challenge. Lack of consensus on communication protocols and standards is to blame for this stumbling point. Disparity in communication protocols, even if a wide variety of devices are connected, complicates and hinders data collecting. This inconsistency in the protocols of connected devices slows down the whole process and hinders the scalability of IoT in health care.

8.9.3 Precision and Excess of Data

As previously stated, data aggregation is difficult because of the use of several communication protocols and standards. On the other side, IoT devices continue to amass enormous amounts of data. Using data acquired by IoT devices, critical choices may be made. It's becoming increasingly difficult for doctors to get insight from all this data, which has an effect on the quality of their decisions.

In addition, this danger is expanding as more devices are connected and collect more data.

8.10 COST

In the challenging parts, you may be startled to see money issues. However, the truth is that IoT has yet to make health care more affordable for the common individual. Anyone concerned about rising health-care costs, especially in wealthier countries, should keep an eye on this trend. In any case, the costs remain a concern as of right now. It is essential for stakeholders to make IoT app development cost-effective in order to make it accessible to everyone except the most affluent people. It is possible to think of an IoT health-care facility as a collection of ubiquitous computing devices that are mostly used for outside activities. A wide variety of patient data is gathered through IoT-based health-care systems, which get input from doctors and medical specialists. Continuous glucose monitoring for insulin pens is the best example.

8.11 CONCLUSION

IoT-enabled devices have enabled remote health-care monitoring. Remote patient monitoring decreases hospital stays and avoids readmissions. IoT capabilities can help reduce health-care costs and improve treatment results. As a result, people's lives have been transformed. Data from devices may help physicians treat patients better.

An IoT sensor that identifies a patient's unusually low heart rate may alert medical staff. Many hospitals use IoT devices to remind patients to wash their hands. The devices may also advise on ways to lessen a patient's risk. The Internet of Things (IoT) links smart devices to health apps. Smart garments, smart spectacles, digital stethoscopes, etc.

Physicians may employ connected medical devices to monitor patient adherence. With IoT in remote patient monitoring, the patient's health is constantly monitored by smart sensors and home devices.

REFERENCES

Farahani, Bahar, Firouzi, Farshad, Chang, Victor, Badaroglu, Mustafa, Constant, Nicholas, & Mankodiya, Kunal. Towards Fog-driven IoT eHealth: Promises and Challenges of IoT in Medicine and Healthcare. *Future Generation Computer Systems*, 78(Part 2), 659–676 (2018).

Goud, Bollipelly PruthviRaj, Prasanth Rao, A., & Sathiyamoorthi, V. Patient Monitoring System Using Internet of Things. In: Sathiyamoorthi, V. & Elci, Atilla. (eds) *Challenges and Applications of Data Analytics in Social Perspectives*, pp. 275–289, 2021. IGI Global, Hershey, PA.

Iranpak, S., Shahbahrami, A., & Shakeri, H. Remote Patient Monitoring and Classifying Usingthe Internet of Things Platform Combined with Cloud Computing. *Journal of Big Data*, 8, 120 (2021). https://doi.org/10.1186/s40537-021-00507-w

Narang, N. Kishor. Mentor's Musings on the Role of Disruptive Technologies and Innovation in Making Healthcare Systems More Sustainable. *IEEE Internet of Things Magazine* 4(3), 80–89 (2021).

Narendar Singh, D., & Pavitra, B. IOT Enabled Patient Health Monitoring and Assistant System. *European Journal of Molecular & Clinical Medicine*, 7(11), 5608–5618 (2021).

Pavitra, B., Singh, D. N., & Sharma, S. K. Predictive and Interactive IOT Diagnosis System with AI and ML Tools: Review. In: Chillarige, R., Distefano, S., & Rawat, S. (eds) *Advances in Computational Intelligence and Informatics. ICACII 2019. Lecture Notes in Networks and Systems*, vol. 119, 2020. Springer, Singapore. https://doi.org/10.1007/978-981-15-3338-9_2.

Rowley, W. R., Bezold, C., Arikan, Y., Byrne, E., & Krohe, S. Diabetes 2030: Insights from Yesterday, Today, and Future Trends. *Population Health Management*, 20(1), 6–12 (2017). https://doi.org/10.1089/pop.2015.0181

Singh, D. N., Hashmi, M. F., & Sharma, S. K. Predictive analytics & modeling for modern health care system for cerebral palsy patients. *Multimedia Tools and Applications*, 79, 10285–10307 (2020). https://doi.org/10.1007/s11042-019-07834-4

9 Applying Machine Learning Logistic Regression Model for Predicting Diabetes in Women

Nitin Jaglal Untwal and Utku Kose

CONTENTS

9.1 INTRODUCTION

Diabetes is a metabolic disease. Diabetes severely impacts or damages various body organs, like the kidneys, eyes, nervous system, dysfunction, etc. The factors which are responsible for diabetes are obesity, poor lifestyle, poor diet, heredity, stress, no exercise. As cited, there are various factors that influence diabetes. Keeping the previous problem in consideration, the researchers have decided to conduct the study titled "Predicting Diabetes in Women by Applying Logistic Regression Model Using Python Programming." For the purpose of conducting the study, secondary data is collected from the Kaggle database regarding diabetes in women. The dependent variable for diabetes in women (Y) is binary class, and there are five independent variables, "pregnancies," "glucose," "blood pressure," "BMI," and "age," which are continuous in nature. The study is based on the objective of creating a predictive

DOI: 10.1201/9781003309451-9

model for diabetes in women. The model used for achieving the objective of the study is the logistic regression model. The study is carried out in six steps.

1. Introduction
2. Data description and creating trail and testing
3. Logistic regression model using Python programming
4. Creating confusion matrix
5. Accuracy statistics
6. Conclusion

9.1.1 Introduction to Multinomial Logistic Regression Model

The supervised learning classification algorithm, the logistic regression model, is used to predict the target variable, which is binary in nature [1, 2]. The dependent variable is dichotomous, which means it has only two classes. Further, we can say that it is in binary form as the outcome is binary in nature, like male/female, yes/no, success/failure. A logistic regression model predicts P(Y = 1) as a function of *x* [3–7]. The logistic regression model is the easiest and simplest algorithm extensively applied in various classification problems [1, 2, 8–11]. The multinomial logistic regression is an extension of the logistic regression model, under which the independent variable has more than two classes of outcomes [12–17]. The target variable has more possible unordered types. That is, the type has no quantitative significance [18–25].

The MLR model for prediction is represented as multinomial logistic regression using a linear predictor function f(k,i) to predict the probability that observation *I* has outcome *k*, of the following form [8–13, 25–31]:

$$f(k,i) = \beta_{0,k} + \beta_{1,k}x_{1,i} + \beta_{2,k}x_{2,i} + \cdots + \beta_{M,k}x_{M,i},$$

9.1.2 Research Methodology

Data Source

Data taken for the study is the Kaggle dataset.

Sample Size

It contains health details of 767 women tested for diabetes.

Software Used for Data Analysis

Python programming.

Model Applied

For the purpose of this study, we had applied multinomial logistic regression model.

```
# importing libraries
import statsmodels.api as sm
import pandas as pd
df = pd.read_csv (r'C:\Users\nitin\Desktop\Diabetes.csv')
print (df)
     Pregnancies  Glucose  BloodPressure  SkinThickness  Insulin   BMI  \
0              6      148             72             35        0  33.6
1              1       85             66             29        0  26.6
2              8      183             64              0        0  23.3
3              1       89             66             23       94  28.1
4              0      137             40             35      168  43.1
..           ...      ...            ...            ...      ...   ...
763           10      101             76             48      180  32.9
764            2      122             70             27        0  36.8
765            5      121             72             23      112  26.2
766            1      126             60              0        0  30.1
767            1       93             70             31        0  30.4

     DiabetesPedigreeFunction  Age  Outcome
0                       0.627   50        1
1                       0.351   31        0
2                       0.672   32        1
3                       0.167   21        0
4                       2.288   33        1
..                        ...  ...      ...
763                     0.171   63        0
764                     0.340   27        0
765                     0.245   30        0
766                     0.349   47        1
767                     0.315   23        0

[768 rows x 9 columns]
```

FIGURE 9.1 Creating data frame.

We are creating a knowledge representation structure by applying Python programming through feature engineering.

Feature engineering is the process of preparing raw data which will be ready for the program to utilize. The data frame is made as per the requirement of the model. The data frame is the first step in making the data structured, which is easy to understand by the program. Once the data frame is created, it is ready to be used by the algorithm. The syntax used for creating a data frame in Python programming is represented in Figure 9.1.

TABLE 9.1
Classes of the Variable Used in Multinomial Logistic Regression Model

Variable	Classes
Diabetes (dependent)	No diabetes = 0
	Diabetes = 1
Glucose	Continuous
Blood pressure	Continuous
Pregnancies	Continuous
BMI	Continuous
Age	Continuous

9.1.3 Data Description and Creating Trail and Testing

For the purpose of conducting the study, secondary data is collected from the Kaggle database regarding diabetes in women. The dependent variable for diabetes in women (Y) is binary class, and there are five independent variables, "pregnancies," "glucose," "blood pressure," "BMI," and "age," which are continuous in nature.

9.2 CREATING TRAIL AND TESTING

Around 80% of data is being considered for the trial, and 20% of data is used for testing purposes in order to evaluate the model. The main objective of the partitioning of the data is to convert data into training and testing datasets. For analysis, we have randomly selected the data to ensure the training dataset and testing dataset are similar. By using similar dataset for training and testing, we can minimize data errors; hence, a better understanding of the model can be achieved.

9.2.1 Intercepts and Coefficient

The first array contains three intercepts, and the second array contains regression coefficients in three sets.

```
# defining the dependent and independent variables
X = df[['Pregnancies', 'Glucose', 'BloodPressure', 'BMI', 'Age']]
y = df[['Outcome']]
```

```
X_train, X_test, y_train, y_test = sklearn.model_selection.train_test_split(X, y,
print(X_train.shape)
print(X_test.shape)
print(y_train.shape)
print(y_test.shape)
```

```
(614, 5)
(154, 5)
(614, 1)
(154, 1)
```

FIGURE 9.2 Python code for trail and testing.

```
log_reg = sm.Logit(y_train, X_train).fit()

# printing the summary table
print(log_reg.summary())
```

```
Optimization terminated successfully.
         Current function value: 0.610146
         Iterations 5
                           Logit Regression Results
==============================================================================
Dep. Variable:                Outcome   No. Observations:                  614
Model:                          Logit   Df Residuals:                      609
Method:                           MLE   Df Model:                            4
Date:                Sat, 11 Sep 2021   Pseudo R-squ.:                 0.05628
Time:                        11:49:26   Log-Likelihood:                -374.63
converged:                       True   LL-Null:                       -396.97
Covariance Type:            nonrobust   LLR p-value:                 4.623e-09
==============================================================================
==
                   coef    std err          z      P>|z|      [0.025      0.97
5]
------------------------------------------------------------------------------
--
Pregnancies      0.1279      0.031      4.080      0.000       0.066       0.1
89
Glucose          0.0129      0.003      4.546      0.000       0.007       0.0
18
BloodPressure   -0.0307      0.005     -5.903      0.000      -0.041      -0.0
21
BMI             -0.0017      0.011     -0.159      0.873      -0.023       0.0
20
Age             -0.0112      0.009     -1.225      0.221      -0.029       0.0
07
==============================================================================
==
```

FIGURE 9.3 Stats models analysis in Python.

9.2.2 Log Odds Ratio Estimates

The odds ratio is the probability of the binary outcome of the target variable. It is the probability of success (A)/failure (A'). Mathematically, it can be expressed in the equation:

$$\text{odd ratio} = P(A)/P(-A)$$

In the interpretation of exponential coefficient for a unit change in predictor variable, here the odds will be multiplied by a factor indicated by the exponent of the beta coefficient. The condition to all other variables remains constant.

9.2.3 The Stats Models Analysis in Python

The six independent variables are "age," "gender," "body temperature," "dry cough," "sore throat," and "breathing," which are binary in nature. Body temperature and breathing problem are the two variables that are found to be significant in class 1 (moderate severity) cases, as the p-values are less than 0.05. Similarly, only the breathing problem variable is found to be significant in class 2 (severe cases), as the p-value is less than 0.05. The aforementioned statistical analysis shows a clear indication of immunity compromise of the severe cases, since the body temperature variable becomes highly insignificant.

```
from sklearn.metrics import (confusion_matrix,
                             accuracy_score)

# confusion matrix
cm = confusion_matrix(ytest, prediction)
print ("Confusion Matrix : \n", cm)

# accuracy score of the model
print('Test accuracy = ', accuracy_score(ytest, prediction))
```

```
Confusion Matrix :
 [[438  62]
 [174  94]]
Test accuracy =  0.6927083333333334
```

FIGURE 9.4 The Python code for the confusion matrix.

9.2.4 Creating Confusion Matrix

Confusion matrix measures model performance. It evaluates the values of actual values and predicted values. It is of the order N*N matrix; here, *N* is for the class of dependent/target variable.

For binary classes, it is a 2 × 2 confusion matrix.
For multiclasses, it is a 3 × 3 confusion matrix.

Calculating False Negative, False Positive, True Negative, True Positive
The confusion matrix for our dataset is as follows:

True Positive: The true positive is represented by cell one.
True Positive = 438
False Negative: The sum of the values apart from the true positive value.
False Negative = 94
False Positive: The total value does not include the true positive value.
False Positive = 174
True Negative: The sum of values of all columns and rows, excluding the values of the class that we are calculating the values for.
True Negative = 62

TABLE 9.2
The Confusion Matrix

	0	1
0	438 (TP)	62 (TN)
1	174 (FP)	94 (FN)

```
#Accuracy statistics

print('Accuracy Score:', metrics.accuracy_score(ytest, prediction))

#Create classification report
class_report=classification_report(ytest, prediction)
print(class_report)
```

```
Accuracy Score: 0.6927083333333334
              precision    recall  f1-score   support

           0       0.72      0.88      0.79       500
           1       0.60      0.35      0.44       268

    accuracy                           0.69       768
   macro avg       0.66      0.61      0.62       768
weighted avg       0.68      0.69      0.67       768
```

FIGURE 9.5 Python code for accuracy statistics.

9.3 ACCURACY STATISTICS

With AccuracyNow, to obtain accuracy from the confusion matrix, we apply the following formulae:

$$\text{Model Accuracy} = \frac{\text{True Positive} + \text{True Negative}}{\text{True Positive} + \text{True Negative} + \text{False Positive} + \text{False Negative}}$$

$$\text{Model Accuracy} = \frac{438+62}{438+62+94+174} = 0.65$$

9.3.1 Recall

Recall measures the ratio of true positive with correctly classified positive examples divided by the total number of positive examples. High recall indicates the class is correctly recognized (a small number of FN).

$$\text{Recall} = \frac{\text{True Positive}}{\text{True Positive} + \text{False Negative}}$$

$$\text{Recall} = \frac{438}{438+94} = 0.82$$

9.3.2 Precision

Precision is the measure of how often it is correct when positive results are predicted.

$$\text{Precision} = \frac{\text{True Positive}}{\text{True Positive} + \text{False Positive}} = \frac{438}{438 + 174} = 0.71$$

9.4 CONCLUSION

The five independent variables "pregnancies," "glucose," "blood pressure," "BMI," and "age" are continuous in nature. "Pregnancies," "glucose," and "blood pressure" are the three variables that are found to be significant, as the p-values are less than 0.05. Similarly, "BMI" and "age" variables are found to be insignificant, as the p-value is more than 0.05. The logistic regression model has a precision of 71%, and overall model accuracy is 0.65%.

REFERENCES

1. Austin, J. T., Yaffee, R. A., and Hinkle, D. E. 1992. Logistic regression for research in higher education. *Higher Education: Handbook of Theory and Research*, 8: 379–410.
2. Cabrera, A. F. 1994. Logistic regression analysis in higher education: An applied perspective. *Higher Education: Handbook of Theory and Research*, 10: 225–256.
3. Chuang, H. L. 1997. High school youth's dropout and re-enrollment behavior. *Economics of Education Review*, 16(2): 171–186.
4. Cleary, P. D., and Angel, R. 1984. The analysis of relationships involving dichotomous dependent variables. *Journal of Health and Social Behavior*, 25: 334–348.
5. Cox, D. R., and Snell, E. J. 1989. *The analysis of binary data*, 2nd ed. London: Chapman and Hall.
6. Demaris, A. 1992. *Logit modeling: Practical applications*. Newbury Park, CA: Sage.
7. Efron, B. 1975. The efficiency of logistic regression compared to normal discriminant analysis. *Journal of the American Statistical Association*, 70: 892–898.
8. Haberman, S. 1978. *Analysis of qualitative data*, Vol. 1. New York: Academic Press.
9. Hosmer, D. W. Jr., and Lemeshow, S. 2000. *Applied logistic regression*, 2nd ed. New York: Wiley.
10. Janik, J., and Kravitz, H. M. 1994. Linking work and domestic problems with police suicide. *Suicide and Life-Threatening Behavior*, 24(3): 267–274.
11. Lawley, D. N., and Maxwell, A. E. 1971. *Factor analysis as a statistical method*. London: Butterworth & Co.
12. Lei, P.-W., and Koehly, L. M. Linear discriminant analysis versus logistic regression: A comparison of classification errors. Paper presented at the annual meeting of the American Educational Research Association. April, New Orleans, LA.
13. Long, J. S. 1997. *Regression models for categorical and limited dependent variables*. Thousand Oaks, CA: Sage.
14. Marascuilo, L. A., and Levin, J. R. 1983. *Multivariate statistics in the social sciences: A researcher's guide*. Monterey, CA: Brooks/Cole.
15. Menard, S. 1995. *Applied logistic regression analysis*. Thousand Oaks, CA: Sage (Sage University Paper Series on Quantitative Applications in the Social Sciences, 07–106).
16. Menard, S. 2000. Coefficients of determination for multiple logistic regression analysis. *The American Statistician*, 54(1): 17–24.

17. Nagelkerke, N. J. D. 1991. A note on a general definition of the coefficient of determination. *Biometrika*, 78: 691–692.
18. Peng, C. Y., Manz, B. D., and Keck, J. 2001. Modeling categorical variables by logistic regression. *American Journal of Health Behavior*, 25(3): 278–284.
19. Peng, C. Y., and So, T. S. H. 2002. Modeling strategies in logistic regression. *Journal of Modern Applied Statistical Methods*, 14: 147–156.
20. Peng, C. Y., and So, T. S. H. 2002. Logistic regression analysis and reporting: A primer. *Understanding Statistics*, 1(1): 31–70.
21. Peng, C. Y., So, T. S., Stage, F. K., and John, St. E. P. 2002. The use and interpretation of logistic regression in higher education journals: 1988–1999. *Research in Higher Education*, 43: 259–293.
22. Peterson, T. 1984. A comment on presenting results from logit and probit models. *American Sociological Review*, 50(1): 130–131.
23. Press, S. J., and Wilson, S. 1978. Choosing between logistic regression and discriminant analysis. *Journal of the American Statistical Association*, 73: 699–705.
24. Ryan, T. P. 1997. *Modern regression methods*. New York: Wiley.
25. SAS Institute Inc. 1999. *SAS/STAT[rgrave] user's guide*, Vol. 2. Cary, NC: Author. Version 8.
26. Schlesselman, J. J. 1982. *Case-control studies: Design, control, analysis*. New York: Oxford University Press.
27. Scott, K. G., Mason, C. A., and Chapman, D. A. 1999. The use of epidemiological methodology as a means of influencing public policy. *Child Development*, 70(5): 1263–1272.
28. Siegel, S., and Castellan, N. J. 1988. *Nonparametric statistics for the behavioral science*, 2nd ed. New York: McGraw-Hill.
29. Tabachnick, B. G., and Fidell, L. S. 1996. *Using multivariate statistics*, 3rd ed. New York: Harper Collins.
30. Tabachnick, B. G., and Fidell, L. S. 2001. *Using multivariate statistics*, 4th ed. Needham Heights, MA: Allyn & Bacon.
31. Tolman, R. M., and Weisz, A. 1995. Coordinated community intervention for domestic violence: The effects of arrest and prosecution on recidivism of woman abuse perpetrators. *Crime and Delinquency*, 41(4): 481–495.

10 Compressive Sensing-Based Medical Imaging Techniques to Detect the Type of Pneumonia in Lungs

Vivek Upadhyaya, Girraj Sharma, Tien Anh Tran, and Mohammad Salim

CONTENTS

10.1 SIGNIFICANCE OF MEDICAL IMAGING IN HEALTH MONITORING

The modern lifestyle creates too many diseases due to mental pressure and unhealthy habits. Early-age diagnosis of these diseases can increase the chances of survival. ECG, MRI, and computed tomography are the tools to efficiently understand and diagnose various diseases. These tools are used to monitor the condition of the patients with the help of processing data associated with patients. That's why the data should be of acceptable quality, so the clinical decision should be correct. Thus,

DOI: 10.1201/9781003309451-10

compressive sensing is a valuable scheme that can enhance the processing speed and maintain the proper grade of medical data. Health monitoring is the technology that can assess the vital parameter of the patients by using wireless sensor networks and wearable devices.

The invention of the X-ray in 1895 was a tremendous breakthrough for image-based patient analysis. This X-ray machine diagnoses various diseases, like cancer, tuberculosis, and verifies bone fractures. A computed tomography scan is the extended version of the X-ray imaging scheme that can be used for high-quality volumetric reconstruction of body parts. CT scan can also be used to diagnose the earlier stages of heart diseases and other infectious diseases. MRI is the technique by which we can ensure high-quality images than CT with the help of large magnets. Acquisition time is an important factor in both MRI and computed tomography. Maintaining the quality of the image with minor exposure to the hazardous impact of equipment is the primary concern in today's medical challenges.

ECG is the process of capturing the electrical activities of the heart. Thus, any action shown by the heart is the immediate change in the electrical signals. By analyzing these electrical signals, the doctor can verify whether the functioning of the heart is correct or not. Similarly, the use of MRI is in visualizing soft tissues, cancer, and tumor in the body. MRI provides exact information related to soft tissues visualization without any invasive method. Computed tomography is also an extended version of traditional X-ray techniques. Still, in CT, we take the images at different angles to acquire much information about the body part on which analysis is conducted.

Further, this chapter will provide various health monitoring techniques and the application of compressive sensing. The prime motive to use compressive sensing and the importance of CS in health monitoring are discussed in this chapter.

10.1.1 Compressive Sensing

The traditional theory of compression, given by Nyquist, has various issues if we move towards higher samples or higher-frequency components. So due to this, the Nyquist sampling rate is impossible to achieve in some specific applications. Medical imaging is the domain in which the data size is too large, like in magnetic resonance imaging, computed tomography, X-Ray, ultrasound, etc. So it is tedious to reconstruct the signal using traditional sampling theory.

So Donoho, Candes, Romberg, and Tao [1–3] proposed a theory that if a signal is sparse on some specific basis, this signal can be reconstructed using a few measurements. This theory is known as the compressive sensing technique. For the compression purpose, we can opt for measurement matrix, and for the representation purpose, we can use the basis matrices. The three important functions that make the compressive sensing theory much favorable than the classical sampling theory are:

- To understand the test signal
- To collect the measurements from the test sample at a specific time
- To identify the reconstruction algorithm that is efficient enough to recover the actual signal

The following section mentions how the sparsity will affect the compression and reconstruction process.

10.1.1.1 Sparsity

Sparsity is the prerequisite for the compressive sensing technique. If a signal contains any information sparsely, this information can be calculated by measuring the sparsity of that signal [4, 5]. Therefore, sparsity is closely attached to the sampling and recovery process rate in compressive sensing. This sparsity concept is entirely different from the Nyquist sampling strategy and is used in traditional sampling theory. In the traditional view, the higher signal frequency value needed many samples to reconstruct the signal. Still, in compressive sensing, when the sparsity is high, then the requirement of samples is very low. So it is crucial to obtain the sparse representation of a signal in the compressive sensing theory [6].

If we can project the natural signals on a suitable basis, these signals can be compressed; these signals are like images, different sound signals, seismic data, etc. The basis is an important parameter; when this parameter is appropriately selected, then many projected values are zero or very small, so we can neglect them. S-nonzero components are considered S-sparse; if the signal is highly sparse, then it is much easier to compress the signal and the base of compressive sensing theory. Two-dimensional wavelets, fractal waveforms, and curvelets are suitable bases for images, spiky data, wavefield propagation, etc.

10.1.1.2 Compressibility

A key assumption used for compressive sensing (CS) is that signals show a degree of structure. Sparsity is one of the structures that has been known to us till now, and it is the number of nonzero values the signal has when represented on an orthonormal basis. Few nonzero values in a signal compared to the whole length of the signal make it sparse. Few structured signals are truly sparse; rather, they are compressible. A signal is compressible if its sorted coefficient magnitudes in Ψ decay rapidly. The mathematical structure is represented next:

Suppose “a” is the compressible signal in the basis Ψ:

$$a = \Psi.\xi \qquad \text{(equation 1.1)}$$

Here are the coefficients of “a” in the basis Ψ. If “a” is compressible, then the magnitude of sorted coefficients observes a power-law decay:

$$a = \Psi.\xi \qquad \text{(equation 1.2)}$$

If a signal obeys the power-law decay, as given in equation 1.2, then it will be compressible. A high value of r indicates the faster decay in magnitude, and the signal is more compressible [7].

10.1.2 Use of Compressive Sensing in Computed Tomography

Computed tomography has so much relevance in today’s clinical era. The reduction in the doses at the time of CT scan is the prime concern in this domain. The amount of medical dose directly affects the clarity of the image. Compressive sensing is the approach that can provide a high-quality reconstructed image by using less dose given to the patients. Some applications involving compressive sensing with computed tomography are mentioned here, so a quick review of the technique can be delivered. Yu et al. first came up with this concept. They obtained very fine results with the help of compressive sensing–based tomography compared to the other global projection method [8], as intensive iterations are required to reconstruct the proper images. Hence, the computation cost is high at that time. Van Sloun et al. proposed a method that was based on ultrasound computed tomography [9]. As a result of this method, the acquisition is much more compact, and the acquisition time as well is also very less, without any bad impact on the high-quality image. Some more precise work and efficient algorithms are proposed by Szczykutowicz and Chen [10–12].

10.1.3 Related Work for Compressive Sensing–Based Computed Tomography

We have discussed various aspects of the compressive sensing. Now we shall elaborate on using the compressive sensing approach for the efficient recovery of computed tomography–based images.

Zhanli Hu et al. mentioned their reviews in [13] related to CS-based CT imaging. Fundamental requirements, sparsity concepts, incoherence, and types of reconstruction algorithms are mentioned in this paper. Various applications associated with the compressive sensing–based computed tomography approach are also discussed in this work. The author's prime concern in this work is to generalize the foundation framework for the CS-based CT scheme.

Dictionary learning–based image reconstruction methods are very much popular nowadays. Conventional dictionary learning–based computed tomography image reconstruction schemes are patch-based, ignoring the overlapped patches. In [14], Peng Bao et al. came up with an idea to apply a convolutional sparse coding approach to compressive sensing–based computed tomography image reconstruction. They did not follow the conventional approach, in which the image is divided into patches; they considered the whole CT image and then applied the proposed method for reconstruction. The proposed method is applied on simulated and real data, and as per the authors, the proposed method is very robust when compared with other approaches.

Computed tomography is an innovative approach that can reduce the effect of radiation on a patient's body without degrading image quality. Compressive sensing is an approach that can further reduce the effect of radiation because it requires fewer samples to reconstruct an image. But traditional methods of compressive sensing are too complex and time-consuming. Sayed Masoud Hashemi proposed an approach to speed up the compressive sensing reconstruction method using pseudo-polar Fourier-based random transform. In [15], CT image reconstruction problem is modeled as a weighted CS optimization problem. According to the authors, by using 78% less exposure to radiation (dose), the CT scan image is reconstructed as the same diagnostic quality as using FBP from full data. A 512 × 512 image was recovered in less than 30 seconds on a desktop computer without numerical optimization.

The work of Oren Barkan et al. in [16] is similar to the previously mentioned [15]. They proposed a mathematical model that can adaptively reconstruct the CT signal by reducing the radiation dose level and better CT image quality. The proposed mathematical model can effectively recover the CT image, and the proposed method is compared with nonadaptive models as well. The exposure to radiation (dose) in a CT scan depends on the machine flux intensity. If the intensity of the machine is low, then the quantity of the dose is also low, but this low intensity will produce the Poisson-type noises in the measurements. So the authors utilized some more realistic models. They defined one additional dimension (dose dimension) in the modeling that can control the intensity of each ray individually that can be used to reconstruct the image. So the overall measure of the dose can be done by using the summation of all the intensities used for the measurements in place of a total number of projected rays.

We know that the compressive sensing–based computed tomography imaging delivers blurring edges in the image. So to overcome this problem, Chia-Jui Hsieh et al. in [17] proposed a modified Canny operator that can be incorporated with a compressive sensing approach. It precisely acquires the object image with its edges and enhances the contrast of these edges in the reconstructed image, enhancing image quality. Double-response edge detection and directional edge tracking are the two methods that the authors also propose. The compressive sensing approach with a modified Canny operator is termed edge-enhanced compressive sensing (EECS). RMSE, PSNR, and UQI are used to analyze the reconstructed image quality. Further comparative analysis shows that the proposed technique is far better than conventional edge detection schemes like Laplacian, Prewitt, and Sobel.

Random matrices enforce some constraints when applying compressive sensing in practical applications like computed tomography, SAR, high-resolution RADAR, and other imaging applications. It is very tedious to make a criterion to design a stable sensing matrix, so most researchers use the trial-and-error method to create sensing matrices. So in [18], the authors propose a constructive approach that can be used to develop a stable measurement matrix that can be further used to establish simple sensing matrices that don't have the mentioned drawbacks. A parameter known as sparsity constant is proposed in work associated with the measurement matrix, and this parameter estimates the improvement in the performance of the reconstruction algorithm. In this paper, the authors considered the MRIP constant and null space property to design a robust measurement matrix to efficiently recover a computed tomography scan.

10.1.4 Basic Framework to Analyze COVID-19 Cases

In the diagram that follows, the basic framework is mentioned, by using which we can quickly investigate COVID-19 cases. The framework is divided into four basic categories, as mentioned. Different aspects that can conclude the COVID-19 cases are also mentioned in Figure 10.1

In the latest COVID-19 cases, we are facing an issue related to the identification of COVID and the variant of the cases.

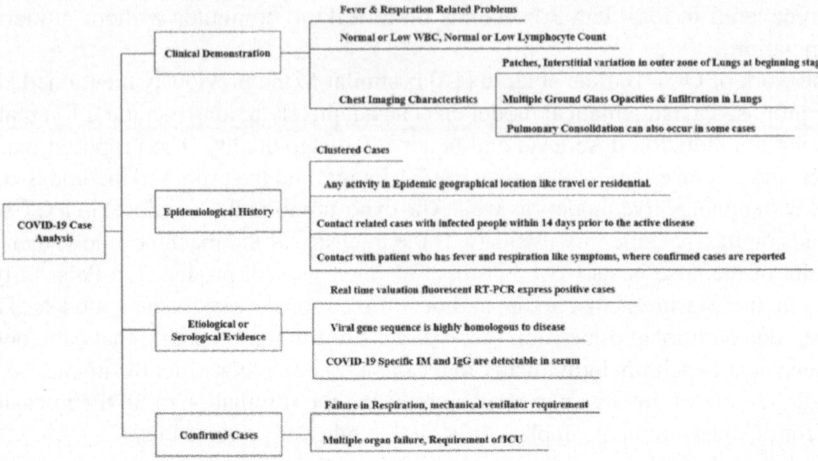

FIGURE 10.1 Different analytics for COVID-19 case study.

10.1.5 Importance of Medical Imaging to Identify Different Types of Pneumonia

Different researchers came up with different types of ideas to identify COVID-19 cases. But it is very much typical to say that the case is confirmed COVID-19 or not, and the mutation makes it much more complicated. So if we can detect the type of pneumonia, we can easily understand the type of pneumonia and its variant. That is why in Figure 10.2, we are discussing various imaging modalities.

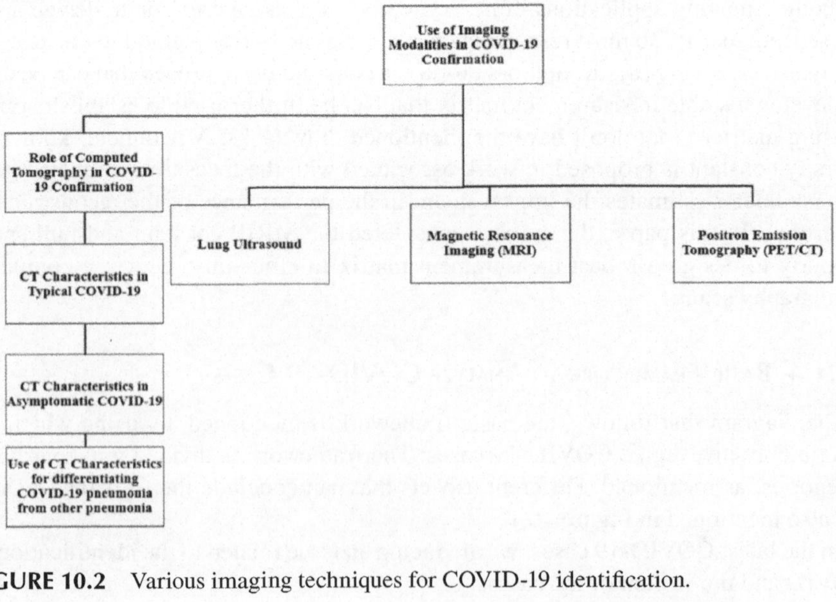

FIGURE 10.2 Various imaging techniques for COVID-19 identification.

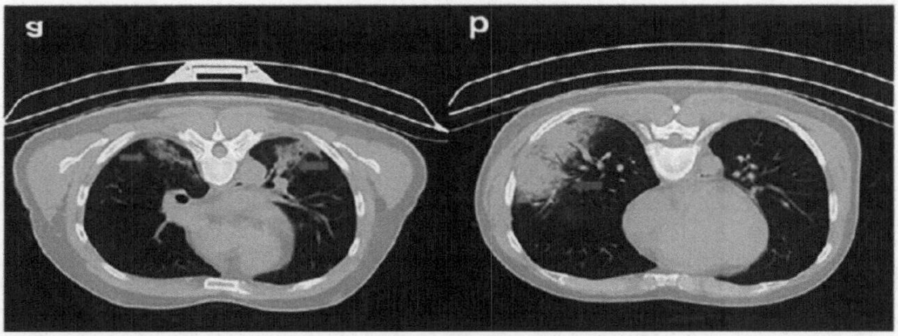

FIGURE 10.3 Sample of CT scan images indicating COVID-19 pneumonia and normal pneumonia: (a) computed tomography image representing bilateral GGOs for COVID-19, with red arrows; (b) computed tomography image representing unilateral GGOs for normal pneumonia, with blue arrows. [19]

A CT scan image for the lungs is given next in Figure 10.3 to classify COVID-19 pneumonia from normal pneumonia.

10.1.6 Proposed Approach

Compressive sensing–based computed tomography scan is already discussed in the literature. Our prime objective is to apply accumulated concepts to identify the type of pneumonia. It is a tedious task to determine pneumonia by analyzing an image. Still, as we know, the structure of normal pneumonia scan and COVID-19 patient scan is slightly different, and the opacities are also different in different scans. So to enhance the sampling rate, we are using the compressive sensing approach here. The reconstruction process requires the basis and sensing matrices. For a generic case, we consider discrete cosine transform as a basis matrix and Gaussian random matrix as a sensing matrix. Our focus is on the section of CT scan where the ground glass opacities and other differences should be easily visible.

10.1.7 Simulation and Associated Results

The computed tomography findings for ordinary pneumonia and COVID-19 pneumonia are too complex to observe due to the large-scale insignificant parts of the lungs. So we cropped the GGO based on the region of interest to overcome the interference with unimportant details. That can help doctors analyze the GGO properly to differentiate ordinary pneumonia from COVID-19 pneumonia. As a large dataset is required to investigate the difference accurately, if we can compress the CT image and only reconstruct the region of interest part of the image, it is helpful to detect COVID-19 pneumonia. We considered a database of CT-based images [20]. This database contains around 4,100 computed tomography scans of approximately 200 patients, out of which about 80 patients are infected with COVID-19, around 75 patients are infected with ordinary pneumonia, and 50 patients are healthy. Around 20 CT scans of every patient are mentioned on average.

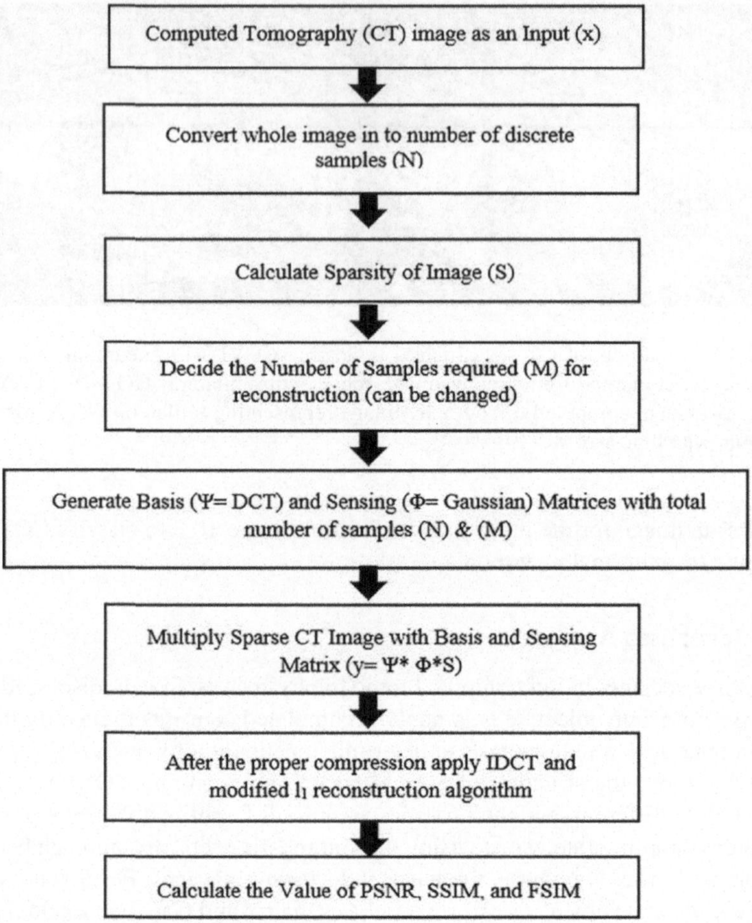

FIGURE 10.4 Steps for proposed approach.

Here we have considered 150 CT lung scan images of a healthy patient, a COVID-19 patients, and a normal pneumonia patient (50 images for each). To calculate the reconstruction quality efficiently, we consider seven different compression ratios. To verify the quality of the reconstruction scheme, we believe in three IQA methods: PSNR, SSIM, FSIM. Each IQA method requires around 1,050 individual simulations to calculate the quality score values for three types of CT scan images. So around 3,150 values we have to represent to show IQA score for each image at different compression ratios. That is why we are not representing tabular data here; in place of that, we represent a plot consisting of three types of patients for each compression ratio.

Figures 10.5 to 10.11 represent the plots for PSNR values for the reconstructed images. In these plots, we have considered 50 CT scan images for each type of

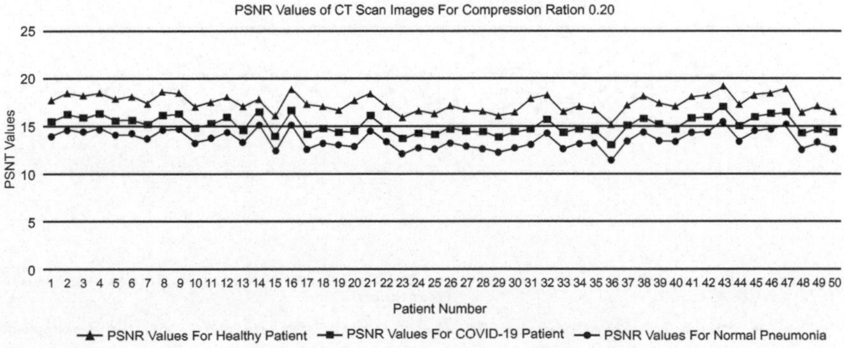

FIGURE 10.5 Plot representing PSNR values for healthy, COVID-19 pneumonia, normal pneumonia CT scan images at 0.20 compression ratio.

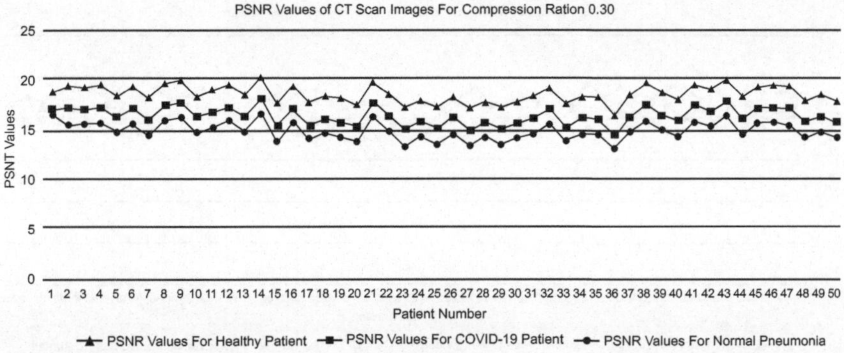

FIGURE 10.6 Plot representing PSNR values for healthy, COVID-19 pneumonia, normal pneumonia CT scan images at 0.30 compression ratio.

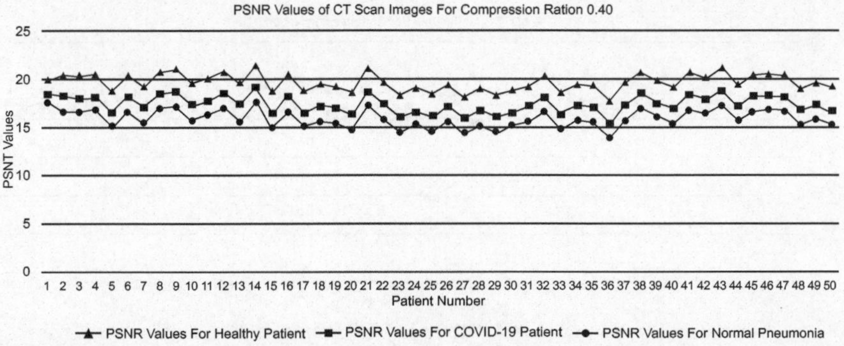

FIGURE 10.7 Plot representing PSNR values for healthy, COVID-19 pneumonia, normal pneumonia CT scan images at 0.40 compression ratio.

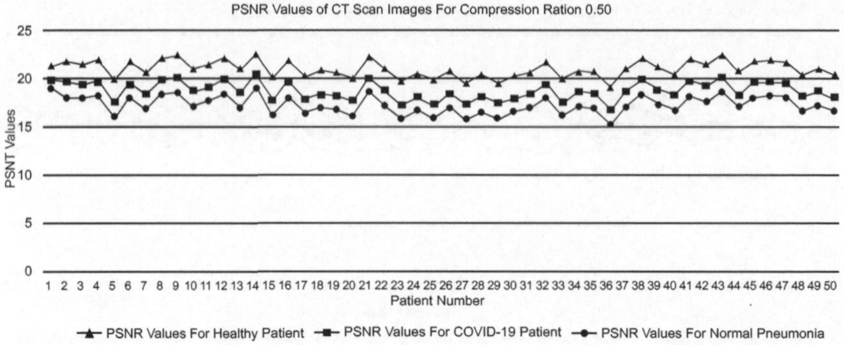

FIGURE 10.8 Plot representing PSNR values for healthy, COVID-19 pneumonia, normal pneumonia CT scan images at 0.50 compression ratio.

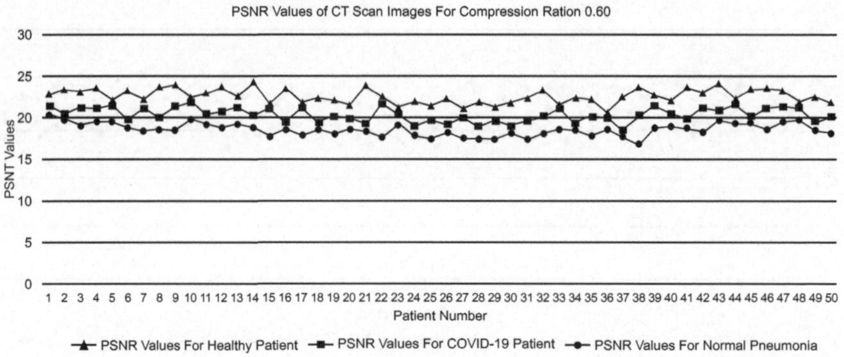

FIGURE 10.9 Plot representing PSNR values for healthy, COVID-19 pneumonia, normal pneumonia CT scan images at 0.60 compression ratio.

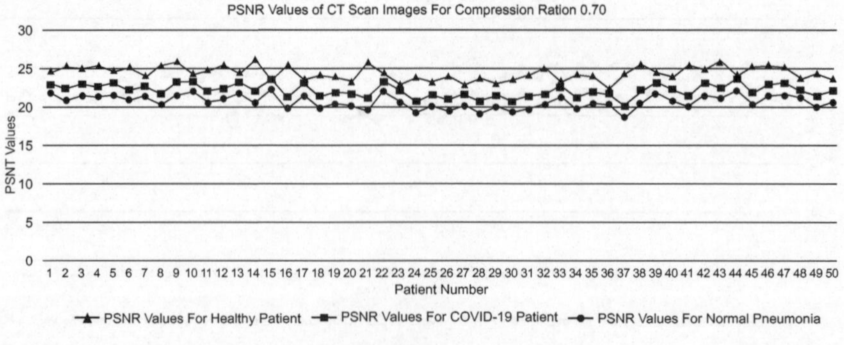

FIGURE 10.10 Plot representing PSNR values for healthy, COVID-19 pneumonia, normal pneumonia CT scan images at 0.70 compression ratio.

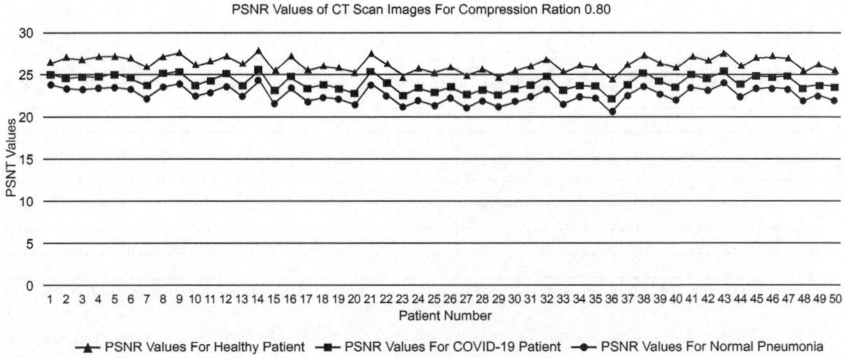

FIGURE 10.11 Plot representing PSNR values for healthy, COVID-19 pneumonia, normal pneumonia CT scan images at 0.80 compression ratio.

patient. Three classifications of patients are taken according to the database. Each plot is drawn for a specific compression ratio value from 0.20 to 0.80. As the compression ratio increases, the value of PSNR also increases because the number of samples taken for the reconstruction also increases. In these plots, we can see that the PSNR values for the healthy patient CT scan image are higher than the COVID-9 patient and the normal pneumonia patient. As PSNR is not the only IQA method, we are moving towards another IQA method called SSIM. So from Figures 10.12 to 10.18, we represent plots drawn for SSIM values.

Figures 10.12 to 10.18 consist of SSIM values for different recovered CT scan images at different compression ratios given. SSIM is used to calculate the structural similarity measure between the two images. It is crucial to check structural aspects for the recovered image to check how much change is at the level of contrast, luminance, and structure concerning the actual image. Here we can see enough variation

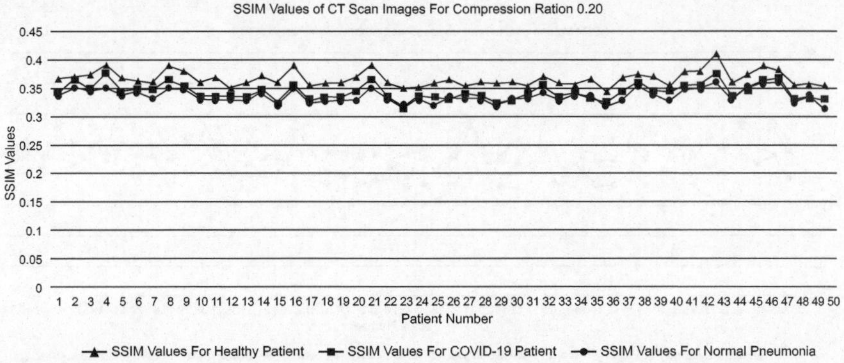

FIGURE 10.12 Plot representing SSIM values for healthy, COVID-19 pneumonia, normal pneumonia CT scan images at 0.20 compression ratio.

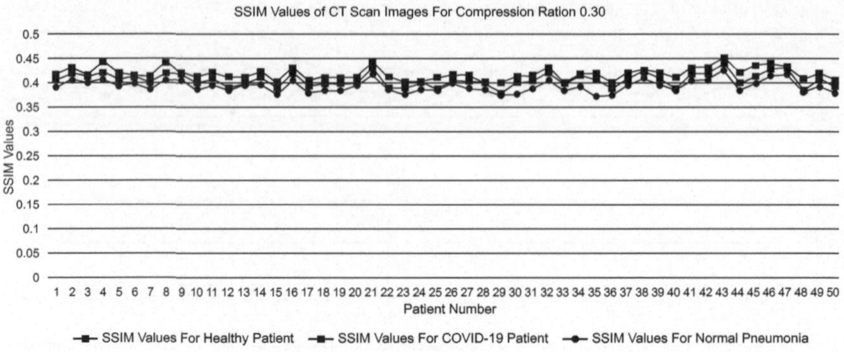

FIGURE 10.13 Plot representing SSIM values for healthy, COVID-19 pneumonia, normal pneumonia CT scan images at 0.30 compression ratio.

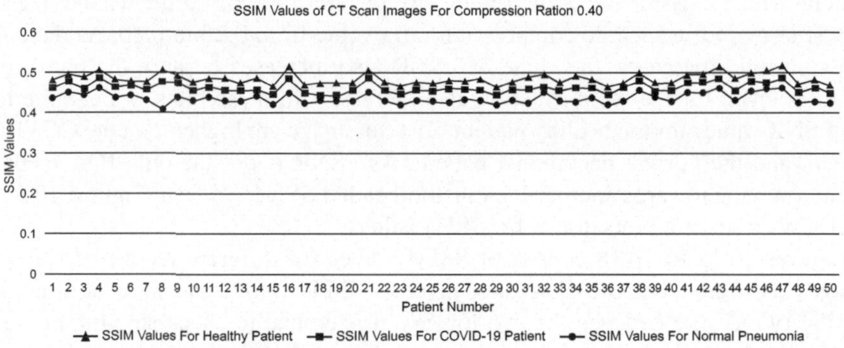

FIGURE 10.14 Plot representing SSIM values for healthy, COVID-19 pneumonia, normal pneumonia CT scan images at 0.40 compression ratio.

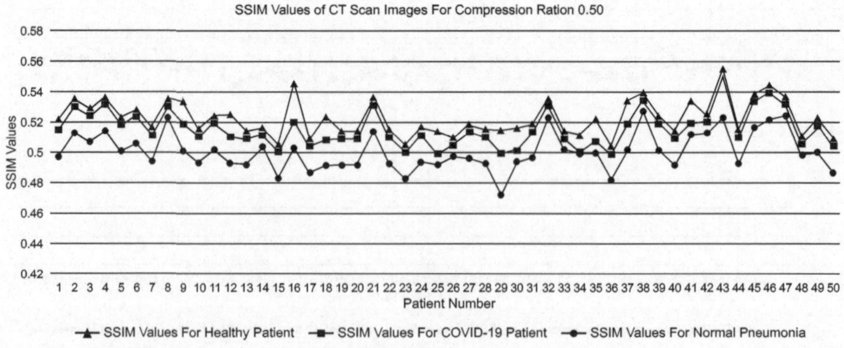

FIGURE 10.15 Plot representing SSIM values for healthy, COVID-19 pneumonia, normal pneumonia CT scan images at 0.50 compression ratio.

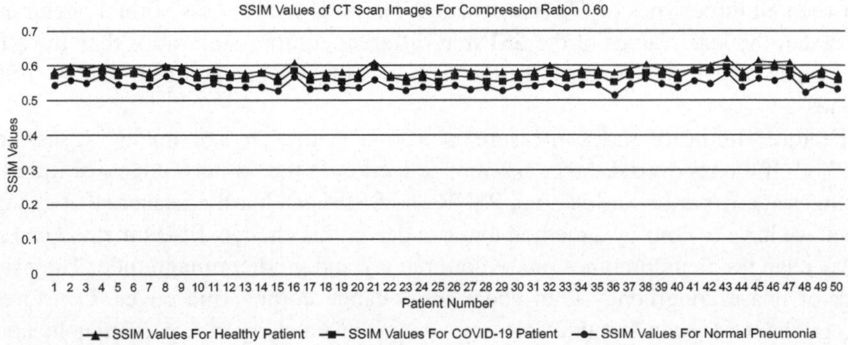

FIGURE 10.16 Plot representing SSIM values for healthy, COVID-19 pneumonia, normal pneumonia CT scan images at 0.60 compression ratio.

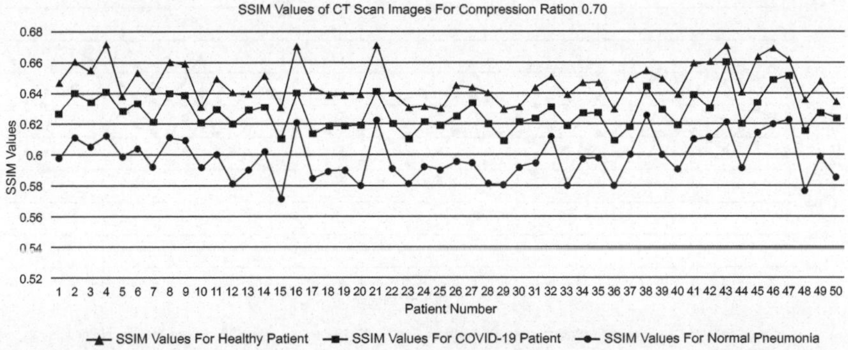

FIGURE 10.17 Plot representing SSIM values for healthy, COVID-19 pneumonia, normal pneumonia CT scan images at 0.70 compression ratio.

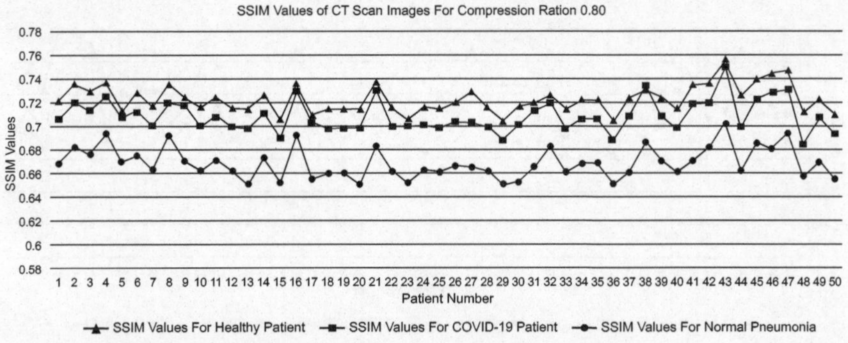

FIGURE 10.18 Plot representing SSIM values for healthy, COVID-19 pneumonia, normal pneumonia CT scan images at 0.80 compression ratio.

between all three types of CT scan images. The SSIM values for normal pneumonia represent the least values of the SSIM at different compression ratios than the other two types of CT scan images. In the next section, we are going to discuss FSIM values.

Feature similarity index measure is also a critical image quality assessment method. If the recovered image has any deflection in the corners, edges, or intensity changes at any particular location, PSNR and SSIM can hardly measure that change. Then we have to employ a method that can detect this change. FSIM method can see these changes as it combines phase congruency and gradient magnitude. The prime task of phase congruency is to check any change corners and edges. Conversely, the gradient magnitude calculates any change in intensity at a particular location. Figures 10.19 to 10.25 show the plots consisting of FSIM values; the FSIM score values for the healthy patients represent higher values than COVID-19 and ordinary pneumonia CT scan images.

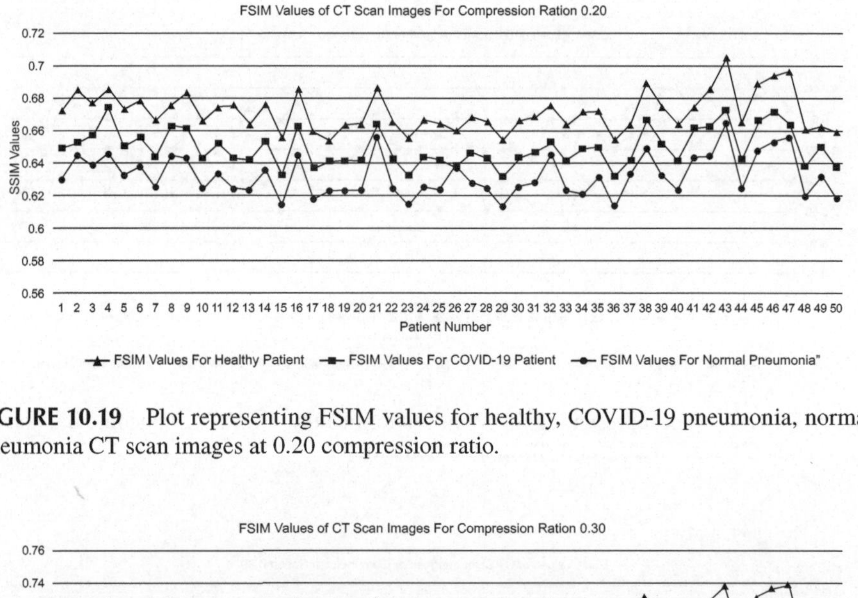

FIGURE 10.19 Plot representing FSIM values for healthy, COVID-19 pneumonia, normal pneumonia CT scan images at 0.20 compression ratio.

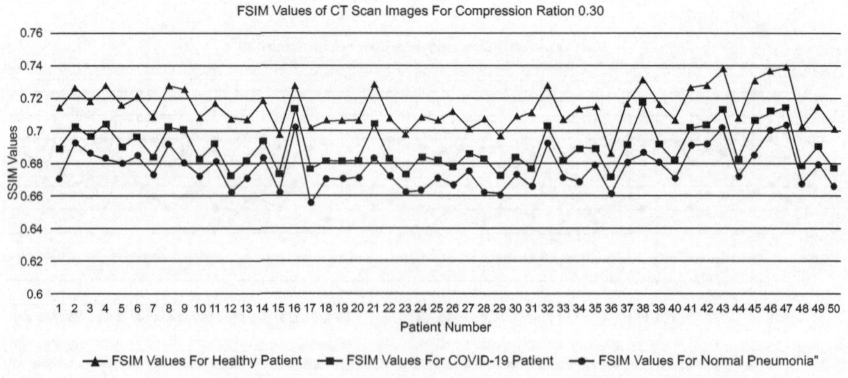

FIGURE 10.20 Plot representing FSIM values for healthy, COVID-19 pneumonia, normal pneumonia CT scan images at 0.30 compression ratio.

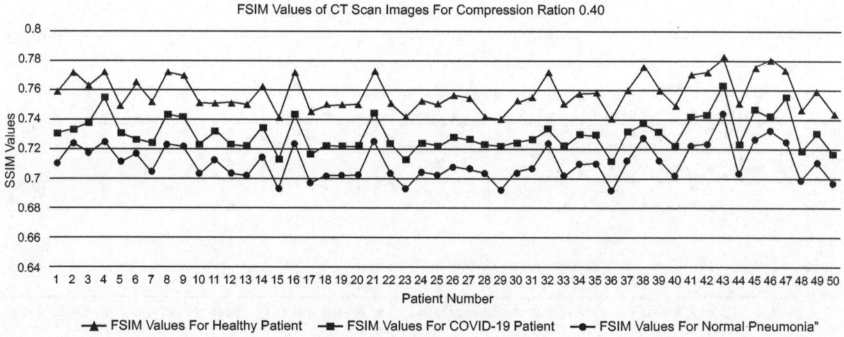

FIGURE 10.21 Plot representing FSIM values for healthy, COVID-19 pneumonia, normal pneumonia CT scan images at 0.40 compression ratio.

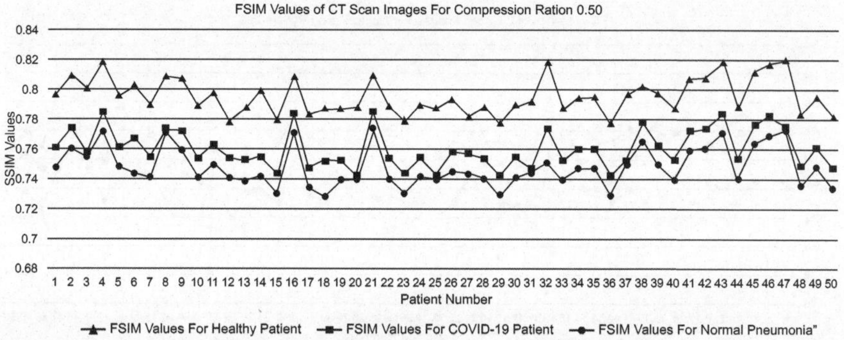

FIGURE 10.22 Plot representing FSIM values for healthy, COVID-19 pneumonia, normal pneumonia CT scan images at 0.50 compression ratio.

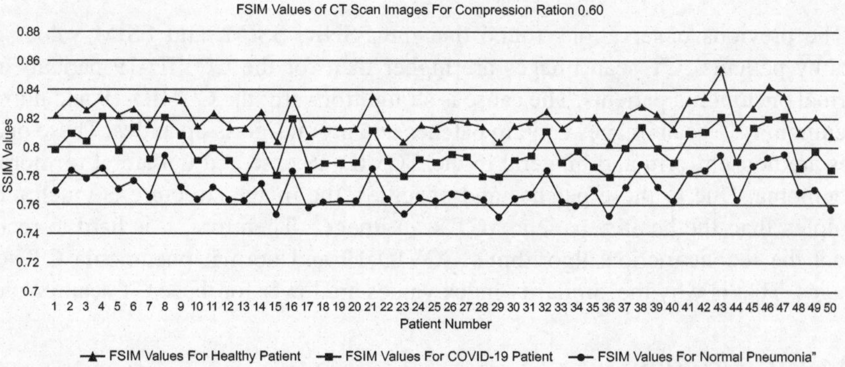

FIGURE 10.23 Plot representing FSIM values for healthy, COVID-19 pneumonia, normal pneumonia CT scan images at 0.60 compression ratio.

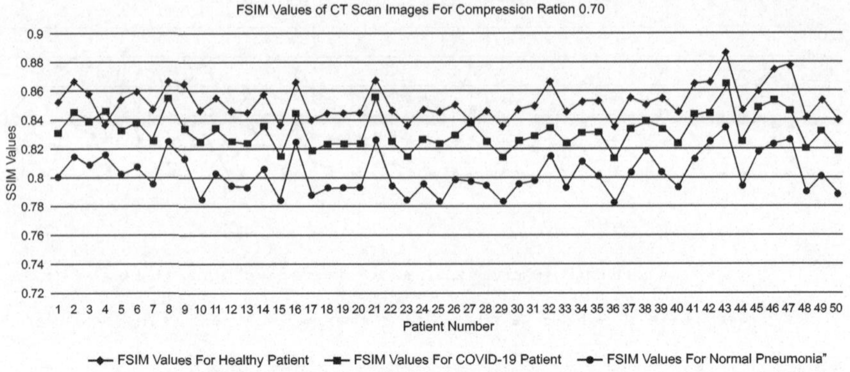

FIGURE 10.24 Plot representing FSIM values for healthy, COVID-19 pneumonia, normal pneumonia CT scan images at 0.70 compression ratio.

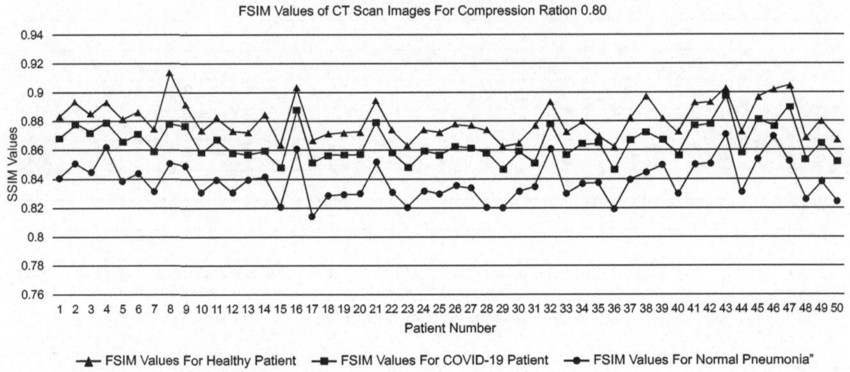

FIGURE 10.25 Plot representing FSIM values for healthy, COVID-19 pneumonia, normal pneumonia CT scan images at 0.80 compression ratio.

The previous observations found that the PSNR, SSIM, and FSIM values for healthy patients' CT scan images are higher than for the COVID-19 patients and normal pneumonia patients. The cause is straightforward: the COVID-19 and normal pneumonia CT scan images contain patches and ground-glass opacities. These opacities are non-uniform and bilateral in the COVID-19 case and unilateral for normal pneumonia. Due to these patches and opacities, the image structure is much more complex than the healthy patients' CT scan images. Therefore, it is hard to reconstruct the reconstruction algorithm's COVID-19 and normal pneumonia CT scan images. That is why the similarity index values are lower for these CT scan images.

10.1.8 Conclusion

We have already discussed the proposed method in the previous section. Now we are going to summarize some of the key findings.

- Our prime focus is to differentiate ordinary pneumonia from COVID-19 pneumonia. In this work, we found that beyond the 60% compression ratio, it is much easier to find out the difference between the different types of pneumonia.
- The comparative study of different plots results that the PSNR, SSIM, and FSIM IQA score values for healthy patients are much higher than in their counterpart CT scan image of COVID-19 and normal pneumonia patients.
- The COVID-19 CT scan images and normal pneumonia images consist of many patches and GGO in the CT scan image, so these images have a complex structure. So it is tedious to recover these images. That is why the values of IQA parameters are lower than in healthy patient CT scan images.

REFERENCES

[1] Donoho, D. "Compressed sensing," *IEEE Transactions on Information Theory*, 52(4): 1289–1306, 2006.

[2] Baraniuk, R. "Compressive sensing," *IEEE Signal Processing Magazine*, 24(4): 118–120, 124, 2007.

[3] Candes, Emmanuel, Nathaniel Braun, and Michael Wakin. "Sparse signal and image recovery from compressive samples." In *Proceedings IEEE International Symposium on Biomedical Imaging: From Nano to Macro*, IEEE, Arlington, VA, April 2007, pp. 976–979.

[4] Gedalyahu, Kfir, Ronen Tur, and Yonina C. Eldar. "Multichannel sampling of pulse streams at the rate of innovation," *IEEE Transactions on Signal Processing*, 59(4): 1491–1504, 2011.

[5] Mishali, M., and Y. C. Eldar. "From theory to practice: Sub-Nyquist sampling of sparse wideband analog signals," *IEEE Journal of Selected Topics in Signal Processing*, 4(2): 375–391, 2010.

[6] Robucci, R., L. Chiu, J. Gray, J. Romberg, P. Hasler, and D. Anderson. "Compressive sensing on a CMOS separable transform image sensor". In *Proceedings of IEEE International Conference on Acoustics, Speech, and Signal Processing* (ICASSP), Las Vegas, NV, April 2008.

[7] Shukla, U. P., N. B. Patel, and A. M. Joshi. "A survey on recent advances in speech compressive sensing." In *Proceedings of IEEE International Multi-Conference on Automation, Computing, Communication, Control and Compressed Sensing (iMac4s)*, Kottayam, India, March 2013, pp. 276–280.

[8] Yu, Hengyong, Ge Wang, Jiang Hsieh, Daniel W. Entrikin, Sandra Ellis, Baodong Liu, and John Jeffrey Carr. "Compressive sensing—Based interior tomography: Preliminary clinical application." *Journal of Computer Assisted Tomography*, 35(6): 762, 2011.

[9] Geethanath, Sairam, Rashmi Reddy, Amaresha Shridhar Konar, Shaikh Imam, Rajagopalan Sundaresan, Ramesh Babu DR, and Ramesh Venkatesan. "Compressed sensing MRI: A review." *Critical Reviews™ in Biomedical Engineering* 41(3), 2013.

[10] van Sloun, Ruud, Ashish Pandharipande, Massimo Mischi, and Libertario Demi. "Compressed sensing for ultrasound computed tomography." *IEEE Transactions on Biomedical Engineering*, 62(6): 1660–1664, 2015.

[11] Szczykutowicz, Timothy P., and Guang-Hong Chen. "Dual energy CT using slow kVp switching acquisition and prior image constrained compressed sensing." *Physics in Medicine & Biology*, 55(21): 6411, 2010.

[12] Leng, Shuai, Jie Tang, Joseph Zambelli, Brian Nett, Ranjini Tolakanahalli, and Guang-Hong Chen. "High temporal resolution and streak-free four-dimensional cone-beam computed tomography." *Physics in Medicine & Biology*, 53(20): 5653, 2008.

[13] Hu, Zhanli, Dong Liang, Dan Xia, and Hairong Zheng. "Compressive sampling in computed tomography: Method and application." *Nuclear Instruments and Methods in Physics Research Section A: Accelerators, Spectrometers, Detectors and Associated Equipment*, 748: 26–32, 2014.

[14] Bao, Peng, Wenjun Xia, Kang Yang, Weiyan Chen, Mianyi Chen, Yan Xi, Shanzhou Niu, et al. "Convolutional sparse coding for compressed sensing CT reconstruction." *IEEE Transactions on Medical Imaging*, 38(11): 2607–2619, 2019.

[15] Hashemi, Sayed Masoud, Soosan Beheshti, Patrick R. Gill, Narinder S. Paul, and Richard S. C. Cobbold. "Accelerated compressed sensing based CT image reconstruction." *Computational and Mathematical Methods in Medicine* 2015 (2015).

[16] Barkan, Oren, Jonathan Weill, Shai Dekel, and Amir Averbuch. "A mathematical model for adaptive computed tomography sensing." *IEEE Transactions on Computational Imaging*, 3(4): 551–565, 2017.

[17] Hsieh, Chia-Jui, Ta-Ko Huang, Tung-Han Hsieh, Guo-Huei Chen, Kun-Long Shih, Zhan-Yu Chen, Jyh-Cheng Chen, et al. "Compressed sensing based CT reconstruction algorithm combined with modified Canny edge detection." *Physics in Medicine & Biology*, 63(15): 155011, 2018.

[18] Shahbazi, Hamid, Mohammad Reza Taban, and Jamshid Abouei. "A new approach to design sensing matrix based on the sparsity constant with applications to computed tomography." *IEEE Access*, 7: 175396–175410, 2019.

[19] Liu, Chenglong, Xiaoyang Wang, Chenbin Liu, Qingfeng Sun, and Wenxian Peng. "Differentiating novel coronavirus pneumonia from general pneumonia based on machine learning." *Biomedical Engineering Online*, 19(1): 1–14, 2020.

[20] Zou, Sijuan, and Xiaohua Zhu. "FDG PET/CT of COVID-19." *Radiology*, 296(2): E118–E118, 2020.

11 Electroencephalogram (EEG) Signal Denoising Using Optimized Wavelet Transform (WT)

A Study

Chaudhuri Manoj Kumar Swain, Ashish Singh, and Indrakanti Raghu

CONTENTS

DOI: 10.1201/9781003309451-11

11.1 INTRODUCTION

Electroencephalography (EEG) is a signal which reflects the activity of the brain in a graphical manner. In other words, it represents the functionality of the brain with respect to time, which helps in diagnosing neurological disorders. The various diseases that can be detected through this include brain tumors, epilepsy, sleep disorders, head injury, dementia, autism, attention disorders, language delay, learning problems, etc. The EEG machine was first developed by Hans Berger in 1929. The mechanism of this machine is that electrical signals generated in different parts of the brain are captured through the electrodes. The different types of techniques which are used to capture EEG signal are magnetoencephalography (MEG), electrical activities from scalp, functional magnetic resonance imaging (FMRI), functional near-infrared spectroscopy (fNIRS) due to changes oxygenation level in blood, etc. Accurate capturing of EEG signals is an important step, as several natural activities, such as muscle activity, eye movements, eye blink, interference of electronic device signals may add noises in the signal, and hence, it affects the reliability in diagnosis process. Hence, noise-removal algorithms, such as adaptive thresholding, filtering, etc., are used during signal processing to remove noises contaminated with the EEG signal. In addition to this, wavelet transform (WT) has been adopted in removing the noise from EEG signal recently due to its inherent property of localization both in time as well as frequency domain, fast computation, and capability of analyzing nonstationary signals. Although WT has been proven as an efficient technique for analyzing EEG signal experimentally, still there are scopes that exist to further improve its performance in denoising through optimization of its parameters. The various denoising parameters of WT are mother wavelet function (Φ), decomposition level (L), thresholding selection rule (λ), thresholding function (β), and rescaling approach (ρ). Different values of these parameters are used for various denoising techniques. The values of these parameters should be chosen optimally for enhancing the performance of denoising. For this, metaheuristic algorithms can be utilized, as these algorithms work on the principle of using learning operator to find out the optimum value of these parameters from search space region.

The organization of this chapter is as follows. Section 2 outlines the background study. Section 3 constitutes the description of optimization techniques. Section 4 explores the conclusion and scope of future work.

11.2 BACKGROUND STUDY

11.2.1 EEG Signal

Electroencephalogram (EEG) signals are the signals captured from the brain of a human which characterizes the activities of the brain. Trillions of synapses exist in the cortical region of the brain. All these generate different types of positive or negative voltage, as a result of which there is a flow of current from one part of the brain to the other part. Hence, there is a shifting electrical pole due to the flow of current which produces waves in EEG. The different types of waves generated due to activities of the brain are (i) beta brainwaves (β waves), (ii) alpha brainwaves (α waves), (iii) theta brainwaves (θ waves), and (iv) delta brainwaves (δ waves).

11.2.1.1 Beta (β) Brainwaves

These waves come under the frequency band of 13–30 Hz. These waves are irregular in shape and are produced at the frontal lobes of the brain. The frequency of these waves is highest among all other brainwaves. These waves are generated in circumstances such as when a person is in a state of conversation, focused on decision-making, anxiety, depression, etc.

11.2.1.2 Alpha (α) Brainwaves

The frequency band under this category is 8–10 Hz. These are synchronized waves and regular in shape. These are produced from the parietal lobe portion or the occipital lobe portion of the brain. The different activities due to which these waves are produced include closing of the eyes, alertness during yoga and meditation, any creativity works, state of relaxation, etc.

11.2.1.3 Theta (θ) Brainwaves

The frequency band that belongs to these waves is 4–7 Hz. The origin of these waves is the hippocampus of the brain. The different activities due to which these waves are generated include disappointment state of mind, before feeling sleepy and waking up from sleep, unconscious state of mind, at the time of recollecting long-forgotten memories, etc.

11.2.1.4 Delta (δ) Brainwaves

These waves come under the frequency band of 2–4 Hz. The frequency of these waves is lowest among all other brainwaves. The different states for this wave include deep sleep state, low state of mind, feeling sad, etc. The various brainwaves, along with their frequency range, amplitude, and recording locations, are outlined in Table 11.1 (Kaur et al. 2018).

11.2.2 Basics of Denoising Process

The common sources of noise are external or environmental sources, such as lightning, AC power lines, an array of electronic equipment, etc. Another general source of noise in EEG include noise generated from physiological activities of the brain. Common examples of such noises are electrocardiogram (ECG) signal, the signal

TABLE 11.1
Recording Locations of Brainwaves (Kaur et al. 2018)

Waves	Frequency (Hz)	Amplitude (μv)	Recording location
β waves	13–30	20	Parietal–frontal region main
α waves	8–10	50–100	Occipital region (more intense), thalamocortical
θ waves	4–7	Above 50	Emotional stress in adults, Parietal–temporal in children and degenerative brain states
δ waves	2–4	Above 50	Deep sleep, infant, serious organic brain disease

generated from muscle contraction (i.e., EMG), signal that originates from movement of the eyeball (i.e., electrooculogram, EOG), etc. These noises when contaminated with EEG signal severely affect the reliability of detection of any disease related to the brain. Hence, denoising is an essential requirement in the processing of EEG signal. In general, the denoising process includes three phases: (i) signal (EEG) decomposition phase, (ii) thresholding phase, and (iii) EEG signal reconstruction phase. The detailed description of these processes are discussed in Section 3.

11.2.3 Wavelet Transform

Wavelet transform (WT) have been utilized for decomposition of EEG signal over Fourier transform due to its properties of multirate filtering, time–frequency localization, and scale–space analysis. It encompasses different window size wrt, the length of EEG signal that permits the wavelet to be expanded or compressed with respect to the frequency of the signal (Adeli et al. 2003). In general, wavelet can be classified into two categories, namely, continuous wavelet transform (CWT) and discrete wavelet transform (DWT). CWT is suitable for applications such as time–frequency analysis, filtering of time localized frequency components, etc. Similarly, DWT is applied in denoising as well as compression of images. Specific to EEG signal, DWT is used for decomposition, whereas inverse discrete wavelet transform (iDWT) is applied for reconstruction of denoisy EEG signal. In general, the decomposition with DWT involves convolution of EEG signal with a bank of low-pass filters as well as high-pass filters. In other words, DWT can be considered as a process of decomposition of EEG signals into different frequency bands (Burrus 1997). There are five wavelet parameters which are used for the denoising process. Those parameters are (i) mother wavelet function (Φ), (ii) decomposition level (L), (iii) thresholding function (β), (iv) threshold selection rules (λ), and (v) threshold rescaling methods (ρ). There exists a range of values of these parameters which can be used in WT. But to achieve the optimality of WT, appropriate values of these parameters are required to be chosen. Hence, requirement of an appropriate optimization algorithm is essential to find out the best values of WT parameters for denoising of EEG signal.

11.2.4 Metaheuristic Algorithms

In general, the optimization algorithms can be classified into two categories, namely, (i) classical technique and (ii) heuristic technique. Linear and nonlinear programming are the optimization algorithms which come under the classical technique. These techniques are not considered as efficient due to the requirement of some conditions, such as differentiability and continuity in the objective function, before applying optimization. Further, metaheuristic algorithms are considered as powerful techniques to solve optimization problems (Siarry 2016). Heuristics techniques are bioinspired optimization algorithms which are widely used in solving optimization problems. Some examples of this technique are genetic algorithm (GA), particle swarm optimization (PSO) algorithm, harmony search algorithm (HMA), flower pollination (FA) algorithm, etc. (Yang 2012). Again, the metaheuristic optimization algorithms can be classified into three categories, namely, (i) evolutionary algorithms (EA), (ii) swarm intelligence (SI) algorithms, (iii) trajectory-based algorithms. Genetic algorithm (GA), harmony search (HM), etc. belong under EA. Similarly, artificial bee colony (ABC), particle swarm optimization (PSO), flower pollination (FA) algorithms, etc. come under SI algorithm category. In addition, tabu search (TS), simulated annealing (SA), β-hill climbing, etc. are examples of trajectory-based optimization algorithm (Al-Betar 2017). The idea of this chapter is that the optimal parameters of WT are chosen with the help of metaheuristic algorithms for denoising of EEG signal. Hence, the efficiency of denoising depends upon two factors, that is, (i) selection of best metaheuristic algorithm and (ii) choosing the optimal values of parameters which can affect the performance of the selected metaheuristic optimization algorithm. Further, these algorithms should be chosen tactically in solving optimization problems related to signal and image processing, wireless communication, embedded system, etc. In the following subsections, some of these optimization algorithms, along with the integral parameters which can affect their performances, are explored.

11.2.4.1 Genetic Algorithm (GA)

Genetic algorithm is a popular evolutionary algorithm which was developed on the principle of "survival of the fittest" (Holland 1975). This algorithm initiates with more than one solution for a problem, in which each solution is represented by a set of decision variables and each of these variables has a certain range of values. The sets of parameters in GA include population size, (P_{size}), number of generations (P_{no}), crossover rate ($P_{crossover}$), and mutation rate ($P_{mutation}$). The values of these parameters affect the efficiency of optimization and hence should be chosen carefully (Burrus 1997). The different processes of GA include (i) initialization of population size, (ii) selection of best individuals from the population, (iii) crossover, and (iv) mutation. The population size reflects the number of solutions of an optimization problem. Each individual solution is associated with a chromosome. Each chromosome has a set of genes which define the characteristics of an individual. Further, each gene is represented as a string of 0s and 1s. Here, each individual is assigned with a fitness value, and a fitness function is used to choose the more suitable individuals on the basis of satisfying some threshold parameters. After selection of best individuals, the crossover process is started, in which a new generation of individuals are formed from the old generations. This process is carried out by selection of a random point in the chromosome and exchange of genes before and after of this point from its old individuals. The resulting chromosomes are known as offspring. Finally, the process of mutation occurs in which genetic diversity is maintained from one generation of a population to the next generations. In this algorithm, the values of parameters such as P_{size}, P_{no}, $P_{crossover}$, $P_{mutation}$, and $P_{mutation}$ should be carefully chosen for improving the reliability of GA.

11.2.4.2 Particle Swarm Optimization (PSO)

Partition swarm optimization (PSO) technique was first studied in (Poli et al. 2007). It comes under the category of swarm intelligence optimization approach. Some important features of this algorithm include (i) improved accuracy, (ii) fast convergence, (iii) computational complexity that is less affected by initial solutions, (iv) less parameters to be tuned for optimization, and (v) performance that is less affected by increase in dimensionality. This algorithm is started with a swarm, that is, a set of candidate solutions. Each of these solutions is known as a particle which moves around the search space in an iterative manner. During each iteration, these particles are affected by the position of best solution, which is acquired in the form of an objective function. The performance of each particle is determined by its objective function, which is searching the local as well as global best solution continuously (Burrus 1997). Further, each particle is associated with these following features: (i) p_i, which represents the position of particle i at current time; (ii) v_i, which reflects the velocity of particle i at current time; (iii) q_i, which indicates the local best of particle I; and (iv) $\hat{q}_i$, which specifies the global best of particle i. In each iteration, these four characteristics of each particle are updated with respect to time t as follows:

$$q_i(t+1) = \begin{cases} q_i(t) & , f\left(p_i(t+1)\right) \geq f\left(q_i(t)\right) \\ p_i(t+1) & , f\left(p_i(t+1)\right) < f\left(q_i(t)\right) \end{cases} \tag{1}$$

$$\hat{q} = \left\{ q_i(t) \middle| t = \underset{i=1,2,\ldots.N}{\arg\min} \; f\left(q_i(t)\right) \right\} \tag{2}$$

Where N represents the total number of particles in a swarm.

$$v_i(t+1) = wv_{i,j}(t) + c_1 r_{i,j}(t)\left(q_{i,j}(t) - p_{i,j}(t)\right) + c_2 r_{2,j}(t)\left(\hat{q}_j(t) - p_{i,j}(t)\right) \tag{3}$$

The updated velocity of each particle is represented in equation (3). It combines factors such as (i) $wv_{i,j}(t)$, where $v_{i,j}$ reflects the previous velocity and its impact is controlled by weight ω; (ii) $q_{i,j}(t) - p_{i,j}(t)$, which represents the extent in which particle i is moved towards local best direction; and (iii) $q_{i,j}(t) - p_{i,j}(t)$, which is responsible for how particle i is forwarded towards global best direction. The other parameters in equation (3), such as c_1, c_2, represent acceleration constants, and r_1, r_2

generate random numbers between 0 and 1 with uniform distribution, that is, r_1, $r_2 \sim U(0, 1)$. Hence, in this algorithm, the parameters such as p_i, v_i, $v_{i,j}$, etc. should be chosen so that its performance is improved.

11.2.4.3 Harmony Search (HS) Algorithm

Harmony search (HS) is an optimization algorithm developed by Geem Z. L. (Geem et al. 2001). The working principle of such algorithm is demonstrated as follows.

11.2.4.3.1 Step 1: Initialization of HS Parameters

In this step, the parameters of this algorithm are initialized. Those parameters include the harmony memory consideration rate (HMCR), the harmony memory size (HMS), the pitch adjustment rate (PAR), the fret width (FW), and the number of improvisations (NI). HMCR determines the selection rate of values from memory, whereas HMS represents the population size. Similarly, PAR reflects the local improvement probability, whereas FW indicates the distance of adjustment. Finally, NI represents the number of iterations.

11.2.4.3.2 Step 2: Harmony Memory (HM) Initialization

In this step, the size of HM is initialized. HM represents a repository of an individual's population, that is, $HM = [h_1, h_2, \ldots\ldots h_{hms}]^T$. Further, these individuals are obtained as follows: $h_i = LB_i + (UB_i - LB_i) \times U(0,1)$, where $I = 1, 2, \ldots, N$ and $U(0, 1)$ produces random numbers between 0 and 1 with uniform distribution. LB_i and UB_i are the lower bound and upper bound values of h_i, respectively.

11.2.4.3.3 Step 3: Generation of New Harmony Memory (HM)

Here, the HM is modified with improved harmony vector, generated as $h_0 = [h_{01}, h_{02}, \ldots\ldots h_{0N}]^T$, on the basis of three operators, namely, (i) random consideration (RC), (ii) pitch adjustment (PA), and (iii) memory consideration (MC). These operators are responsible for assigning values to each of these decision variables in new HM.

11.2.4.3.4 Step 4: Updation of Harmony Memory (HM)

In this step, the HM is updated by comparing the new harmony vector with the worst harmony vector. If it is found that the new harmony vector, that is, $h_0 = [h_{01}, h_{02}, \ldots\ldots h_{0N}]$, is better, then it replaces the worst harmony vector stored in HM.

11.2.4.3.5 Step 5: Check Stop Criteria

In this step, the algorithm stops its execution after satisfying the stop criteria.

From the procedure of this algorithm, it is realized that parameters such as HM, RC, PA, and MC should be chosen to get the optimal performance of harmony search optimization algorithm.

11.2.5 EEG Dataset

The scope in EEG research depends upon the available dataset collected with different activities of the human brain. The datasets can be of various categories, such as (i) motor imagery EEG, (ii) emotion-recognition EEG, (iii) visually evoked potentials (VEPs) EEG, (iv) slow-cortical potentials (SCPs) EEG, (v) resting state EEG, (vi) eye blinks/movements EEG, and (vii) clinical EEG.

11.2.6 Performance Parameters

The efficacy of the proposed methods is evaluated on the basis of some parameters, such as root mean square error (RMSE), percentage root mean square difference (PRD), signal-to-noise ratio (SNR), etc. (Nagar and Kumar 2021). The formulation of these parameters is as follows:

$$RMSE = \sqrt{\frac{1}{N}\sum_{n=1}^{N}\left[x(n) - \hat{x}(n)\right]^2} \tag{4}$$

$$PRD = 100 * \sqrt{\frac{\sum_{n=1}^{N}\left[x(n) - \hat{x}(n)\right]^2}{\sum_{n=1}^{N}\left[x(n)\right]^2}} \tag{5}$$

$$SNR_{out} = 10 * \log_{10}\left[\frac{\sum_{n=1}^{N}\left[x(n)\right]^2}{\sum_{n=1}^{N}\left[x(n) - \hat{x}(n)\right]^2}\right] \tag{6}$$

Where $x(n)$ and $\hat{x}(n)$ represent the original EEG signal and denoised EEG signal, respectively.

11.3 OPTIMIZATION OF WAVELET TRANSFORM PARAMETERS FOR EEG SIGNAL DENOISING WITH METAHEURISTIC ALGORITHMS

In this section, the procedure of optimizing wavelet parameters using metaheuristics algorithm for EEG denoising is elaborated. This process is executed in four steps. The first phase includes initialization of parameters related to EEG, noise, wavelet transform, and metaheuristics algorithm. In the second step, the optimal value of wavelet parameters is evaluated using metaheuristics algorithms. In the third step, the denoising of EEG signal is done by using optimized DWT. Finally, in the final step, the denoisy EEG signal is reconstructed using iDWT. Figure 11.1 illustrates the steps of optimized wavelet transform in denoising EEG signal. The discussion of each of these steps follows.

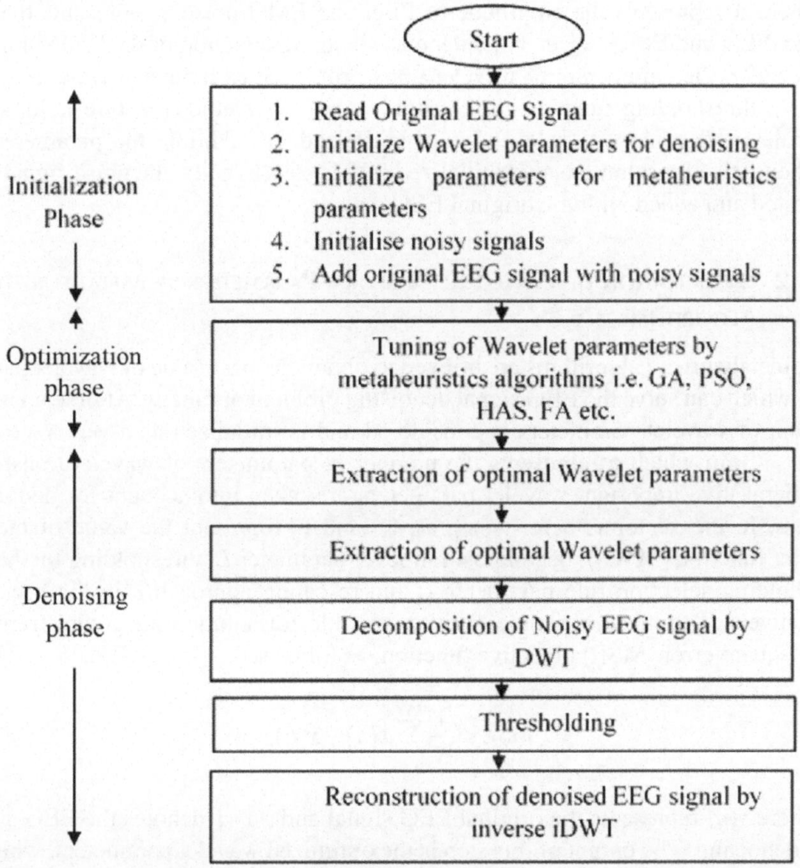

FIGURE 11.1 Steps of EEG denoising with optimal wavelet parameters.

11.3.1 Initialization

In this step, first, an original EEG signal is generated. Further, the noises, such as electromyogram (EMG), power line noise (PLN), white Gaussian noise (WGN), etc., are initialized with the corresponding parameters and added with the original EEG signal. The reason of adding the noise is that these noises are generated during the dataset generation due to eye movement, electro signal distortion, eye blink, etc. The mathematical models for these noises, such as PLN, EMG, and WGN, are represented by equation (7), equation (8), and equation (9), respectively.

$$Noise_{PLN} = A * \sin(2\Pi f\, n) \tag{7}$$

$$Noise_{EMG} = A * rand(n) \tag{8}$$

$$Noise_{WSN} = x(n) + \sigma \tag{9}$$

Where A represents the amplitude of PLN and EMG noise, f represents the frequency PLN and EMG noise, and σ denotes standard deviation of the WGN noise.

Secondly, the parameters of wavelet transform, such as mother wavelet function (MWF), thresholding function β, threshold re-scaling methods ρ, threshold selection rules λ, decomposition level L, are initialized. In addition, the parameters of metaheuristic algorithm are also initialized in this step. Finally, the noise signals are generated and added with the original EEG signal.

11.3.2 Optimization of Wavelet Transform Parameters by Metaheuristic Algorithm

Here, metaheuristic algorithms are utilized to obtain the best value of wavelet parameters, which can solve the EEG signal denoising problem optimally. At first, an initial solution of wavelet parameters is considered and is initialized as a vector $v = (v_1, v_2, \ldots v_n)$, in which n represents the number of parameters of wavelet transform. In general, five important wavelet parameters are taken into account for denoising process. In the vector v, here, v_1, v_2, v_3, v_4, and v_5 represent the value of mother wavelet function (MWF), decomposition level parameter L, thresholding method β, thresholding selection rule parameter λ, and rescaling approach ρ in the vector x, respectively. The metaheuristic algorithm suitable for solution are found from the mean square error (MSE) objective function, as follows.

$$MSE = \frac{1}{N}\sum_{n=1}^{N} x(n) - \hat{x}(n) \tag{10}$$

Where $x(n)$ represents the original EEG signal and $\hat{x}(n)$ denotes the EEG signal after denoising. The output of this step is the optimized wavelet parameters, which is represented by $v_{0opt} = (v_{01}, v_{02}, \ldots v_{0n})$.

11.3.3 EEG Denoising Using Optimized Parameter of Wavelet

In this step, the denoising of EEG is performed. This process involves three main steps, namely, (a) EEG signal decomposition, (b) application of thresholding for noise removal, and (c) reconstruction of EEG signal. The EEG signal decomposition and reconstruction can be executed by DWT and *i*DWT, respectively. Here, decomposition of EEG signal by DWT is done by passing the signal through a group of low-pass filters (LPF) and high-pass filters (HPF) to generate approximation coefficients (*cA*) and details coefficients (*cD*), respectively. The primary purpose of using DWT for EEG denoising is to decompose the noisy EEG signal into various coefficient levels for evaluating its high-frequency components because it assumes that artifacts may have high amplitudes in the corresponding frequency bands. In thresholding step, the magnitude of wavelet coefficients is compared with some threshold level. Further, in signal reconstruction phase, the iDWT is used in the reconstruction of denoisy EEG signal. Figure 11.2 shows the block diagram of EEG signal denoising process (Alyasseri et al. 2018). The procedure of each of the steps is discussed in the following subsections.

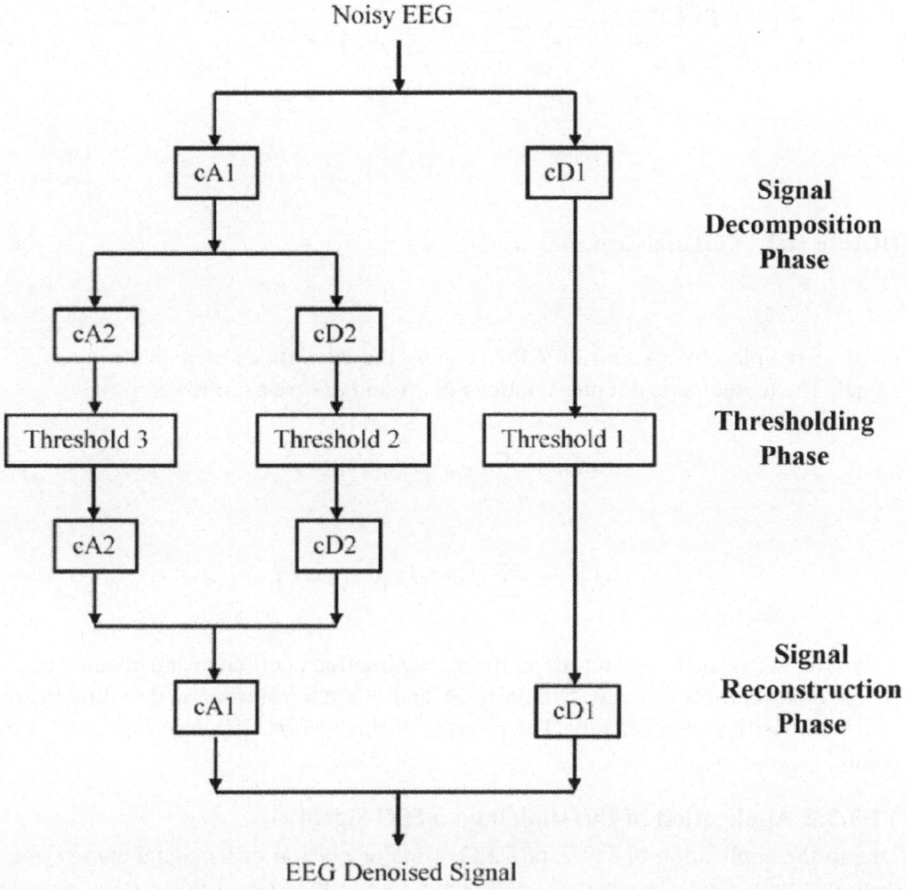

FIGURE 11.2 Block diagram of EEG signal denoising process.

11.3.3.1 EEG Signal Decomposition Using Discrete Wavelet Transform (DWT)

Here, DWT is used for decomposition of noisy EEG signals *x*(n), which can be comparable to passing the signal through a set of low-pass filters as well as high-pass filters, simultaneously. The outputs of these low-pass filters and high-pass filters are denoted as approximation coefficients (*cA*) and detail coefficients (*cD*), respectively. This low-pass filter as well as high-pass filter bank used in decomposition process is called analysis filter bank. Similarly, the low-pass filters as well as high-pass filters used in signal reconstruction phase are known as synthesis filter bank. The decomposition level can be taken as 2, 3, 4, etc. Each level is decomposed into two different parts, that is, *cA* and *cD*. A high-pass filter is used to process *cA*, whereas a low-pass filter is utilized to process *cD*. The separation of EEG signals into *cA* and *cD* is dependent upon their frequencies and amplitudes. Further, the decomposed signal

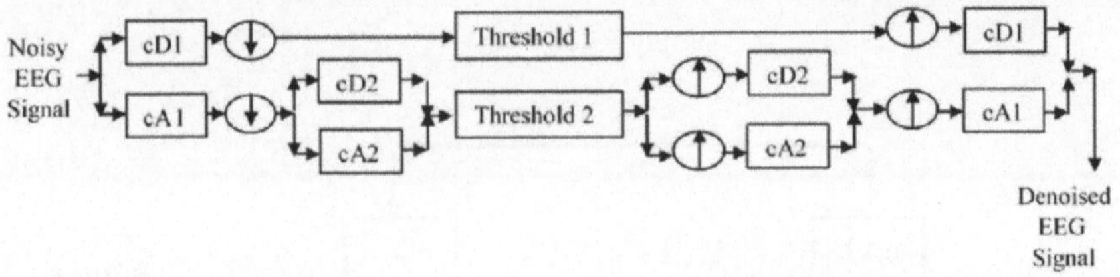

FIGURE 11.3 EEG denoising process.

is down-sampled by a factor of 2 for keeping the even index elements in the EEG signal. The mathematical representations of *cA* and *cD* are as follows.

$$cA_i(n) = \sum_{k=-\infty}^{\infty} cA_{i-1}(k)\phi_i(n-k) \tag{11}$$

$$cD_i(n) = \sum_{k=-\infty}^{\infty} cD_{i-1}(k)\psi_i(n-k) \tag{12}$$

Where $cA_i(n)$ and $cD_i(n)$ represent the approximation coefficient and detailed coefficient on level *i*, respectively. Similarly, ϕ and ψ are the shifting and scaling factor of decomposition, respectively. The process of this step is shown in Figure 11.3 in detail.

11.3.3.2 Application of Thresholding on EEG Signal

Due to the application of DWT on EEG, a major portion of its signal energy concentrates with fewer number of coefficients, while the spread noise is on a large number of coefficients. Hence, the process of noise removal involves comparison of magnitude of wavelet coefficients with a threshold value λ. This process is called thresholding. The coefficients having large magnitude are selected, whereas coefficients having smaller magnitude are set to zero. This is because the coefficients with smaller magnitude are considered as noise. For calculation of threshold level λ, the standard deviation σ of the noise is considered. Although there are many techniques present to evaluate λ, the method proposed by Donoho and Johnstone is widely used. It depends on the median absolute deviation (MAD) value of detail coefficients and represented by the following formula (Choudhry et al. 2016).

$$\sigma = median\left(\frac{\left|x - \hat{x}\right|}{0.6745}\right) \tag{13}$$

Here, the value 0.6745 is considered for scaling purpose in case of normally distributed random data.

The threshold level is determined by (Choudhry et al. 2016):

$$\lambda = \sigma\sqrt{2\log(k)} \tag{14}$$

Where k represents the length of the signal.

The thresholding process is applied on the detailed coefficients only, as approximation coefficients contain lower-frequency components; hence, these are least influenced by noise. There exist different thresholding techniques which are used for denoising. Some of these are as follows.

11.3.3.2.1 Hard Thresholding

The shrinkage or threshold function of this technique is represented as (Choudhry et al. 2016):

$$S^{\lambda}(d) = \begin{cases} d, & |d| > \lambda \\ 0, & |d| < \lambda \end{cases} \tag{15}$$

Where $S^{\lambda}(.)$ represents the shrinkage or threshold function and d denotes the single detail coefficient.

11.3.3.2.2 Soft Thresholding

The shrinkage or threshold function of this technique is represented as (Choudhry et al. 2016):

$$S^{\lambda}(d) = \begin{cases} Sgn(d)(|d| - \lambda), & |d| \geq \lambda \\ 0, & |d| < \lambda \end{cases} \tag{16}$$

11.3.3.3 Reconstruction of Denoisy EEG Signal by Inverse DWT (iDWT)

The reconstructed EEG signal is obtained by applying inverse discrete wavelet transform (iDWT) on the decomposed EEG signal $x_{decomposed}(n)$ as follows:

$$\begin{aligned} \hat{x}(n) &= iDWT\left(x_{decomposed}(n)\right) \\ &= \sum_{k=-\infty}^{\infty} cA_L(k)\phi_i'(n-k) + \sum_{i=1}^{L}\sum_{k=-\infty}^{\infty} cD_{i+1}(k)\psi_i'(n-k) \end{aligned} \tag{17}$$

Further, the EEG signal is up-sampled by a factor of 2 (↑2) for inserting zeros at even indexed elements of EEG signals.

11.4 CONCLUSION AND SCOPE OF FUTURE WORK

In this chapter, a detailed discussion of combined impact of metaheuristic algorithm and wavelet transform on the process of denoising of EEG signal is demonstrated. In this technique, at first, the parameters of wavelet transform are optimized by using metaheuristic algorithm. Further, with the optimized wavelet parameters, EEG signal denoising is performed. The denoising process includes EEG signal decomposition, thresholding, and signal reconstruction. Although this method can be applied for denoising of EEG signal efficiently, its application in real time is limited by the convergence time metaheuristic algorithm. This research work can also be extended in determining energy degradation of EEG signal after application of this technique, as most of the research methodology reduces the energy content of the signal after denoising.

REFERENCES

Adeli, Hojjat, Ziqin Zhou, and Nahid Dadmehr. 2003. "Analysis of EEG Records in an Epileptic Patient Using Wavelet Transform." *Journal of Neuroscience Methods* 123 (1): 69–87.

Al-Betar, Mohammed Azmi. 2017. "β-Hill Climbing: An Exploratory Local Search." *Neural Computing and Applications* 28 (1): 153–168.

Alyasseri, Zaid Abdi Alkareem, Ahamad Tajudin Khader, Mohammed Azmi Al-Betar, Joao P. Papa, and Osama Ahmad Alomari. 2018. "EEG Feature Extraction for Person Identification Using Wavelet Decomposition and Multi-Objective Flower Pollination Algorithm." *IEEE Access* 6: 76007–76024.

Burrus, C. Sidney. 1997. *Introduction to Wavelets and Wavelet Transforms: A Primer*. Englewood Cliffs. Prentice Hall PTR.

Choudhry, Mahipal Singh, Rajiv Kapoor, Anuj Gupta, and Bhaskar Bharat. 2016. "A Survey on Different Discrete Wavelet Transforms and Thresholding Techniques for EEG Denoising." In *2016 International Conference on Computing, Communication and Automation (ICCCA)*, Greater Noida, India, 1048–1053. IEEE.

Geem, Zong Woo, Joong Hoon Kim, and Gobichettipalayam Vasudevan Loganathan. 2001. "A New Heuristic Optimization Algorithm: Harmony Search." *Simulation* 76 (2): 60–68.

Holland, J. H. 1975. "An Introductory Analysis with Applications to Biology, Control, and Artificial Intelligence." In *Adaptation in Natural and Artificial Systems. First Edition*. The University of Michigan.

Kaur, Bhavneet, Meenakshi Sharma, Mamta Mittal, Amit Verma, Lalit Mohan Goyal, and D. Jude Hemanth. 2018. "An Improved Salient Object Detection Algorithm Combining Background and Foreground Connectivity for Brain Image Analysis." *Computers &Electrical Engineering* 71: 692–703.

Nagar, Subham, and Ahlad Kumar. 2021. "Orthogonal Features Based EEG Signals Denoising Using Fractional and Compressed One-Dimensional CNN AutoEncoder." In *IEEE Transactions on Neural Systems and Rehabilitation Engineering*, vol. 30, 2474–2485.

Poli, Riccardo, James Kennedy, and Tim Blackwell. 2007. "Particle Swarm Optimization." *Swarm Intelligence* 1 (1): 33–57.

Siarry, Patrick. 2016. *Metaheuristics*. Springer.

Yang, Xin-She. 2012. "Flower Pollination Algorithm for Global Optimization." In *International Conference on Unconventional Computing and Natural Computation*, 240–249. Springer.

12 Predicting Diabetes in Women by Applying the Support Vector Machine (SVM) Model

Nitin Jaglal Untwal and Utku Kose

CONTENTS

12.1 INTRODUCTION

In terms of ML efforts, classification has been always a remarkable solution methodology for real world problems [1–7]. Support vector machine is a supervised ML model. It was developed in 1960, and its improved version was launched in the 1990s [8–10]. The model became known and got tremendous recognition for achieving accurate results. Support vector machine is different from other machine learning models since it minimizes the misclassification errors. As we closely analyze the machine learning models, we will get to know about the accuracy of the support

DOI: 10.1201/9781003309451-12

vector machine in comparison to other machine learning models. We can find that the support vector machine algorithm correctly differentiates the data point. The support vector machine algorithm optimizes the classification, which is an important feature of the support vector machine, and hence, it is more reliable to depend on it. The complex mathematical problems can be resolved with the support vector machine algorithm [11–15]. From the previous discussion, it is clear that the support vector machine algorithm is superior to other machine learning models, and hence, we have decided to apply the support vector machine algorithm in the study titled "Predicting Diabetes in Women by Applying the Support Vector Machine (SVM) Model Using Python Programming."

Support vector machine algorithm is the most used and applied machine learning classification algorithm. The support vector machine algorithm is used since it has the upper hand over other machine learning classification models, like the logistic regression model and naive Bayes model. The support vector machine algorithm gives more optimum solutions than any other machine learning classification model. The support vector machine algorithm is also known for the accuracy which it provides [9–12]. The main utility of the support vector machine algorithm is to find out a hyperplane (N-dimensional) that creates a difference in data points (refer to Figure 12.1). To bifurcate the data as per classes, many hyperplanes can be drawn. The best optimum hyperplane, which has the maximum margin, is considered to be the best (refer to Figure 12.1). The decision boundaries are hyperplanes that help classify the data points.

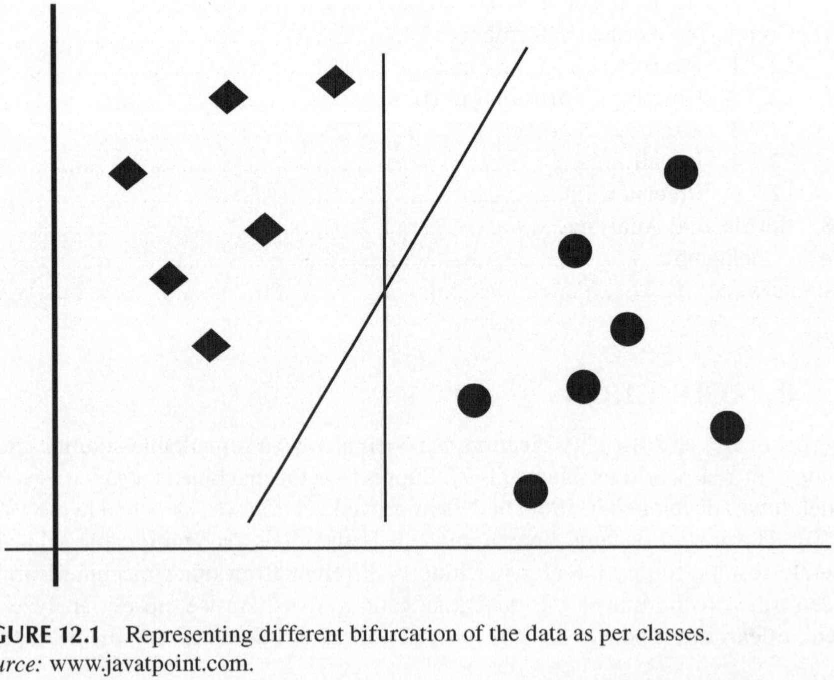

FIGURE 12.1 Representing different bifurcation of the data as per classes.
Source: www.javatpoint.com.

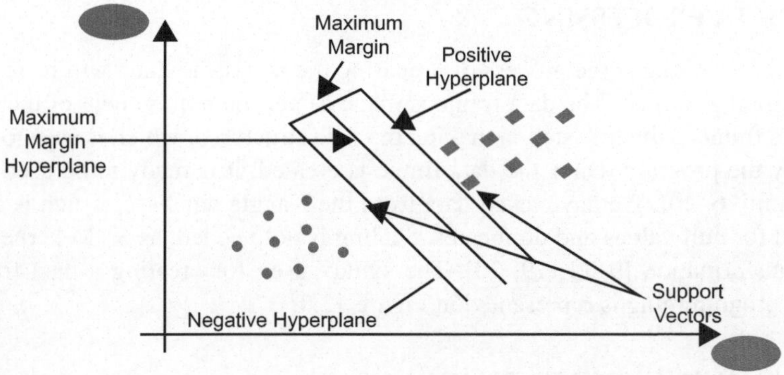

FIGURE 12.2 Representing different bifurcation of the data with maximum margin hyperplane.
Source: www.javatpoint.com.

12.2 RESEARCH METHODOLOGY

Data Source

Data used for the study is from Kaggle database.

Sample Size

It contains health details of 767 women tested for diabetes.

Software Used for Data Analysis

Python programming.

Model Applied

For the purpose of this study, we have applied the support vector machine algorithm.

Limitation of the Study

The study is restricted to only predicting diabetes in women with a given dataset.

Future Scope of the Study

In the future, the study can be extended by comparing the support vector machine algorithm with other models with increased sample size/comparative study with male sample data.

12.3 METHODOLOGY

For creating a predictive model, we have selected and applied the support vector machine algorithm. The research is carried out in five steps.

1. **Data processing**
2. **Training and testing**
3. **Making a prediction kernel support vector machine**
4. **Comparing kernel performance**
5. **Results and analysis**

12.4 DATA PROCESSING

Feature engineering is the process of preparing the raw data which will be ready for the program to utilize. The data frame is made as per the requirement of the model. The data frame is the first step in making the data structured, which is easy to understand by the program. Once the data frame is created, it is ready to be used by the algorithm [16–20]. We have taken data from the Kaggle database, which is already checked for null values and duplicates. Scaling is not needed, as SVM kernels have data transformation [9–11, 19, 20]. The syntax used for creating a data frame in Python programming is represented in Figure 12.3.

12.5 TRAINING AND TESTING

For the purpose of conducting the study, secondary data is collected from the Kaggle database regarding diabetes in women. The dependent variable for diabetes in women (Y) is binary class (Figure 12.4), and there are five independent variables, "pregnancies," "glucose," "blood pressure," "BMI," and "age," which are continuous in nature [9].

```
# importing libraries
import statsmodels.api as sm
import pandas as pd
df = pd.read_csv (r'C:\Users\nitin\Desktop\Diabetes.csv')
print (df)
```

```
     Pregnancies  Glucose  BloodPressure  SkinThickness  Insulin   BMI
0              6      148             72             35        0  33.6
1              1       85             66             29        0  26.6
2              8      183             64              0        0  23.3
3              1       89             66             23       94  28.1
4              0      137             40             35      168  43.1
..           ...      ...            ...            ...      ...   ...
763           10      101             76             48      180  32.9
764            2      122             70             27        0  36.8
765            5      121             72             23      112  26.2
766            1      126             60              0        0  30.1
767            1       93             70             31        0  30.4

     DiabetesPedigreeFunction  Age  Outcome
0                       0.627   50        1
1                       0.351   31        0
2                       0.672   32        1
3                       0.167   21        0
4                       2.288   33        1
..                        ...  ...      ...
763                     0.171   63        0
764                     0.340   27        0
765                     0.245   30        0
766                     0.349   47        1
767                     0.315   23        0

[768 rows x 9 columns]
```

FIGURE 12.3 Creating data frame.

```
# defining the dependent and independent variables
X = df[['Pregnancies', 'Glucose', 'BloodPressure', 'BMI', 'Age']]
y = df[['Outcome']]
```

FIGURE 12.4 Defining the dependent and independent variables.

```
X_train, X_test, y_train, y_test = sklearn.model_selection.train_test_split(X, y,
print(X_train.shape)
print(X_test.shape)
print(y_train.shape)
print(y_test.shape)
```

```
(614, 5)
(154, 5)
(614, 1)
(154, 1)
```

FIGURE 12.5 Representing the Python code for training and testing.

```
from sklearn.svm import SVC
from sklearn.metrics import accuracy_score

clf = SVC(kernel='linear')
clf.fit(X_train,y_train)
y_pred = clf.predict(X_test)
print(accuracy_score(y_test,y_pred))
```

```
C:\Users\nitin\anaconda3\lib\site-packages\sklearn\utils\validation.py:72: Data
ConversionWarning: A column-vector y was passed when a 1d array was expected. P
lease change the shape of y to (n_samples, ), for example using ravel().
  return f(**kwargs)

0.7987012987012987
```

```
from sklearn.svm import SVC
svclassifier = SVC(kernel='poly', degree=8)
svclassifier.fit(X_train, y_train)
```

```
C:\Users\nitin\anaconda3\lib\site-packages\sklearn\utils\validation.py:72: Data
ConversionWarning: A column-vector y was passed when a 1d array was expected. P
lease change the shape of y to (n_samples, ), for example using ravel().
  return f(**kwargs)

SVC(degree=8, kernel='poly')
```

```
y_pred = svclassifier.predict(X_test)
```

FIGURE 12.6 Representing the Python code for SVC linear kernel.

Around 80% of data is being considered for trial, and 20% of data is used for testing purposes to evaluate the model. The main objective of the partitioning of the data is to convert data into training and testing datasets. For analysis, we have randomly selected the data to ensure the training and testing datasets are similar. By using similar datasets for training and testing, we can minimize data errors, and hence, a better understanding of the model can be achieved.

12.6 PREDICTING WITH KERNEL SUPPORT VECTOR MACHINE

Kernel function in support vector machine is considered to be an important element in overall support vector machine. It is a set of mathematical functions. The kernel takes data input and converts it into required transformation.

12.6.1 Polynomial Kernel Function

Support vector machines represent the similarity of vectors (training and testing samples) in a feature space over polynomials of the original variables, allowing learning of nonlinear models.

It is represented by:

$$K\left(x_i, x_j\right) = \left(x_i, x_j\right)^d \qquad \text{Eq (1)}$$

Where *d* is the degree of polynomial and *xi, xj* are input space vectors.

12.6.2 RBF Kernel

RBF is the radial basis function. This is used when there is no prior knowledge about the data. It's represented as:

$$K(xi, xj) = \exp(-\gamma \parallel xi - xj \parallel)2 \qquad \text{Equation (2)}$$

```
from sklearn.metrics import classification_report, confusion_matrix
print(confusion_matrix(y_test, y_pred))
print(classification_report(y_test, y_pred))
```

```
[[92  8]
 [25 29]]
              precision    recall  f1-score   support

           0       0.79      0.92      0.85       100
           1       0.78      0.54      0.64        54

    accuracy                           0.79       154
   macro avg       0.79      0.73      0.74       154
weighted avg       0.79      0.79      0.77       154
```

FIGURE 12.7 Polynomial kernel results for SVM.

```
from sklearn.svm import SVC
svclassifier = SVC(kernel='rbf')
svclassifier.fit(X_train, y_train)
```

```
C:\Users\nitin\anaconda3\lib\site-packages\sklearn\utils\validation.py:72: Data
ConversionWarning: A column-vector y was passed when a 1d array was expected. P
lease change the shape of y to (n_samples, ), for example using ravel().
  return f(**kwargs)

SVC()
```

```
y_pred = svclassifier.predict(X_test)
```

```
from sklearn.metrics import classification_report, confusion_matrix
print(confusion_matrix(y_test, y_pred))
print(classification_report(y_test, y_pred))
```

```
[[90 10]
 [25 29]]
              precision    recall  f1-score   support

           0       0.78      0.90      0.84       100
           1       0.74      0.54      0.62        54

    accuracy                           0.77       154
   macro avg       0.76      0.72      0.73       154
weighted avg       0.77      0.77      0.76       154
```

FIGURE 12.8 RBF kernel results.

12.6.3 Sigmoid Kernel Function

It is similar to the two-layer neural network, which is used as an activation function for artificial neurons. This is mainly used in neural networks. It is represented by:

$$K(xi, xj) = \tanh(\alpha xay + c) \quad \text{Equation (3)}$$

12.7 COMPARING KERNEL PERFORMANCE

12.7.1 Precision

Precision only considers the pertinent factors. It is the ratio of correct classification to false-positive classification.

$$\text{Precision} = \frac{\text{True Positive}}{\text{True Positive } + \text{False Positive}}$$

After comparing the three kernel support vector machine, the polynomial kernel is considered to be the best, with an accuracy of 79%. The precision of diabetes in women is 78%.

TABLE 12.1
Showing Comparative Analysis for Precision

	Linear	Polynomial	RBF	Sigmoid
Precision (0)	0.82	0.79	0.78	0.49
Precision (1)	0.72	0.78	0.74	0.06

```
y_pred = svclassifier.predict(X_test)
```

```
from sklearn.metrics import classification_report, confusion_matrix
print(confusion_matrix(y_test, y_pred))
print(classification_report(y_test, y_pred))
```

```
[[49 51]
 [51  3]]
              precision    recall  f1-score   support

           0       0.49      0.49      0.49       100
           1       0.06      0.06      0.06        54

    accuracy                           0.34       154
   macro avg       0.27      0.27      0.27       154
weighted avg       0.34      0.34      0.34       154
```

FIGURE 12.9 Sigmoid kernel results.

12.7.2 Creating Confusion Matrix

Confusion matrix measures model performance. It evaluates the values of actual values and predicted values. It is of the order N*N matrix; here, *N* is for the class of dependent/target variable.

For binary classes, it is 2 × 2 confusion matrix.

Calculating False Negative, False Positive, True Negative, True Positive

The confusion matrix for our dataset is as follows:

12.7.3 Accuracy Statistics

It measures the overall accuracy of the model by analyzing the output predicted about incorrect predictions.

To obtain the accuracy of the model, we apply the following formulae:

$$\text{Accuracy} = \frac{\text{True Positive} + \text{True Negative}}{\text{True Positive} + \text{True Negative} + \text{False Positive} + \text{False Negative}}$$

$$\text{Accuracy} = \frac{92 + 8}{92+8+25+29} = 0.65$$

Accuracy for the overall model is 0.65.

TABLE 12.2
The Confusion Matrix

	0	1
0	92 (TP)	8 (TN)
1	25 (FP)	29(FN)

12.7.4 Recall

It is the ratio of true positive predictions divided by the total number of true positive predictions and false-negative predictions. Higher recall implies more correct prediction (a small number of FN).

$$\text{Recall} = \frac{\text{True Positive}}{\text{True Positive} + \text{False Negative}}$$

$$\text{Recall} = \frac{92}{92+29} = 0.76$$

Recall for the overall model is 0.76.

12.7.5 Precision

Precision measures how correctly we have predicted the true positive prediction. It is the qualitative analysis of correctly predicted values.

$$\text{Precision} = \frac{\text{True Positive}}{\text{True Positive} + \text{False Positive}} = \frac{92}{92+25} = 0.78$$

Precision for the overall model is 0.78.

12.8 RESULTS AND ANALYSIS

The table that follows shows the comparative analysis of the confusion matrix, which indicates that poly kernel's performance is notable, as it has a higher rate of accuracy in predicting positive values in the confusion matrix. The matrix also includes the logistic regression model (LRM) for comparison.

12.9 CONCLUSION

After comparing the confusion matrix for four kernels, the polynomial kernel is considered to be the best, with an accuracy of 78%. The precision of predicting diabetes in women is 78%; further, the comparative analysis of the confusion matrix indicates

TABLE 12.3
Comparative Analysis of Confusion Matrix

	TP	TN	FN	FP
Linear	88	12	19	35
Poly	92	8	25	29
RBF	90	10	25	29
Sigmoid	49	51	51	3
Logistic regression model	438	62	174	94

that poly kernel's performance is notable, as it has a higher rate of accuracy in predicting positive values in the confusion matrix.

REFERENCES

1. Allwein, E. L., Schapire, R. E. and Singer, Y. 2001. Reducing multiclass to binary: A unifying approach for margin classifiers. *Journal of Machine Learning Research*, 1: 113–141.
2. Bottou, L., Cortes, C., Denker, J. S., Drucker, H., Guyon, I., Jackel, L. D., LeCun, Y., Müller, U. A., Sackinger, E., Simard, P. and Vapnik, V. 1994. Comparison of classifier methods: A case study in handwriting digit recognition. *Proceedings of the 12th IAPR International Conference on Pattern Recognition, Vol. 3-Conference C: Signal Processing (Cat. No. 94CH3440-5)*, 2, pp. 77–82. IEEE.
3. Crammer, K. and Singer, Y. 2002. On the learnability and design of output codes for multiclass problems. *Machine Learning*, 47: 201–233.
4. Dietterich, T. G. and Bakiri, G. 1995. Solving multi-class learning problems via error-correcting output codes. *Journal of Artificial Intelligence Research*, 2: 263–286.
5. Fürnkranz, J. 2002. Round robin classification. *Journal of Machine Learning Research*, 2: 721–747.
6. Hastie, T. J. and Tibshirani, R. J. 1998. Classification by pairwise coupling. In *Advances in Neural Information Processing Systems*, Edited by Jordan, M. I., Kearns, M. J. and Solla, S. A., Vol. 10, 507–513. Cambridge, MA: MIT Press.
7. Javatpoint. www.javatpoint.com (assessed on September 2021).
8. Kreßel, U. 1999. Pairwise classification and support vector machines. In *Advances in Kernel Methods: Support Vector Learning*, Edited by Schölkopf, B., Burges, C. J.C. and Smola, A. J., 255–268. Cambridge, MA: MIT Press.
9. UCI Machine Learning. 2017. Pima Indian Diebetes dataset. Version.1. www.kaggle.com/uciml/pima-indians-diabetes-database.
10. Vapnik, V. 1998. *Statistical Learning Theory*. New York: Wiley.
11. Bennett, K. P. 1999. Combining support vector and mathematical programming methods for classification. In *Advances in Kernel Methods: Support Vector Learning*, Edited by Schölkopf, B., Burges, C. J.C. and Smola, A. J., 307–326. Cambridge, MA: MIT Press.
12. Crammer, K. and Singer, Y. 2002. On the algorithmic implementation of multiclass kernel-based vector machines. *Journal of Machine Learning Research*, 2: 265–292.
13. Lee, Y., Lin, Y. and Wahba, G. 2001. Multicategory support vector machines. *Computing Science and Statistics*, 33: 498–512.
14. Weston, J. and Watkins, C. 1998. *Multi-class Support Vector Machines. CSD-TR-98–04 Royal Holloway*. Egham: University of London.
15. Hsu, C. W. and Lin, C. J. 2002. A comparison of methods for multiclass support vector machines. *IEEE Transactions on Neural Networks*, 13: 415–425.
16. Rifkin, R. and Klautau, A. 2004. In defense of one-vs-all classification. *Journal of Machine Learning Research*, 5: 101–141.
17. Platt, J., Cristianini, N. and Shawe-Taylor, J. 2000. Large margin DAGs for multiclass classification. In *Advances in Neural Information Processing Systems*, Edited by Solla, S. A., Leen, T. K. and Müller, K.-R., Vol. 12, 547–553. Cambridge, MA: MIT Press.
18. Moreira, M. and Mayor, E. 1998. Improved pairwise coupling classification with correcting classifiers. *Lecture Notes in Computer Science: Proceedings of the 10th European Conference on Machine Learning*, 160–171. Berlin, Heidelberg: Springer.
19. Platt, J. 1999. Probabilistic outputs for support vector machines and comparisons to regularized likelihood methods. In *Advances in Large Margin Classifiers*, Edited by Smola, A. J., Bartlett, P., Schölkopf, B. and Schuurmans, D., 61–74. Cambridge, MA: MIT Press.
20. Angulo, C., Parra, X. and Català, A. 2003. K-SVCR. A support vector machine for multi-class classification. *Neurocomputing*, 55: 57–77.

13 Data Mining Approaches on EHR System

A Survey

Thilagavathy R, Veeramani T, Deebalakshmi R, and Sundaravadivazhagan B

CONTENTS

13.1 INTRODUCTION

The broad scope of new data gathering approaches in the field of data analysis has initiated a colossal advancement in the measure of data and extended the multifaceted nature of the data structure, which changes human services as an information-escalated field that can improve clinical quality and diminish social protection costs through gigantic data examination. As shown by the ongoing examination between 2007 and 2013 in *Health Affairs*, the rate of tremendous practices that assembled data on quality measures nearly increased; that activity extended significantly more in little and medium-sized practices someplace in the scope of 2009 and 2013. In any case, the use of electronic vaults to perceive the disparity in patient care and the reaction of execution data to specialists stayed somewhat to dimensions of practices.

DOI: 10.1201/9781003309451-13

Presently, the headway of advancement which engaged the field of human consideration to accumulate massive data and examine the quality thought and cost-suitability has instigated a noteworthy stream of research, which is information mining in electronic medical/health records (EMR/EHR). Despite the fact that electronic health records (EHRs) were first delineated more than 50 years back, as of late, EHR has become unavoidable. Remarkably, the Unites States saw EHR determination triple someplace in the scope of 2009 and 2013; gathering among kids' facilities brought from 21% up in 2008 to 59% in 2011. The advancement found over the span of the most recent five years is likely going to be dynamic and prudent, as, with time, countries, for instance, Norway, the Netherlands, New Zealand, and the United Kingdom, have achieved close all-inclusive EHR confirmation. As contribution with EHRs has extended clinicians additionally, researchers have begun to see the EHR from a substitute viewpoint, as an enormous database, likewise and interventional instrument at the intersection purpose of specialists, patients, and care conveyance. In the previous strategy, inquiry about databases, devices, and records has been disconnected from clinical databases, instruments, and records; EHRs cloud these refinements and association of these storage facilities.

With satisfactory EHR data and a motivation for examination, what is required is a channel for the disclosures. The learning social insurance system, a thought that has gotten a handle on by the Institute of Medicine, is an adaptable structure, with "best practices reliably embedded in the movement method and new learning got as a fundamental symptom of the transport involvement." In this particular condition, advanced examination, clinical data mining included, will expect a key occupation in expelling new getting the hang of, estimating the effects of changes in consideration transport, and possibly dealing with the plan of best-practice rules.[1] EHRs include persistent information, for instance, financial, drugs, and looking into office test results, discovering codes, and strategies. EHRs have organized information relating to disease diagnosis for a large number of patients. So mining them could result in an improvement in patient well-being across the field.

Clinical data mining using EHR data yields results that are comparable to clinical testing. Because EHRs monitor people across extended time allotments, they have considerable test sizes and potentially longer follow-up times, allowing for the testing of a large number of hypotheses. Using clinical data mining approaches like classification, clustering, anomaly detection, etc. on EHR data can give profitable information to energize clinical decision help, to separate condition-unequivocal clinical procedure results, and to improve assembly-based mass care, past the customary clinical encounters.[2] We predict that the progress in clinical information mining determination will accelerate now that all three components of information mining success—information, motivation, and channels for results—have come together.[4] The fundamental job of this investigation is to give the basic establishment and a survey of the current writing of mining that composed EHR data. We foresee that our study will be a valuable apparatus for breaking down the current information mining approaches in an EHR framework.

13.2 DATA MINING METHODS

Analysts discover that prediction and description are the two critical objectives of information mining. To start with, prediction is conceivable by the utilization of

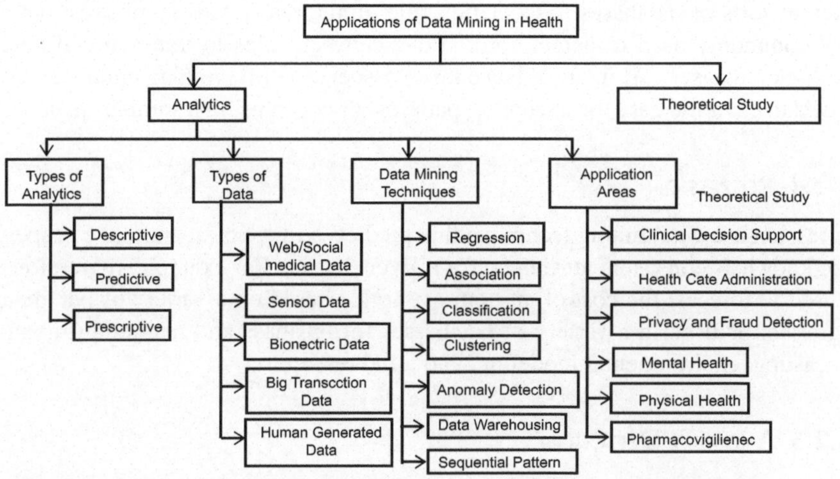

FIGURE 13.1 Characterization of data mining applications in health care.

existing factors in the database so as to anticipate obscure or future estimations of intrigue. Second, the description essentially centers on discovering designs depicting information on the consequent introduction for client translation. The general accentuation of both prediction and description contrasts concerning the hidden application and strategy.[1,3] Figure 13.1 illustrates the characterization of data mining approaches in the field of health care.

13.2.1 Classification

Classification[9] is the procedure of foreseeing the class of another object from the arrangement of predefined classes. To construct a classification model, the dataset is separated in two sections. One section is assigned as preparing dataset, and the other section as testing dataset. The entire model is devised by utilizing preparing dataset, and afterward, its exactness is estimated by approving the classes in which it characterizes the test dataset examples.

13.2.2 Clustering

Clustering is a kind of unsupervised learning procedure in which classes are obscure.[8] It works by finding the similitude/distance among two objects. On the off chance that the similitude/distance measure is within the threshold value, items will have a place with the same class—generally not. The items that have a place with the same class will together shape a cluster; thus, this procedure is named clustering.

13.2.3 Association Rule Mining

Association rules are a set of if–then statements that assist to demonstrate the likelihood of connections between the data items in substantial informer indexes in different sorts of databases. Association rule mining has a variety of uses, but it is most commonly used to detect correlations between sales in transactional data or health-care datasets. At its most basic level, association rule mining entails using AI models to examine data for intriguing patterns, or coevents, in a sample space.

13.2.4 Regression

Regression is a data mining technique that predicts a range of quantitative properties (also known as consistent attributes) for a given dataset. For example, regression can be used to forecast the cost of an item or service based on a variety of parameters. Regression is used in a variety of businesses for business and marketing, financial forecasting, environmental modeling, and trend research.

13.2.5 Anomaly Detection

Anomaly discovery is a system used to distinguish abnormal patterns that don't comply with anticipated conduct, known as outliers. It has numerous applications in business, from interruption location (distinguishing odd examples in system traffic that could flag a hack) to the health monitoring framework (recognizing a harmful tumor in an MRI check), and from extortion recognition in visa exchanges to blame discovery in operational environments.[8]

Finding peculiarity is used in finding the most enormous changes in informational collection. Bo Lie et al. had used three unmistakable irregularity disclosure system, standard help vector data delineation, thickness-actuated help vector data portrayal, and Gaussian blend to evaluate the exactness of the abnormality area on a sketchy dataset of the liver issue dataset which is gotten from the UCI. The procedure is surveyed using AUC precision. The results gained for a sensible informational index by ordinary was 93.59%, while the ordinary standard deviation gained from the equal informational index is 2.63. The uncertain dataset is slated to be open in all datasets; the eccentricity disclosure would be a not-too-bad technique to settle this issue at any rate, since there is only a solitary paper discussing this system, and we can't comment much on the feasibility of the procedure.

13.3 DATA MINING APPROACHES IN EHR SYSTEMS

The EHR information mining discipline remains at the convergence of the study of biostatistics, disease transmission, and data mining. From the study of biostatistics and disease transmission, health-care information mining has acquired an application framework, the system that enables us to arrange EHR information into a framework that is amiable to the utilization of information mining calculations and can address important clinical inquiries accurately. Likewise, it has acquired essential methodologies from biostatistics and the study of disease transmission to address the difficulty present in EHR information, including controlling, examination of unpredictable time–series information, and approaches to causal inference. With the achievement that information mining has accomplished in numerous business applications, the desire from information mining in the medical space is high, and its commitments are getting to be evident.[5,7] In this current section, we analyze about these contributions.

13.3.1 Discriminant Analysis

Discriminant examination is a strategy for gathering a ton of discernments into predescribed classes. The explanation behind existing is to choose the class of discernment reliant on a great deal of elements known as markers or information factors. The model is built subject to a great deal of discernments for which the classes are known. This game plan of discernments is every so often suggested as the training set.[9] In light of the training set, the technique builds up a great deal of straight components of the pointers, known as discriminant capacities, with the true objective that:

$$L = b_1 x_1 + b_2 x_2 + \ldots + b_n x_n + c \quad (3.1)$$

Where c is a consistent, the x's are the data elements or pointers, and the b's are discriminant coefficients.

These discriminant limits are used to foresee the class of discernment with darkens class. For a k class issue, k discriminant limits are manufactured. Given discernment, all the k discriminant limits are surveyed, and the observation is doled out to class I if the ith discriminant work has the most significant regard.

13.3.1.1 Clustering Methodologies in EHR

Data recovery, classification, and summarizing of documents that can be used to support patients' care and therapeutic research require distinguishing comparative clinical archives in the electronic health record (EHR). *Similarity* can be defined as a percentage of how close information items' relationships are (i.e., amount of indistinguishable information objects are). The way to deal with deciding similitude can be seen as changing the information into a similitude space and, after that, playing out the investigation of the connections among this information. In this situation, likeness is the proportion of how much (at least two) medical records share content, which is not quite the same, as an accurate match isn't required.

The K-means clustering approach, a widely used grouping strategy, divides all data into K groups, with each data point assigned to the group that is closest to it. Even though K-means defines a centroid to explain a cluster's mean, it never matches to a genuine information point. Furthermore, the group lacks inalienable, predefined semantic information (e.g., it isn't limited to collecting archives from the same report type or originator).[6]

Todd Lingren et al. (2016)[16] clustered comorbidities based on the disease codes of the automatically mined autism spectrum disorder (ASD) patient cohort to automate the cohort selection in large-scale EHR system. To identify similarity among free-text clinical documents, Chunlei Tang et al. (2018)[10] proposed a variation of the k-means method which first detects centroids through producing a one-level partitioning among all fingerprints made by Charikar's SimHash and thereafter upgrades the centroids successively through identifying each point to its nearest centroid and recomputing the centroid of each cluster. Thilagavathy et al. (2014)[13, 14] proposed a semantics-based suffix tree clustering algorithm which effectively clusters the document contents in a meaningful way with reduced search space.

Kirstine Rosenbeck Goeg et al. (2015),[12] utilized semantic comparability for bunching clinical models from neighborhood electronic well-being records. In this investigation, Lin similitude evaluations and Sokal and Sneath closeness gauges were utilized together with two accumulation systems (normal and best-coordinate normal separately), bringing about a sum of four strategies. The closeness estimations are utilized to progressively bunch layouts. The test material comprises of formats from Danish and Swedish EHR frameworks. Kalankesh et al. (2013),[17] noticed that speaking to the restorative condition of a patient with conclusion codes can prompt scanty bunches since EHRs regularly contain an extensive number (i.e., thousands) of finding codes. To beat this issue, they utilized principal component analysis (PCA) (Dunteman 1989)[25] to decrease dimensionality, consequently making the structure increasingly amiable for representation and bunching.

13.3.1.2 Association Analysis

Association rule mining results will be increasingly precise if the information of all neighborhood EHR frameworks is coordinated and then performs association rule mining. Incorporation of nearby EHR frameworks requires the sharing of neighborhood EHR information, but sharing of patient records abuses the protection of patients. Subsequently, medicinal research is centered on the issue of mining association rules without the sharing of nearby private EHR information. To understand this issue, Nikunj Domadiya et al. (2019)[18] proposed a privacy-preserving distributed association rule mining (PPDARM) by mining the association rules while safeguarding the security of patients. Dingcheng Li et al. (2013)[19] proposed a productive AI system—distributional association rule mining (ARM)—for self-loader demonstrating of affiliation calculations. ARM gives a very productive and vigorous system for finding the most prescient arrangement of phenotype definition criteria and standards from extensive informational collections.

In a similar line, the investigation directed by Stephen M. Kang'ethe et al. (2014)[20] demonstrated that affiliation rule mining cannot exclusively be utilized to affirm what is, as of now, known from health information as comorbidity designs, yet additionally produce some exceptionally fascinating disease diagnosis associations that can give a decent beginning stage and space for further investigation through examinations by restorative specialists to clarify the patterns that are apparently obscure or curious in the concerned populaces.

13.3.2 Predictive Analysis in EHR

Predictive analysis can support medical service providers to precisely expect and react to patient needs. It gives the capacity to settle on monetary and clinical choices depending on the predictions made by the framework.[11]

13.3.2.1 Effect of Classification Algorithms in EHR

Predictive analysis can be done on different elements with statistic information, medical clinic parameters, persistent previous history, and different indicators for a

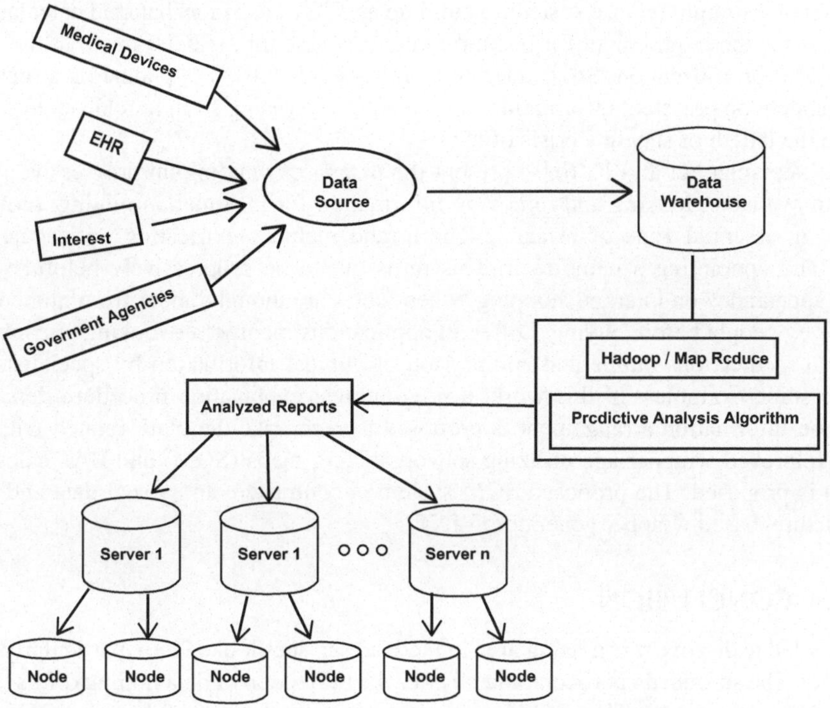

FIGURE 13.2 Predictive analysis framework in the health-care system.

particular ailment. However, recognizing the solid indicator for exact prediction is a dispute.[11] From the components being considered for predictive analysis, different models and algorithms should be examined. Classification methods like naive Bayes, linear regression, random forest, logistic regression, hidden Markov models, and generalized additive models must be considered for building up a predictive system. Colin G. Walsh et al. (2017),[21] applied random forest classification method on HER to predict the suicide attempt more accurately than the traditional approaches. Vyshali S. Parsania et al. (2014),[22] compared the effect of classification algorithms like BayesNet, naive Bayes, ZeroR, JRip, OneR, and PART with EHR database of rheumatoid arthritis to discover the unknown patterns.

13.3.2.2 Use of Regression Techniques in EHR

Regression analysis is a dominant statistical technique that enables us to inspect the connection between at least two factors of intrigue. It yields accurate knowledge that can be used for the additional improvement of products and services.[9]

Rui Duan et al. (2020),[23] utilized a one-shot appropriated calculation to perform strategic relapses on electronic well-being records from the University of Pennsylvania well-being framework to assess the dangers of fetal misfortune because of different medicine exposures. Haijun Zhai et al. (2014),[24] proposed a novel AI calculation utilizing a straight relapse system to build up an EHR-based mechanized calculation to foresee the requirement for pediatric intensive care unit (PICU) move in the initial 24 h of affirmation. So, Catherine Combes et al. (2014)[26] planned a structure dependent on prescient information mining approach (using straight relapse) to evaluate the length of stay in a crisis office.

T. Veeramani et al. (2016),[15] present the databases winding up progressive program available, looking and removing information for information mining, turning into an essential zone of research. The errand includes extricating critical qualities and appointing a name to it. This turns out to be progressively helpful with the appearance on internet shopping, where costs are thought about from numerous sites by people before buying. Different applications incorporate looking at conventional medications online and comparison of clinical information by specialists to give some examples. In this work, a programmed explanation procedure dependent on information arrangement is proposed to recognize the mark related with it. An improved wrapper age utilizing self-organizing maps (SOM) and BAT calculation is proposed. The proposed SOM system is contrasted, and visual data and tag structure–based wrapper generator (VINT).

13.4 CONCLUSION

Knowledge discovery is a rising area in medical services looked into, particularly for EHRs. The shrouded data accumulated after the utilization of data mining offers help to therapeutic clients. The work done to date on the usefulness of data mining in the health-care industry yields significant improvement on interoperability and decision support systems. In this study, we investigated the advancement of the EHR framework by using data mining approaches. The fundamental target of the examination was to comprehend the EHR inquiry about patterns and characterize a worldwide viewpoint on the past, present, and eventual fate of human services.

REFERENCES

[1] Abdel Nasser H. Zaied, Mohammed Elmogy, and Seham Abd Elkader, "Electronic Health Records: Applications, Techniques and Challenges", *International Journal of Computer Applications*, 119, no. 14 (2015): 38–49.

[2] Tapas Ranjan Baitharua and Subhendu Kumar Panib, "Analysis of Data Mining Techniques for Healthcare Decision Support System Using Liver Disorder Dataset", *Procedia Computer Science*, 85 (2016): 862–870.

[3] Shivani Batra, Shelly Sachdeva, Hem Jyotsana Parashar, et al., "Applying Data Mining Techniques to Standardized Electronic Health Records for Decision Support", 6th International Conference on Contemporary Computing, IC3 (2013): 510–515. DOI: 10.1109/IC3.2013.6612249.

[4] Najwashehadeh Ahmad Altabeiri, "Applications of Data Mining Approaches in Healthcare", *IOSR Journal of Engineering*, 6, no. 10 (2016): 16–21.

[5] Elizabeth A. Madigan and Olivier Louis Curet, "A Data Mining Approach in Home Healthcare: Outcomes and Service Use", *BMC Health Services Research*, (2006): 6-18.

[6] Yong-Mi Kim, "Medical Informatics Research Trend Analysis: A Text Mining Approach", *Health Informatics Journal*, 24, no. 4 (2016): 432–452. DOI: 10.1177/1460458216678443.

[7] Jingfeng Chen, Wei Wei, Chonghui Guo, et al. "Textual Analysis and Visualization of Research Trends in Data Mining for Electronic Health Records", *Health Policy and Technology*, 6, no. 4 (2017): 389–400. DOI: 10.1016/j.hlpt.2017.10.003.

[8] Neesha Jothi, Nur' Aini Abdul Rashid, and Wahidah Husain, "Data Mining in Healthcare—A Review", *Procedia Computer Science*, 72 (2015): 306–313.

[9] Pranjul Yadav, Michael Steinbach, Vipin Kumar, and Simon G, "Mining Electronic Health Records (EHRs): A Survey", *ACM Computing Surveys*, 50, no. 6 (2018): Article 85. DOI: 10.1145/3127881

[10] Chunlei Tang, Joseph M. Plasekd, Yun Xiong, and Li Zhou, "Clustering Similar Clinical Documents in Electronic Health Records", *International Conference on Data Science*, no. 6 (2018).

[11] Kavakiotis Ioannis, Tsave Olga, et al. "Machine Learning and Data Mining Methods in Diabetes Research", *Computational and Structural Biotechnology Journal*, (2017). DOI: 10.1016/j.csbj.2016.12.005.

[12] Kirstine Rosenbeck Gøeg, Ronald Cornet, and Stig Kjaer Andersen, "Clustering Clinical Models from Local Electronic Health Records based on Semantic Similarity", *Journal of Biomedical Informatics*, 54 (April 2015): 294–304. DOI: 10.1016/j.jbi.2014.12.015.

[13] R. Thilagavathy and R. Sabitha, "Enhancing the Quality of Search Results by a Novel Semantic Model", *Asian Journal of Information Technology*, 15, no. 6 (2014): 996–1004.

[14] R. Thilagavathy and R. Sabitha, "A Novel Approach to Aggregation of Web Documents by Semantic Suffix Tree and Self Organizing Feature Map Methods", *Third International Conference on Science Technology Engineering & Management (ICONSTEM)*, (2017): 87–92. DOI: 10.1109/ICONSTEM.2017.8261262.

[15] T. Veeramani and R. Nedunchelian, "An Improved Wrapper Generation Using Self Organizing Maps and Meta Heuristic Technique for web Based Biomedical Data Mining", *Journal of Medical Imaging and Health Informatics*, 6 (2016): 1–4.

[16] T. Lingren, P. Chen, J. Bochenek, F. Doshi-Velez, P. Manning-Courtney, and Julie Bickel, "Electronic Health Record Based Algorithm to Identify Patients with Autism Spectrum Disorder", *PLoS One*, 11, no. 7 (2016): DOI: 10.1371/journal.pone.0159621.

[17] Leila R. Kalankesh, James Weatherall, Thamer Ba Dhafari, and Iain Edward Buchan., "Taming EHR Data: Using Semantic Similarity to Reduce Dimensionality", *Studies in Health Technology and Informatics*, 192, no. 1 (2013): 52–56.

[18] Nikunj Domadiya and Udai Pratap Rao, "Privacy Preserving Distributed Association Rule Mining Approach on Vertically Partitioned Healthcare Data", *Procedia Computer Science*, 148 (2019): 303–312.

[19] Dingcheng Li, Gyorgy Simon, and Christopher Chute, and Jyotishman Pathak, "Using Association Rule Mining for Phenotype Extraction from Electronic Health Records". *AMIA Summits on Translational Science proceedings AMIA Summit on Translational Science*, (2013): 142–146.

[20] Stephen M. Kang'ethe and Peter Waiganjo Wagacha, "Extracting Diagnosis Patterns in Electronic Medical Records Using Association Rule Mining", *International Journal of Computer Applications*, 108, no. 15 (2014).

[21] Colin G. Walsh, Jessica D. Ribeiro, and Joseph C. Franklin, "Predicting Risk of Suicide Attempts Over Time through Machine Learning", *Clinical Psychological Science*, 5, no. 3 (2017): 457–469.

[22] Vaishali S. Parsania, N. N. Jani, and Navneet H. Bhalodiya, "Applying Naïve Bayes, BayesNet, PART, JRip and OneR Algorithms on Hypothyroid Database for Comparative Analysis", *International Journal of Darshan Institute on Engineering Research & Emerging Technologies*, 3, no. 1 (2014): 60–64.

[23] Rui Duan, Mary Regina Boland, Zixuan Liu, Yue Liu, Howard H. Chang, Hua Xu, Haitao Chu, Christopher H. Schmid, Christopher B. Forrest, John H. Holmes, Martijn J. Schuemie,

Jesse A. Berlin, Jason H. Moore, and Yong Chen, "Learning from Electronic Health Records Across Multiple Sites: A Communication-efficient and Privacy-preserving Distributed Algorithm", *Journal of the American Medical Informatics Association*, 27, no. 3 (March 2020): 376–385. https://doi.org/10.1093/jamia/ocz199.

[24] Haijun Zhai, Patrick Brady, Qi Li, Todd Lingren, Yizhao Ni, Derek Wheeler, and Imre Solti, "Developing and Evaluating a Machine Learning Based Algorithm to Predict the Need of Pediatric Intensive Care Unit Transfer for Newly Hospitalized Children", *Resuscitation*, 85, no. 8 (2014). DOI: 10.1016/j.resuscitation.2014.04.009.

[25] George H. Dunteman, *Principal Components Analysis*. Newbury Park, CA: SAGE Publications, Inc., 1989. DOI:10.4135/9781412985475.

[26] Catherine Combes, Farid Kadri and Sondes Chaabane, "Predicting Hospital Length of Stay Using Regression Models: Application to Emergency Department", 10ème Conférence Francophone de Modélisation, Optimisation et Simulation- MOSIM'14, Nov 2014, Nancy, France. <hal-01081557>

14 Chest Tumor Identification in Mammograms by Selected Features Employing SVM

Shivaprasad More, Pallavi Gholap, Rongxing Lu, and Sitendra Tamrakar

CONTENTS

DOI: 10.1201/9781003309451-14

14.1 FOUNDATION

Breast tumor is the common cancer in ladies, as well as the major reason of mortality from cancer. Early identification of breast cancer can thus assist to minimize the disease's morbidity and mortality rates. One of the most trustworthy sources in imaging modalities for early detection of cancers and their characteristics has been discovered to be digital mammography. According to studies, screening is crucial in discovering malignancies at an early stage. For early diagnosis of breast cancer, both traditional film mammography and digital mammography are routinely utilized. Digital mammography is seen to be superior to film mammography, especially because it can be implemented in conjunction with CAD (computer-aided detection or computer-aided diagnostic) systems. The identification of breast tumor, the division of tumor borders, and the categorization of lumps based on their look and appearance aspects are all part of this manual mammography interpretation process. This manual study of breast lesions from mammograms reveals a lot of variation in radiologists' interpretations. Diagnostic systems which are handled by computer help oncologists during the inspection of chest lumps and can help reduce this variability. However, in order for any diagnostic system to be helpful in any medical field, it should be able to differentiate between cancerous and noncancerous lumps.

Digital mammography has been discovered to be the most dependable imaging modality for prior identification of malignancies and their features. The physical examination of mammography lesions indicates a wide range of radiologists' judgments. Systems which can diagnose with the help of computer can help reduce this variability by helping the radiologist to an alternative reader while examining the tumor.

The identification, segmentation of interested region, and categorization of breast anomalies are the essential stages of completely preprogrammed systems that can interpret mammograms. Current techniques have a considerable number of contradictory results and miss a significant number of accurate regions during the tumor identification process. Most tumor segmentation algorithms use ACM and GB techniques, which show the wide variety of form and structure dissimilarity present in breast tumor. Ultimately, tumor identification is typically accomplished using ML algorithms, such as support vector machines, linear discriminate analysis, and artificial neural networks.

This technique describes a new technique for developing a completely mechanized diagnostic system for identification, division, and categorization of lumps from mammographic pictures, based on recently established machine learning models. The four stages of our suggested solution to the mass detection challenge are as follows: (1) segmentation of the original input mammographic image's region of interest (ROI); (2) geometric and texture features extracting information from divided picture; (3) choosing features from the entire number of photos retrieved; and (4) classification of mass (tumor) as benign or malignant.

14.1.1 Review of the Literature

The structures of the breast are revealed by mammography, which allows them to be classed as benign or malignant. The quality of the image, the radiologist's level of experience, and the huge volume of cases all influence the accuracy of screening mammography interpretation. Radiologists benefit from computer-aided diagnostic (CAD) [1] tools and systems that act as a second reader, reminding them to revisit areas of a mammogram that are being considered. As a result, an effective CAD scheme is necessary to investigate the usage of computer output by radiologists in their diagnosis.

Breast tumor is detected by mammography, which detects anomalies like lumps and deposits of calcium. Because of the convolution of the chest anatomy, the low number of cases of the disease and irregularities are not noticed. According to [2], radiologists overlooked between 12% and 27% of suspected instances found in mammographic images.

Double reading, on the other hand, is both costly and time-consuming. As a result, academics and radiologists are interested in computer-aided cancer-detection technology [3]. The use of a diagnostic system can lower the overhead of specialists while also increasing the rate of early cancer diagnosis [4, 5]. According to a recent study [4], the use of a diagnostic system increases the identification of breast cancer by 8.57%. Despite the fact that the deployment of diagnostic systems has received a lot of attention [5], more research into associated core technologies is needed. Identification of calcium deposits, calcification, tumor identification, and lump categorization are some of the main technologies.

[6] looked into the use of gradients for mass classification. This suggested method shows that a tumor border was divided into hollowed-out and rounded-outward lump. Attributes that count the range of the postulated character of the borderline and the level of spicule temperedness were extracted for lump categorization.

[7, 8] employed morphological traits with manually defined limits for lump categorization. Pohlman et al. [8] used a flexible area growth strategy to divide a tumor. Eight variables assessing bulk structure and border harshness were drawn out and used for categorization. [9] looked at lump categorization based on machine-driven mass division. The findings show that features extracted from automated contours can aid in the detection of breast masses. It's also critical to consider feature choice. As all attributes are not required for categorization, attribute election is difficult. In general, using the attributes found by the procedure can upgrade categorization precision. In this presentation, we will pivot attribute election for tumor categorization.

One of the better techniques, support vector machine recursive feature elimination, combines recursive feature elimination with the support vector machine classifier [10]. It has many upper hands, but it also has some drawbacks, such as unnecessary points between the attributes used and its being highly expensive. As a result, there have been a few improvements [11, 12].

Machine learning approaches are garnering a lot of interest from medical academics and physicians after a number of visible triumphs on a variety of predicting tasks [13]. We direct the call for the evolution in this domain by presenting a visionary initiation to ML and also a direction in evolvement and trail of ML algorithms using open-source software and openly available information [14].

Artificial neural network, conventional extreme learning machine, support vector machine (SVM), and K-nearest neighbor are some of the methods used. The dataset was obtained from the University of California–Irvine (UCI) library. Age, BMI, glucose, insulin, homeostatic model assessment (HOMA), leptin, adiponectin, resistin, and chemokine monocyte chemo attractant protein 1 (MCP1) were all employed in this dataset. Four different machine learning algorithms were used to find the parameters with the best accuracy values. The hyperparameter optimization method was employed for this. Finally, the findings were compared and discussed in[15].

Despite the fact that various studies have been published on mass classification, only a handful of the aforementioned studies have looked into feature selection in this method. Feature selection is critical, as all attributes are not suitable for categorization issue. Attribute election has been used in a variety of PR, including body and thumbprint identification. The usage of the attributes chosen by the ways can increase categorization precision in general.

14.2 PROPOSED WORK

14.2.1 Issue Statement

Using a feature selection method like SRN, classify a tumor mass as malignant or benign based on specified geometrical and textural features (SVM-RFE-NMIFS).

14.2.2 Methodology

Refer to Figure 14.1 for the proposed system architecture.

14.2.2.1 Mammograms from Dataset

Here, pictures from the Mammographic Image Analysis Society (MIAS) or the Digital Database for Screening Mammography (DDSM) are used. It contains a large number of mammography cases from various medical institutions. The original photographs are pgm files with a dimension of 1,024 × 1,024 pixels. In these photographs, there is some background noise.

Low contrast, irregular illumination, and noise are all major disadvantages of mammograms. Written labeling and other artifacts in mammograms must be removed. As a result, mammography feature extraction and object detection may be problematic. Image smoothing and contrast enhancement are two crucial phases in preprocessing.

14.2.2.2 Extraction of Interested Area

The first photos are really huge, and the tumor image in that mammogram is extremely little. As a result, it's critical to process the original image. That is, only the region in which we are interested is utilized. The ROI that has been cropped has a poor contrast.

14.2.2.3 Segmentation

This module seeks to extract photos in order to search only the areas suspected of having abnormalities rather than the entire image. As a result, segmentation is the process of separating the ROI from the backdrop, such as pectoral tissues. Local thresholding, K-means clustering, Otsu, and other standard approaches can be applied.

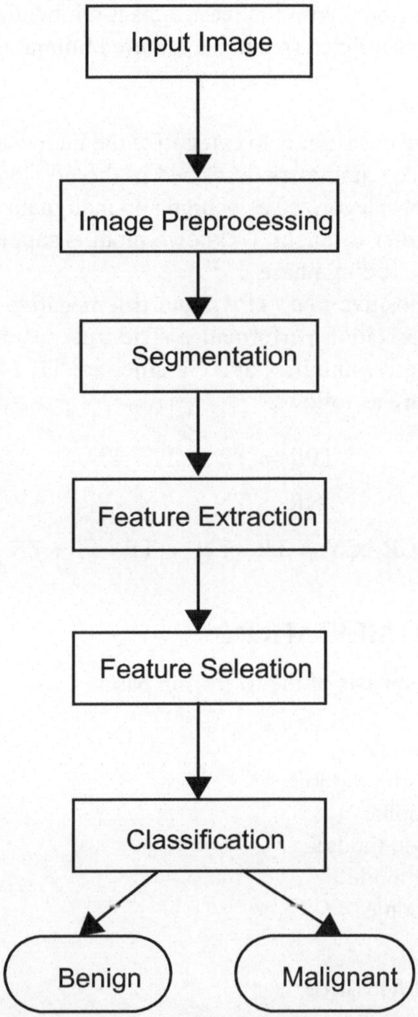

FIGURE 14.1 System architecture.

The border of a benign lesion is irregular, uneven, and hazy, but the boundary of a malignant tumor is usually round, smooth, and well-circumscribed. As a result, a boundary analysis can be used to classify the masses as benign or malignant. After that, we'll look at both the geometry and texture elements.

14.2.2.4 Feature Selection

Basic support vector machine (SVM). Using a sequential elimination strategy that starts with all feature variables and removes one at a time, nested subsets of features are chosen. At each step, the coefficients of the linear SVM's weight w vector are used to calculate the feature ranking score. The feature with the lowest rating score is deleted, as is the feature with the lowest ranking score. Using a ranking criterion entails removing the attribute with the least impact on the aim function. As the aim function, SVM-RFE was utilized (recursive feature elimination).

14.2.2.5 Categorization

The important job of the classifier is to categorize the interested area using the mathematical and appearance parameters described in the previous sections. During the training phase, a dataset categorized as benign and malignant mass is supplied to the classifier, and the classifier is taught. Unknown input is supplied to the classifier for classification during the testing phase.

The accuracy, true positive rate (TPR), and true negative rate (TNR) can all be used to evaluate categorization performance. The true positive, false positive, true negative, and false negative numbers of a classifier are TP, FP, TN, and FN, respectively. The definitions are as follows:

$$\text{TPR} = \text{TP} \div \text{TP} + \text{FN}$$

$$\text{TNR} = \text{TN} \div \text{TN} + \text{FP}$$

$$\text{ACCURACY} = \text{TP} + \text{TN} \div \text{TP} + \text{FP} + \text{TN} + \text{FN}$$

14.3 SYSTEM IMPLEMENTATION

The presented system consists of the following parts:

1. Input pictures module
2. Image preprocessing module
3. Segmentation module
4. Features extraction module
5. Feature selection module
6. Classification module

14.3.1 Input Images Module

We'll use photos from the standard machine learning and breast screening datasets in this module. DDSM has a significant number of mammography cases from various medical facilities. It covers more than 2,500 mammography cases from various medical facilities in the United States. There are 322 digitized films in the database. It also provides "truth" markings on the locations of any anomalies that may be present by the radiologist. The original images are of 1,024×1,024 pixels in. pgm format (portable gray map). These images contain some background noise. Refer to Figure 14.2 for the first mammogram picture from the dataset.

14.3.2 Image Preprocessing Module

Low contrast, irregular illumination, and noise are all major disadvantages of mammograms. Written labeling and other artifacts in mammograms must be removed. Image smoothing and contrast enhancement are two crucial phases in preprocessing.

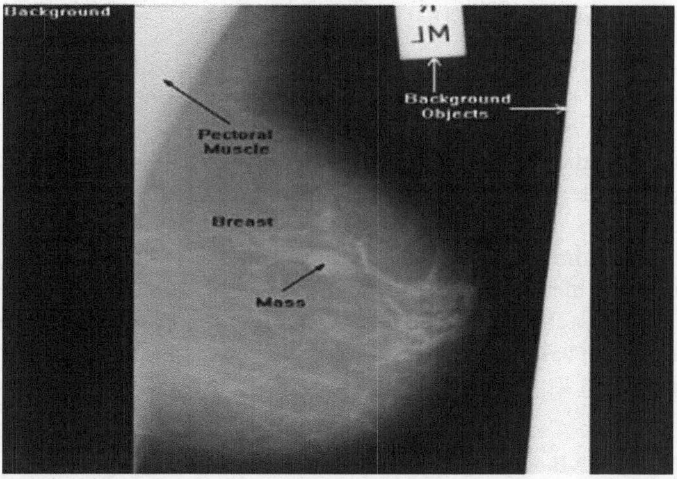

FIGURE 14.2 Original mammogram.

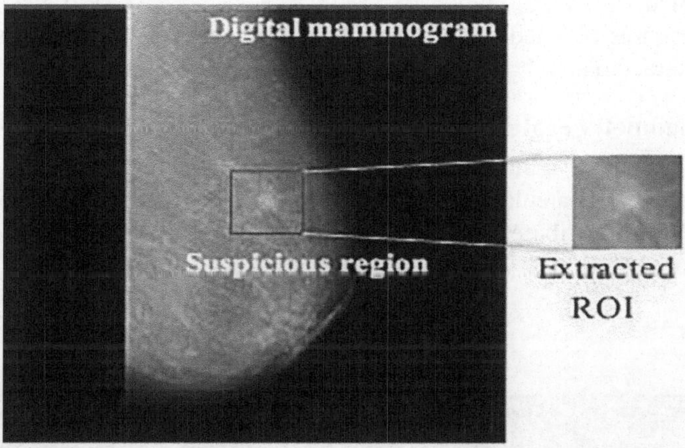

FIGURE 14.3 Region of interest (ROI).

Adaptive histogram equalization (AHE) and other contrast enhancement techniques can be used. It improves image contrast by altering the image's intensity levels.

One of the reasons for the challenges in interpreting mammograms is that the dense tissues that surround the suspicious region mask any lump, calcification, or other types of cancer. The original photographs are enormous, and the tumor image in that mammogram is minuscule. The ROIs that contain masses in the original mammogram will be manually deleted.

Refer to Figure 14.3 for the extracted area of interest of the actual mammogram.

14.3.3 Segmentation Module

To extract features from the abnormal regions, we have to separate the suspicious regions from the background parenchyma, and this is the goal of the segmentation step.

Segmentation is the process of dividing a picture into homogeneous portions based on a set of criteria. It extracts regions from photographs so that it may search only the areas suspected of containing anomalies rather than the full image. A label is applied to each pixel in an image, and pixels with the same label share certain characteristics. It's one of the methods for recognizing pixels based on their color or gray-level intensity similarities. By enabling the concept of partial membership, which allows each image pixel to belong to numerous clusters, fuzzy set theory has improved this process. Take a look at Figure 14.4.

14.3.4 Feature Extraction Module

The qualities of ROI are defined as features. Following the segmentation of the ROI's mass, a set of characteristics linked to the geometry and texture of the boundary and its neighboring regions is computed. The boundary of a benign tumor is usually irregular, rough, and hazy, but the border of a malignant lesion is round, smooth, and well-delineated.

Refer to Figures 14.5 and 14.6 for the schematic diagram of some mass shapes and boundary characteristics.

14.3.4.1 Geometry Features

The contour shape of a mass is represented by the geometry features. After segmentation, they are calculated from the boundary pixels. Brevity, NDM, Fourier attributes, NRLB attributes, and RGOB attributes are all examples of geometry features.

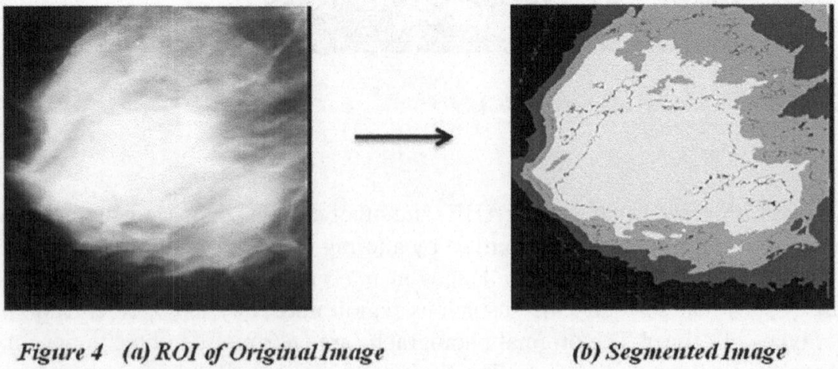

FIGURE 14.4 ROI to segmentation.

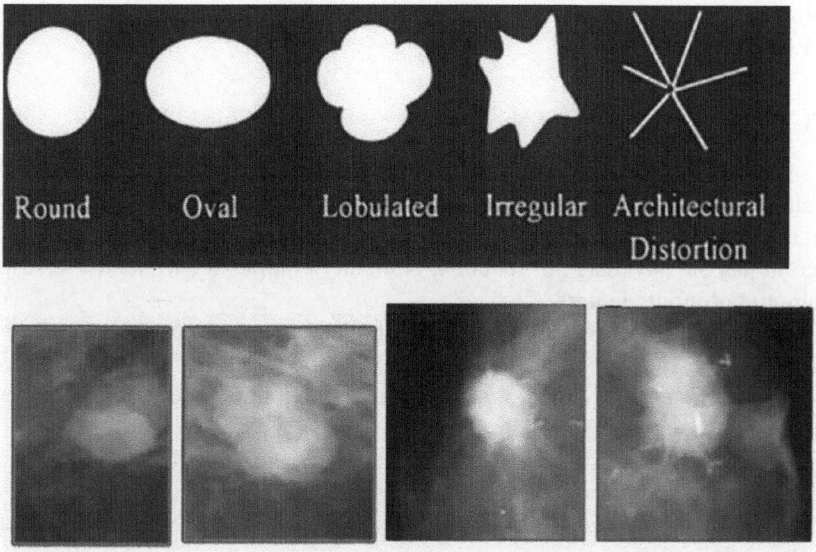

FIGURE 14.5 Types of shape of mass and ROI mass.

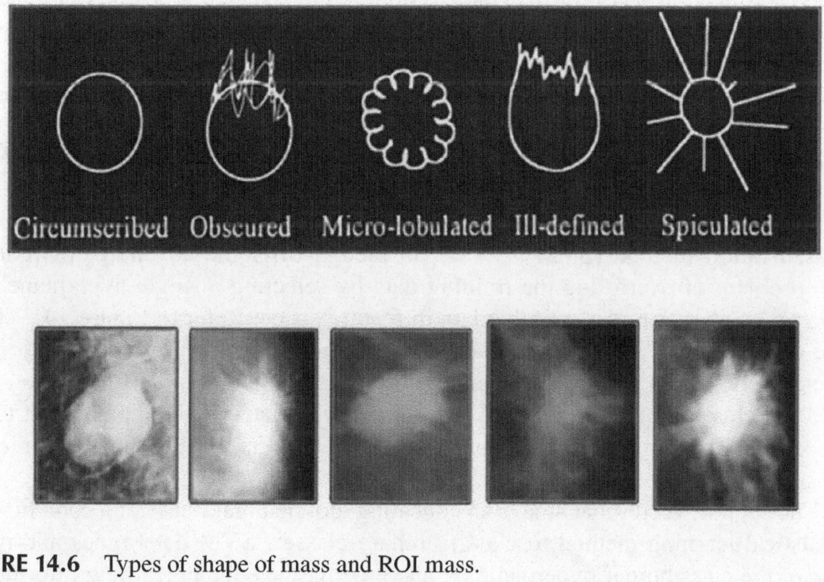

FIGURE 14.6 Types of shape of mass and ROI mass.

14.3.4.2 Texture Features

Mathematically created texture features are difficult to extract visually and are not visible to the naked eye. The texture feature is computed using the grey-level co-occurrence matrix (GLCM), which is a statistical method that evaluates the spatial association of pixels in the gray-level co-occurrence matrix. Correlation, homogeneity, sum average, cluster prominence, and uniformity are some examples of GLCM features.

14.3.5 Feature Selection Module

The selection of characteristics is an important part of the breast cancer diagnosis and classification process. When the traits are extracted, it is observed that not all of them are effective in discriminating between normal and abnormal patterns. Feature selection has the advantage of reducing computer complexity while also restricting the amount of input features.

Methods of feature selection in the past. Feature selection is a technique for extracting the most appropriate subset of information for classification. The aim is to search an attribute subgroup with size D which maximizes categorization precision given an attribute collection with size D.

The retrieved feature space is extremely large and complex due to the wide spectrum of normal tissues and diseases. Excessive features cause the so-called "curse of dimensionality," which reduces the classifier's performance and increases its complexity.

14.3.6 Classification Module

The classification phase of the CAD system is the last step. The classifier uses the robust significant characteristics identified in the feature selection step as input to classify the aberrant lesion as benign or malignant in this step. During the training phase, the dataset identified as benign and malignant tumor is sent to the categorizer as tutoring information, and the categorizer is taught. During the trail phase, unrevealed information (mammogram picture) is submitted to the categorizer for actual categorization.

For breast tumor diagnosis, the support vector machine is a commonly utilized classification method (SVM). SVMs, in theory, offer the advantage of reducing the problem of overfitting the training data by selecting a single hyperplane from a large number that can split the data in feature space. Refer to Figure 14.7. There are many categorizers that can make discrete the information in this situation, but there is one which optimizes the edge (optimizes the length betwixt the categorizer and the closest dot). The optimal separating hyperplane is the name given to this linear classifier. In contrast to other alternative borders, this boundary is intended to generalize well.

The support vector machine is an educating tool that makes use of a contemporary statistical learning method to classify binary classes. SVM determines and makes use of the class border hyperplane by maximizing the edge in learning information. The learning information specimens that run along the hyperplanes at the class border are called support vectors. Using support vectors, SVM identifies an appropriate hyperplane to partition the sets. Patients owned to one group remain on a single side of the plane after separation, while cases belonging to another category remain on the opposite side of the plane.

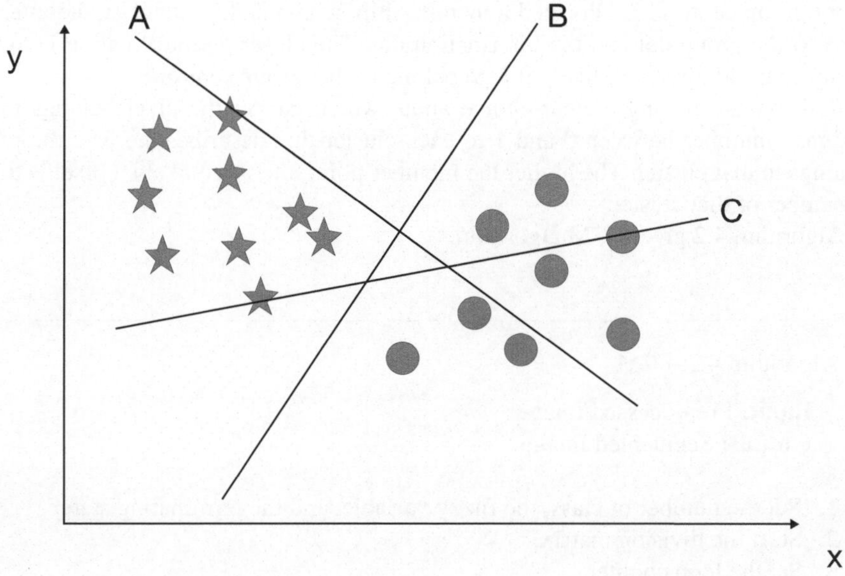

FIGURE 14.7 Support vectors.

14.4 EXECUTION DETAILS

The details of development of different modules to get near the proposed system are user interface (GUI), NR and finding of interested area, division, attributes extraction, features choice, and categorization.

14.4.1 Image Preprocessing

Background noise, pectoral tissue, and other noises can be found in pictures taken from typical mammographic databases. To obtain precise findings, this disturbance must be taken out by prefiltering, which is accomplished by cutting the image's unneeded region.

14.4.2 Segmentation

Following the creation of the required ROI, the image is segmented to produce the segmented image. The fuzzy C-means clustering algorithm is used to segment the data in the proposed system. Based on their similarity, a clustering algorithm separates things into groups. Rather than deciding on absolute membership, it assesses the chance of a data point belonging to a specific cluster. Each object can relate to numerous groups (thus the word "fuzzy"), with the degree of membership determined by a probability distribution across the clusters. Every dot correlating with every group centroid is allocated a membership, which is dependent on the interval betwixt the group dot and the information dot. The closer the information is to the group centroid, the more likely it is to belong to that group centroid.

Rather than making a clear choice about which cluster the pixel belongs to, it assigns a number between 0 and 1 to each cluster that describes the way the pixel belongs to that cluster. The higher the member point, the more likely a pixel is to be a member of that cluster.

Algorithm 4.2 gives FCM algorithm.

Algorithm 4.2: FCM

Input: Preprocessed Image
Output: Segmented Image

1. Put the number of class, the fuzzy variable, and the terminating state.
2. Start the division matrix.
3. Set the loop counter.
4. Determine the group centroid and find the impartial number.
5. For every pixel, for each group, find the member number in the matrix.
6. If the K between consecutive petitions is less than the terminating state, then break.

14.4.3 Features Extraction

Once segmentation is done, geometry and texture features can be extracted. The definitions and formulae for the specified geometry and texture features are aforementioned. To discover the features from the library areas, take note of the following.

dei.lmuz.ifik.dbss.j-featurelib.getfeatures

dei.lmuz.ifik.dbss.j-featurelib.shapeFeatures.Compactness

dei.lmuz.dbss.ifi.j-featurelib.showexamples

14.4.4 Feature Election

We extracted a total of 31 geometry and texture characteristics in the feature extraction module. To determine the optimal amount of characteristics for the classification process, we used the WEKA library.

Attribute selection in WEKA examines all potential combinations of elements in the details to determine which subgroup of elements is most useful for prognostication. It makes use of two things: an attribute tester and a enxplore procedure. To find the grade any union of these variables, judge-and-explore algorithms can be used.

Waikato Environment for Knowledge Analysis uses a subgroup of attributes and instructs the model using them. It either puts in or takes out the attributes from the subgroup.

Its method is as follows:

1. Firstly, it puts uncertainty to the specified information set by making mixed duplicates of all variables.
2. It teaches an RFC on the information and applies an FI measure to find out the importance of each attribute where upper shows more principle.
3. It analyzes whether a real feature is more important than the best of its shadow feature with each iteration and removes features that are judged irrelevant.
4. Lastly, the algorithm breaks when all the attributes get validated or declined or it goes to a given end of RFC.

Then there are the qualities. The WEKA library accepts a. csv file as input. The WEKA method employs a subset evaluation technique. Features are recorded during the execution of the algorithm and will be used for categorization later.

Working. By generating scrambled duplicates of all features, unpredictability is attached to the provided information group. It uses a prime variable measure to test the significance of every variable after training a random forest classifier on the enlarged dataset. It analyzes if a genuine feature is more important than the best of its shadow feature at each iteration and removes features that are regarded highly unimportant until all features are confirmed or rejected.

According to [10], the performance of a classifier is determined by the features chosen. Finding the optimal number of characteristics for a classifier is tough. For the final classification, we employed SVM. In fact, the optimum features for different categorization algorithms for the identical information set may be varying. We discovered that the support vector machine can attain a higher degree with the same number of features in most cases. As a result, we predicated the number of attributes for every election type on the SVM classification presentation.

Using LOO (leave one out) validation scheme:

For 29 attributes

Correct predictions: 17 (80.95%)

Wrong predictions: 4

For 30 attributes

Correct predictions: 20 (95.23%)

Wrong predictions: 1

For 31 attributes

Correct predictions: 20(95.23%)

Wrong predictions: 1

From the previous results, putting all the 33 attributes gives good results. So we have used 33 features as input for classification criteria.

14.4.5 Classification

The fundamental aim of the classifier is to characterize the ROI as benign or malignant based on the geometrical and textural parameters outlined before. The dataset classified as nondangerous and dangerous tumor is submitted to the categorizer as teaching dataset during the training phase, and the categorizer is trained. The unrevealed information (mammographic picture) is sent to the categorizer for actual categorization during the testing phase.

The algorithm used to classify the image is called a support vector machine. The algorithmic steps are outlined in the following sections. The library LibSVM is being loaded.

The two main types of machine learning methods are supervised and unsupervised learning. UL algorithm is based on learn-by-example, and SL is based on acquire-knowledge-by-outcome. The input for supervised learning is a set of instruction information. The SVM is an SL technology that analyzes information to find patterns that may be used to classify it. It takes a set of inputs, reads them, and outputs the specified output form for each input. *Classification* is a term used to describe this type of procedure.

The SVM algorithm translates an input vector into a higher-dimensional feature space in order to construct the most separated hyperplanes possible. The phrase "feature space" refers to an input space set aside for utilizing the kernel function to calculate similarity. It's a high-dimensional space with significantly easier linear separation than input space. This method converts raw data into sample vectors of a specific length. *Feature values* and *feature vectors* are the two terminologies utilized in feature space. Feature vectors are how these feature values are presented to the computer in a vector, and feature values are the attributes of an image.

The kernel function is used in the kernel technique to perform classification and clustering operations on numerous types of data, including text documents, sequences and vectors, collections of points, pictures, and graphs. It converts the incoming data into a higher-dimensional feature space, allowing it to be more easily isolated and organized. The kernel we use in the SVM is most primary, and we take the help of the radial basis function since it works well in a wide range of applications.

14.4.6 Support Vector Machine Algorithm

Algorithm 4.4: SVM

Input: Training set in form of attributes.csv

Output: Image classified as benign or malignant

Begin

1. Load a training dataset as attributes. csv.
2. Construct a LibSVM classifier with default settings.
3. Initialize counters for correct and wrong predictions.
4. Classify all instances and check with the correct class values.
5. For each instance do store inst in predicted class value after classification pass the function inst class value and store is result in real class value. If values of predicted class value and real class value come the same, then we get correct predictions; else, wrong.
6. End for.

14.5 EXPERIMENTAL RESULTS

The experiment's photos come from the Digital Database for Screening Mammography (DDSM), the Mammographic Image Analysis Society (MIAS), and the UCI Machine Learning Repository. More than 2,500 mammography cases from various medical institutes in the United States are included in the DDSM. With a Joyce-Loebl scanning microdensitometer, MIAS films from the UK National Breast Screening Program were digitized to 50 micron pixel edge. There are 322 digitized films in the database. It also provides true readings on the spots of any anomalies that may be there by the radiologist.

14.5.1 Outcome Examination

The examinations were conducted on the complete interested area (n = 786) information group, where all the 33 features were procured without attribute election. The categorization presentation is calculated by the TPR, TNR, and accuracy.

We have taken 531 benign ROIs and 255 malignant ROIs as training dataset and also tested them to find out the accuracy from the experimental results:

From430 Benign ROIs:
TP: Image is benign, and result shows benign=**390.**
FN: Image is benign, and result shows malignant=**40.**
From 240 Malignant ROIs:
TN: Image is malignant, and result shows malignant=**207.**
FP: Image is malignant and result shows benign=**43.**

From the previous calculations of TP, TN, FN, and FP:
TPR = 0.9088
TNR = 0.8632
ACCURACY = 0.8989

These results have been cross-checked on Waikato Environment for Knowledge Analysis discoverer using NB algorithm to test the correctness of the presented support vector machine technique. The relative categorization presentation is shown in the following table.

TABLE 14.1
Relative Categorization Presentation

Precision	NB	Support vector machine
True +	0.9186(395/430)	0.9186(395/430)
True-	0.8916(214/240)	0.8835(212/240)
Prec	0.9(603/670)	0.9055(600/670)

TABLE 14.2
Categorization

Attributes	Linear discriminant analysis True + True-Prec			Support vector machine True +True- Prec		
Shape (16)	0.86 (356/412)	0.80 (341/422)	0.81 (681/834)	0.85 (346/408)	0.87 (362/418)	0.86 (708/826)
Texture (22)	0.75 (314/418)	0.75 (317/422)	0.77 (644/831)	0.83 (345/413)	0.76 (321/422)	0.79 (656/833)
All (38)	0.86 (367/422)	0.86 (363/422)	0.81 (682/834)	0.90 (371/412)	0.90 (373/414)	0.91 (729/802)

TABLE 14.3
Comparative Evaluation with Existing Systems

Live categorizers	Precision
Hippocrates-mst	44.32%
BLC	67.84%
C4.5	87.02%
RBFN	82.60%
PNN	86.10%
ANN	87.88%
NB	88.99%
Support vector machine	91.00%

14.6 CONCLUSION

Breast tumor is the most frequent tumor in ladies and the largest cause of mortality. Breast cancer can be detected and diagnosed early, which increases treatment options and increases the chances of a cure. The JDK environment is used to develop a breast mass classification system for cancerous–noncancerous lesion categorization. In addition, including all 33 features in the classification stage improves the classification outcomes. The suggested work is done using the JDK environment and the JAVA programming language. WEKA explorer was used to cross-check test cases, which yielded substantial findings. Standard datasets are employed, such as DDSM, MIAS, and the UCI machine learning repository. The testing results demonstrate that the suggested method achieves an accuracy of 89.10% on 670 photos; 33 photos were truly suspicious. However, they were flagged as nonsuspicious. The quality of the mammographic image in the dataset has an impact on classification as well. The higher the image quality, the greater the proportion of correctness. Also refer to Figure 14.8 for comparison with existing classifiers.

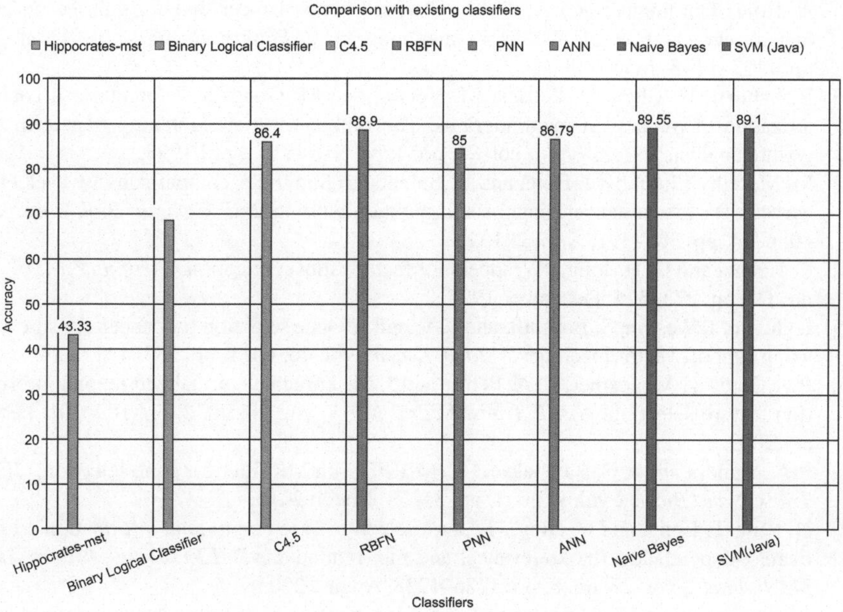

FIGURE 14.8 Comparison with existing classifiers.

14.6.1 Future Work

The suggested system is based on three standard datasets, each of which has an average number of records. This can be expanded in the future to accommodate large amounts of data. The system extracts the image's geometry and texture properties. We can try out more different types of features in the future. Depending on the classifier method to be employed, other feature selection algorithms may be used in the future.

REFERENCES

[1] R. Rangayyan, N. Mudigonda, and J. Desautels, "Boundary modeling and shape analysis methods for classification of mammographic masses," *Med. Biol. Eng. Comput.*, vol. 38, no. 5, pp. 487–496, 2000.

[2] N. Mudigonda, R. Rangayyan, and J. Desautels, "Gradient and texture analysis for the classification of mammographic masses," *IEEE Trans. Med. Imag.*, vol. 19, no. 10, pp. 1032–1043, October 2000.

[3] J. Kilday, F. Palmieri, and M. Fox, "Classifying mammographic lesions using computerized image analysis," *IEEE Trans. Med. Imag.*, vol. 12, no. 4, pp. 664–669, December 1993.

[4] S. Pohlman, K. Powell, N. Obuchowski, W. Chilcote, and S. Grundfest-Broniatowski, "Quantitative classification of breast tumors in digitized mammograms," *Med. Phys.*, vol. 23, no. 8, pp. 1337–1345, August 1996.

[5] A. Rojas Dominguez and A. Nandi, "Toward breast cancer diagnosis based on auto mated segmentation of masses in mammograms," *Pattern Recog.*, vol. 42, no. 6, pp. 1138–1148, June 2009.

[6] B. Sahiner, H. Chan, N. Petrick, M. Helvie, and M. Goodsitt, "Comuterized characterization of masses on mammograms: The rubber band straightening transform and texture analysis," *Med. Phys.*, vol. 25, no. 4, pp. 516–526, April 1998.

[7] M. Meselhy Eltoukhy, I. Faye, and B. Belhaouari Samir, "A comparison of wavelet and curvelet for breast cancer diagnosis in digital mammogram," *Comput. Biol. Med.*, vol. 40, no. 4, pp. 384–391, April 2010.

[8] R. Kohavi and G. H. John, "Wrappers for feature subset selection," *Artif. Intell.*, vol. 97, no. 1/2, pp. 273–324, December 1997.

[9] I. Guyon, J. Weston, S. Barnhill, and V. Vapnik, "Gene selection for cancer classification using support vector machines," *Mach. Learn.*, vol. 46, no. 1, pp. 389–422, 2002.

[10] P. A. Estévez, M. Tesmer, C. A. Perez, and J. M. Zurada, "Normalized mutual information feature selection," *IEEE Trans. Neural Netw.*, vol. 20, no. 2, pp. 189–201, February 2009.

[11] P. A. Mundra and J. C. Rajapakse, "SVM-RFE with MR filter for gene selection," *IEEE Trans. Nano Biosci.*, vol. 9, no. 1, pp. 31–37, March 2010.

[12] H. Peng, F. Long, and C. Ding, "Feature selection based on mutual information criteria of max-dependency, max-relevance, and min-redundancy," *IEEE Trans. Pattern Anal. Mach. Intell.*, vol. 27, no. 8, pp. 1226–1238, August 2005.

[13] Xiaoming Liu and Jinshan Tang, "Mass classification in mammograms using selected geometry and texture features, and a new SVM-based feature selection," *IEEE Syst. J.*, vol. 8, no. 3, September 2014.

[14] J. Sidey-Gibbons and C. Sidey-Gibbons, "Machine learning in medicine: A practical introduction," *BMC Med Res Methodol*, vol. 19, no. 64, 2019.

[15] Anji Reddy Vaka, Badal Soni, and Sudheer Reddy, "Breast cancer detection by leveraging Machine Learning," *ICT Express*, vol. 6, no. 4, pp. 320–324, December 2020.

15 A Novel Optimum Clustering Method Using Variant of NOA

Ravi Kumar Saidala, Nagaraju Devarakonda, Thirumala Rao B, Jabez Syam, and Sujith Kumar

CONTENTS

15.1 INTRODUCTION

Clustering analysis focuses on grouping a set of data points into non-overlapping and homogeneous groups called clusters, and it is essential in solving many real-world clustering problems [1]. Over the past decades, the research community has developed so many clustering methods to solve various real-world clustering problems [2]. A clustering algorithm, regardless of the principle on which it is based, must possess a higher degree of closeness among data points of each cluster and low degree of closeness among data points of the other clusters. Even though clustering is a general task, obtaining optimal clusters is the crucial step in many real-world clustering applications [3, 4]. Obtaining optimal clusters is still a big challenge for the research community. A clustering algorithm divides the data into a set of homogeneous clusters by maximizing intercluster dissimilarity and minimizing intracluster dissimilarity. The optimality of obtained clusters can be measured with the quantization error. The mathematical expression of quantization error (J_e) is presented in equation (1), where $|C_{ij}|$ indicates the number of clusters and d denotes the Euclidean distance function, and it is expressed mathematically as shown in equation (2).

$$J_e = \sum_{j=1}^{N_c} \left[\sum_{\forall Z_p \in C_{ij}} d\left(Z_p, m_j\right) \Big/ \left|C_{ij}\right| \right] \tag{1}$$

$$d\left(D_T, Z\right) = \sum_{i=1}^{n} \sum_{j=1}^{k} \left\| D_{Tij} - Z_j \right\|^2 \tag{2}$$

DOI: 10.1201/9781003309451-15

D_T denotes data point, Z denotes cluster, and $d(D_T, Z)$ represents the distance between D_T and Z. Traditional clustering algorithms are ineffective or infeasible in solving complex problems for real-world challenges. *Computational intelligence (soft computing)* is the one destination for many algorithms that are used to solve many design and engineering optimization problems.

Computational Intelligence is the destination for most algorithms, which exhibits higher optimization performance and requires less computing capacity and time. Recently, many optimization algorithms have been designed and developed for solving many complex real-world problems. The research community has identified the superiority of metaheuristics over other algorithms based on their capabilities and promising results in solving complex design and engineering problems. The key ideas behind the development of metaheuristics are from nature. All popular metaheuristics were developed by taking inspiration from evolution theories, human behavioral psychology, nervous system, animal intelligence, etc. For instance, the key idea behind the development of tornadogenesis optimization algorithm (TOA) is the tornado formations at Oklahoma State, USA. Mathematical modeling of cool air masses and warm air masses led to the development of TOA [5]. The foraging behavioral aspect of humpback whales is fundamental to the development of the whale optimization algorithm (WOA) [6]. The WOA has been formulated from the unique prey searching and encircling mechanism and the bubble-net foraging mechanism [7–9]. In view of authors Wolpert and Macready, there is no single optimizer that is equipped well for tackling all optimization problems, and at the same time, there is no single algorithm superior to all optimization algorithms [10]. Hence, one optimizer can be better in higher performance, requiring less computing capacity and time than other optimization algorithms on various optimization problems, but not on all optimization problems. Thus, there are still many opportunities to develop new optimization algorithms as well as to improve one of the available optimization algorithms. The NFL theorem has motivated many researchers to develop many optimization algorithms. Table 15.1 lists out the some of the widespread heuristic and metaheuristic algorithns, along with the inspiration and their year of development. Although metaheuristics have different distinctive motivation behind their development, they have identical exploration and exploitation phases in the search gradation [11]. The first phase, the exploration, refers to the process of searching for a solution space as extensively, randomly, and globally as possible. The second one, the exploitation phase, refers to the ability of the algorithm to search more accurately in the space obtained by the exploration phase, and its randomness decreases as its accuracy increases. Balancing the exploration and exploitation phases plays a key role in obtaining better optimization performance. Dominance of any of these two phases impacts the performance of the algorithm. If the dominance of the exploration phase occurs in the algorithm, it can search the solution space more extensively and randomly and generate more differential solution sets to fleetly converge. If the dominance of the exploitation phase occurs in the algorithm, it can search the solution space more locally to improve the quality and accuracy of the optimal solution sets. Over the past years, metaheuristics were successfully used in dealing numerous clustering application problems. Many researchers have kept their special interests on designing and developing new metaheuristics and study the applications

TABLE 15.1
List of Popular Heuristic and Metaheuristic Algorithms

S. No	Algorithm	Inspiration	Reference
1	PSO	Bird flock	[Eberhart and Kennedy, 1995]
2	MHBO	Honeybees	[Abbass, 2001]
3	AFSA	Fish swarm	[Li, 2003]
4	TA	Termite colony	[Roth, 2005]
5	ABC	Honeybee	[Basturk, 2006]
6	WPSA	Wolf herd	[Yang et al., 2007]
7	DPO	Dolphin	[Shiqin et al., 2009]
8	CS	Cuckoo	[Yang and Deb, 2009]
9	GSO	Animal searching	[He et al., 2009]
10	HS	Group of animals	[Oftadeh et al., 2010]
11	FA	Firefly	[Yang, 2010a]
12	BIA	Bat herd	[Yang, 2010b]
13	TLBO	Teaching learning	[Rao et al., 2011]
14	FOA	Fruit fly	[Pan, 2012]
15	KH	Krill herd	[Gandomi and Alavi, 2012]
16	BMO	Bird mating	[Askarzadeh and Rezazadeh, 2013]
17	DE	Dolphin	[Kaveh and Farhoudi, 2013]
18	MBA	Mine bomb explosion	[Sadollah et al., 2013]
19	GWO	Gray wolf foraging	[Mirjalili et al., 2014]
20	DA	Dragonfly swarm	[Mirjalili, 2016a]
21	MFOA	Moths	[Mirjalili, 2015]
22	ESA	Elephant herds	[Deb et al., 2015]
23	LOA	Lion's lifestyle	[Yazdani and Jolai, 2016]
24	WOA	Whale's foraging	[Mirjalili and Lewis, 2016]
25	GOA	Grasshopper swarm	[Saremi et al., 2017]
26	SSA	Salp swarm	[Mirjalili et al., 2017]
27	CTO	Students	[Das et al., 2018]

of metaheuristics in solving various clustering problems [12–19]. Clustering plays a major role in the discovery of useful hidden knowledge from data. Obtaining optimal clusters in data analysis and data mining applications is identified as an NP-hard problem. While obtaining optimal clusters in real-world data is becoming a challenging task for many researchers, this research work mainly focuses on developing optimal clustering method. Metaheuristics starts with some initial candidate solution (individuals), then continues to improve all the solutions towards preferred global optimal solution. Iteratively, this process will continue and finishes when encounters predefined stopping criteria. The desired solution may be trapped in local optima without attaining the global optima if premature convergence takes place. To avoid too early convergence and maintain balance between exploration and exploitation phases, this work embeds a new evolutionary strategy called space transformation search (STS) into standard NOA. This research presents implementation of a new optimal clustering method using *K-means* clustering algorithm empowered by *CNOA* for data clustering problems.

The rest of the sections of this chapter are systematized as follows: Standard NOA is illustrated in Section 2. The proposed approach along with pseudocode and flowchart are presented in Section 3. Section 4 summarizes the experimental simulations for variant of NOA and proposed optimum clustering method, and this section also deals with the results analysis and discussions. The last section concludes the whole chapter and outlines the future work.

15.2 STANDARD NOA

Northern bald ibis optimization algorithm (NOA) is a new optimization technique [14] designed to solve numerous kinds of complex optimization problems. It mimics the migrating behavior of ibis birds. Ibises belong to the species *Threskiornithina*, and they live in large and loose flocks. They hover in the sky while they migrate from one place to another and tend to fall in skein formation (either form J-formation or V-formation). Researchers have studied the heart rate of ibises flying in skein formation and identified that they have a characteristic of energy-saving pattern behind the V-formation. In the V-formation flying pattern, ibises are arranged like in a leader–follower manner. For instance, if one ibis bird (leader) is at the front, all the other ibises (followers) will fly around one meter behind and one meter off the side of the leading ibis bird. Likewise, all the ibis birds will fall in V-formation pattern and travel the distance. Each and every ibis benefits from hovering in this manner because when an ibis bird pushes down its wings with certain pressure, then the higher amount of air pressure will come from the conflicting direction, and it would be rather higher than the pressure initiated by the ibis bird. This air pressure varies. All ibises behind that front ibis bird would hover very effortlessly without using much energy. During the relocation, one ibis alone cannot lead the overall journey, because its energy drops out for flipping its wings uninterruptedly for miles. During the overall journey, ibises will rotate themselves from "front to rear" and vice versa to successfully and tirelessly reach their destination. Based on this migrating behavior, the authors of [14] modeled the energy-saving flying pattern in a mathematical form and tested it in benchmark optimization problems.

Algorithm 1. NOA

Input: Initial n Ibis population, t, max_iter
Output: Best solution, best optimal values of objective function

1. Generate initial n ibis population X_i $(i = 1, 2, . . . , n)$, max_iter, t.
2. Compute fitness of each Ibis of the flock X_i.
3. Find X^* = the best ibis position and update position of leader as well as follower position according to STS technique.
4. **While** stopping criterian not met, **do**.
5.   **For** each Ibis in population, **do**.

```
Algorithm 1. NOA
     If i = 1.
   Update the leader position by equation (3).
     Else.
   Update the follower position by equation (4), where C1 and C2
   parameters are replaced with logistic and iterative chaotic
   maps.
     End if.
     End for.
   If there is a better solution, update best position X*.
   Check if any ibis of the flock goes beyond the search space,
   then amend it.
   End while.
   Return X* the best ibis position.
```

NOA starts by dividing the ibis population into two assemblies—that is to say, the leader and the followers. The front ibis bird of the flock is called the leader, and the other ibises are termed as the followers. The position of each ibis is determined in *n*-dimensions, which denote the search space of a problem (or simply problem space), and *n* denotes the problem's variables. These ibises migrate to a place which indicates the destination of the flock. The position of each ibis in the skein formation group should be updated frequently, so the following equation (3) is used to accomplish this action to the ibis leader:

$$X_j^1 = 0.5 * \tanh\left(1.2 * \cos(30)\right) + \left(1 - {l}/{\pi}\right) + 0.5 - X_j^* \tag{3}$$

Where X_j^1 indicates the position of the leader of the ibis flock in the jth dimension. The parameter l represents the current iteration, while X_j^* is the best position in the jth dimension. In this algorithm, authors assumed X_j^* as a food source or destination of the flock in the search space. After updating the ibis flock leader's position, the NOA starts to update the followers' position using the following equation (4):

$$X_j^i = C_1 * \left(e^{-\left(\sum_{i=1}^{3} X_j^i / 3\right)} \right) + C_2 * \left(e^{-\left(\sum_{i=1}^{3} X_j^i / 3\right)} \right) \tag{4}$$

Where $i \geq 2$, X_j^i designates i^{th} follower position of the ibis flock in j^{th} dimension, and parameters c_1 and c_2 designate the random numbers between the interval $[0,1]$.

15.3 VARIANT OF NOA

In this section, the structure of the CNOA, which is a variant of standard NOA, is explained. The CNOA combines the STS, logistic chaotic map, and the NOA algorithms. The basic structure of the NOA algorithm is enhanced by two steps. In the first step of the enhancement of NOA, candidate solutions are calculated in two search spaces—current search space and transformed search space. Newer search space is obtained using STA model. STA can be represented with equation (5).

$$y^* = r(m+n) - y \tag{5}$$

Where r is a real number and m, n denote lower and upper boundaries of search area. Depending upon the random value r, the STS model can be categorized into four types. If $r = 0$, then $y^* = -y$, which is the first type, called symmetrical solutions in STS. If $r = 0.5$, then $y^* = \left(\frac{m+n}{2}\right) - y$, which is the second type, called asymmetrical interval in STS. If $r = 1$, then

$y^* = (m+n) - y$, which is the third type, called STS opposition–based learning. If r is a range between 0 and 1, then $y^* = r(m+n) - y$, which is the fourth type, called random STS. In the second step of the enhancement of NOA, the parameters C1 and C2 are replaced with logistic and iterative chaotic maps, respectively. These two chaotic maps are represented in equations (6) and (7).

$$x^{(n+1)} = c\,x^{(n)}\left(1 - x^{(n)}\right) \quad for\ 0 < c \leq 4 \tag{6}$$

$$x^{(n+1)} = \sin\left(\frac{c.\pi}{x^{(n)}}\right) \tag{7}$$

The integration of STS and chaotic maps adds more flexibility to the NOA in exploring, ensures the diversity of population, and moreover, enables one to arrive at the optimal value quickly.

15.4 EXPERIMENTAL ANALYSIS AND DISCUSSIONS

This section illustrates the summary of investigations carried out on the proposed chaotic NOA algorithm using a comprehensive set of complex benchmark optimization problems. In the experimental simulations, seven unimodal functions (Table 15.2), six multimodal benchmark functions (Table 15.3), and ten benchmark clustering problems have been considered. The benchmark clustering problems are iris, wine, *E. coli*, Hayes-Roth, zoo, glass, breast, Pima Indian diabetes, Haberman's survival, and vowel, which are taken from the UCI data repository (*https://archive.ics.uci.edu/ml/datasets.html*). In order to achieve an unbiased comparison of the achieved results, each and every experiment has been carried out on the same system. Convergence curves obtained during experimental simulations of CNOA are graphically presented in the last portion of each function plot along with the standard NOA, as shown in Figure 15.1 The statistical results, mean ("Average"), median, standard deviation ("Std. Dev."), minimum ("Best"), and Maximum ("Worst"), are also considered in performance comparison. Statistical analysis of achieved solution cost for unimodal and multimodal benchmark optimization problems are presented in Tables 15.4 and 15.5. The popular optimization algorithms (PSO, WOA, TOA, NOA) and other versions of them (HWOA, HTOA) are considered in the result comparisons.

Convergence curves obtained for NOA and CNOA, along with other qualitative results, are presented in Figure 15.1. From the statistical results portrayed in Table 15.4

TABLE 15.2
Functional Description of Unimodal Function

Function description	Range	fmin
$F_1(Dim=30)=\sum_{i=1}^{n} x_i^2$	$[-100, 100]$	0
$F_2(Dim=30)=\sum_{i=1}^{n} \mid x_i \mid + \prod_{i=1}^{n} \lvert x_i \rvert$	$[10, 10]$	0
$F_3(Dim=30)=\sum_{i=1}^{n}(\sum_{j-1}^{i} x_j)^2$	$[-100, 100]$	0
$F_4(Dim=30)=max_i\{\lvert x_i \rvert, 1 \le i \le n\}$	$[-100, 100]$	0
$F_5(Dim=30)=\sum_{i=1}^{n-1}\left[100\left(x_{i+1}-x_i^{2}\right)^2+(x_i-1)^2\right]$	$[30, 30]$	0
$F_6(Dim=30)=\sum_{i=1}^{n}\left(\left[x_i-0.5\right]\right)^2$	$[-100, 100]$	0
$F_7(Dim=30)=\sum_{i=1}^{n} ix_i^{4}+\text{random}\left[0,1\right)$	$[-1.28, 1.28]$	0

and 15.5 and the graphical results presented in Figure 15.1, it is clear that majority of the unimodal and multimodal functions exhibit the improved performance when compared to other optimization algorithms. For the remaining benchmark problems, CNOA gives the closest optimization results to the other algorithm. Several investigations were also carried out on a proposed CNOA-based clustering method for solving ten UCI clustering problems. The performance of the proposed clustering method is evaluated by considering statistical measures like mean and standard deviation. Table 15.6 shows the comparison of CNOA with other popular optimum clustering methods for all the ten UCI clustering problems. From the obtained result, it is obvious that CNOA works better than standard NOA and other optimum clustering methods. This significant improvement achieved by proposed CNOA-based optimum clustering method can be contributed to the STS model and the chaotic maps. From Table 15.6 and Figure 15.3, it is shown that the lower solution cost was achieved using the proposed CNOA clustering method. The obtained mean and standard deviation for iris dataset is 0.2503 and 8.2716e-004, respectively. For glass dataset also, the proposed method achieved mean 0.0400 and standard deviation 0.0015, which are optimal solution cost compared to other methods. For breast dataset also, CNOA achieved low solution cost, and the obtained mean and standard deviation are 0.0509 and 0.0294, respectively. For diabetes dataset, CNOA-based clustering method achieved mean 13.4909 and standard

TABLE 15.3
Functional Description of Multimodal Benchmark Functions

Function description	Range	fmin
$F_8(Dim = 30) = \sum_{i=1}^{n} -x_i^2 \sin(\sqrt{\lvert x_i \rvert})$	$[-500, 500]$	-418.9829*5
$F_9(Dim = 30) = \sum_{i=1}^{n} [x_i^2 - 10\cos(2\pi x_i) + 10]$	$[-5.12, 5.12]$	0
$F_{10}(Dim = 30) = -20e^{\left(-0.2\sqrt{\frac{1}{n}\sum_{i=1}^{n} x_i^2}\right)} - e^{\left(\frac{1}{n}\sum_{i=1}^{n}\cos(2\pi x_i)\right)} + 20 + e$	$[-32, 32]$	0
$F_{11}(Dim = 30) = \frac{1}{4000}\sum_{i=1}^{n} x_i^2 - \prod_{i=1}^{n}\cos(\frac{x_i}{\sqrt{i}}) + 1$	$[-600, 600]$	0
$F_{12}(Dim = 30) = \frac{\pi}{n}\{10\sin(\pi y_1) + \sum_{i=1}^{n-1}(y_i - 1)^2\left[1 + 10\sin \pi y_{i+1}{}^2\right] + (y_n - 1)^2\} + \sum_{i-1}^{n} u(x_i, 10, 100, 4)$ $y_i = 1 + \frac{x_i + 1}{4} \quad u(x_i, a, k, m) = \begin{cases} k(x_i - a)^m & x_i > a \\ 0 & -a < x_i < a \\ k(-x_i - a)^m & x_i < -a \end{cases}$	$[-50, 50]$	0
$F_{13}(Dim = 30) = 0.1\{\sin(3\pi x_1)^2 + \sum_{i=1}^{n}(x_i - 1)^2\left[1 + \sin(3\pi x_i + 1)^2\right] + (x_n - 1)^2\left[1 + \sin 2\pi x_n)^2\right] + \sum_{i-1}^{n} u(x_i, 5, 100, 4)\}$	$[-50, 50]$	0

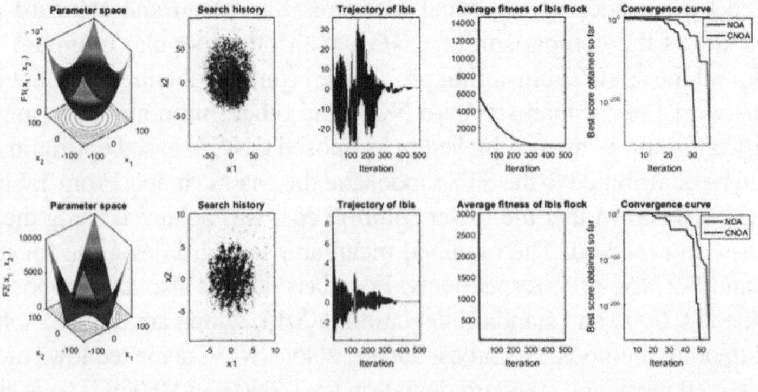

FIGURE 15.1 Convergence curves of CNOA along with other qualitative results.

TABLE 15.4
Statistical Analysis of Acquired Solution Cost for Unimodal Benchmark Problems

Alg.	*Average*	*Median*	*Std. Dev.*	*Best*	*Worst*	*Average*	*Median*	*Std. Dev.*	*Best*	*Worst*
	F1					***F2***				
PSO	115.4e +01	1.45e-29	8.03e+03	1.24e-87	6.83e+04	4.22e+07	2.20e-09	8.74e+08	3.58e-47	7.26e+10
WOA	976.0925	1.49e-29	6.10e+03	1.07e-81	6.24e+04	2.92e+07	1.70e-18	4.61e+08	3.42e-52	7.30e+09
HWOA	133.7060	3.15e-94	2.97e+03	0	6.64e+04	4.13e+06	1.06e-49	9.24e+07	5.58e-318	2.07e+09
TOA	143.2307	1.80e-154	2.63e+03	1.9e-316	5.80e+04	3.94e+07	4.45e-82	8.82e+08	1.18e-165	1.97e+10
HTOA	142.3801	7.61e-164	2.18e+03	0	6.62e+04	1.48e+06	3.61e-81	7.15e+07	1.31e-16	2.43e+09
NOA	30.0501	0	949.8967	0	3.00e+04	0.0140	0	0.1391	0	2.8708
CNOA	40.0011	0	1.26e+03	0	4.00e+04	0.0035	0	0.0370	0	0.4665
	F3					***F4***				
PSO	4.75e+05	4.21e+04	3.14e+05	2.25e+04	1.3549e+05	1.2461	2.68e-36	5.3245	6.57e-49	100.246
WOA	5.80e+04	5.30e+04	3.43e+04	1.70e+04	1.2689e+05	1.4157	2.21e-35	5.2690	6.35e-38	90.7160
HWOA	635.1985	2.68e-47	8.17e+03	9.05e-12	1.5907e+05	0.2516	1.45e-76	4.1991	0	87.8658
TOA	233.5637	2.80e-132	3.89e+03	0	8.3096e+04	0.3498	4.47e-67	4.3485	2.53e-135	82.9751
HTOA	224.6014	2.48e-34	3.96e+03	0	8.1454e+04	0.2498	4.15e-71	4.1126	0	82.4402
NOA	1.23e-08	0	8.83e-08	0	6.4675e-07	0.0036	0	0.0895	0	2.8252
CNOA	10.6617	0	336.371	0	1.0637e+04	4.21e-08	0	3.41e-07	0	2.80e-06
	F5					***F6***				
PSO	2.84e+06	28.2816	1.75e+8	29.1575	2.64e+08	635.9215	0.3843	4.74e+03	0.4637	5.49e+04
WOA	2.61e+06	27.9527	1.94e+7	27.9516	2.54e+08	629.8011	0.3204	4.80e+03	0.3204	5.53e+04
HWOA	5.67e+05	28.4372	1.27e+7	28.4371	2.84e+08	158.9282	0.1634	2.78e+03	0.1633	6.03e+04
TOA	6.70e+05	26.9503	3.20e+06	28.9404	2.12e+08	107.5311	0.4654	2.39e+03	0.0187	6.29e+04
HTOA	5.14e+05	26.0428	3.07e+6	28.1665	2.22e+08	131.0568	0.3454	3.46e+03	0.2715	5.65e+25
NOA	0.1195	6.72e-04	2.1182	0	49.0935	0.1960	0.1866	0.1463	0	2.5000
CNOA	0.1252	2.90e-05	0.9906	0	9.0315	21.0920	0.8718	151.7579	0	1.01e+01
	F7					***F00***				
PSO	1.6845	0.8634	12.6521	0.0654	148.5281	000	000	000	000	000
WOA	1.3233	0.0057	11.1443	0.0052	139.6725	000	000	000	000	000
HWOA	0.1892	1.39e-04	4.2067	1.39e-04	94.0648	000	000	000	000	000
TOA	0.2595	4.46e-04	5.4104	9.01e-05	120.7298	000	000	000	000	000
HTOA	0.1862	2.65e-04	4.1831	4.66e-05	120.5418	000	000	000	000	000
NOA	0.0042	0.0014	0.0359	0	0.7924	000	000	000	000	000
CNOA	0.0028	2.31e-04	0.0160	0	0.2416	000	000	000	000	000

deviation 0.0981, which are optimal solution cost compared to other methods. For *E. coli* dataset also, CNOA achieved optimal results, and the obtained mean and standard deviation of the solution cost are 0.0519 and 0.0012, respectively. For the remaining datasets, the obtained solution costs are almost close to those of the other popular clustering methods. From the obtained statistical results and convergence curves portrayed in Table 15.6 and Figure 15.3, the proposed clustering method, CNOA, is demonstrated to be superior compared to other optimum clustering methods.

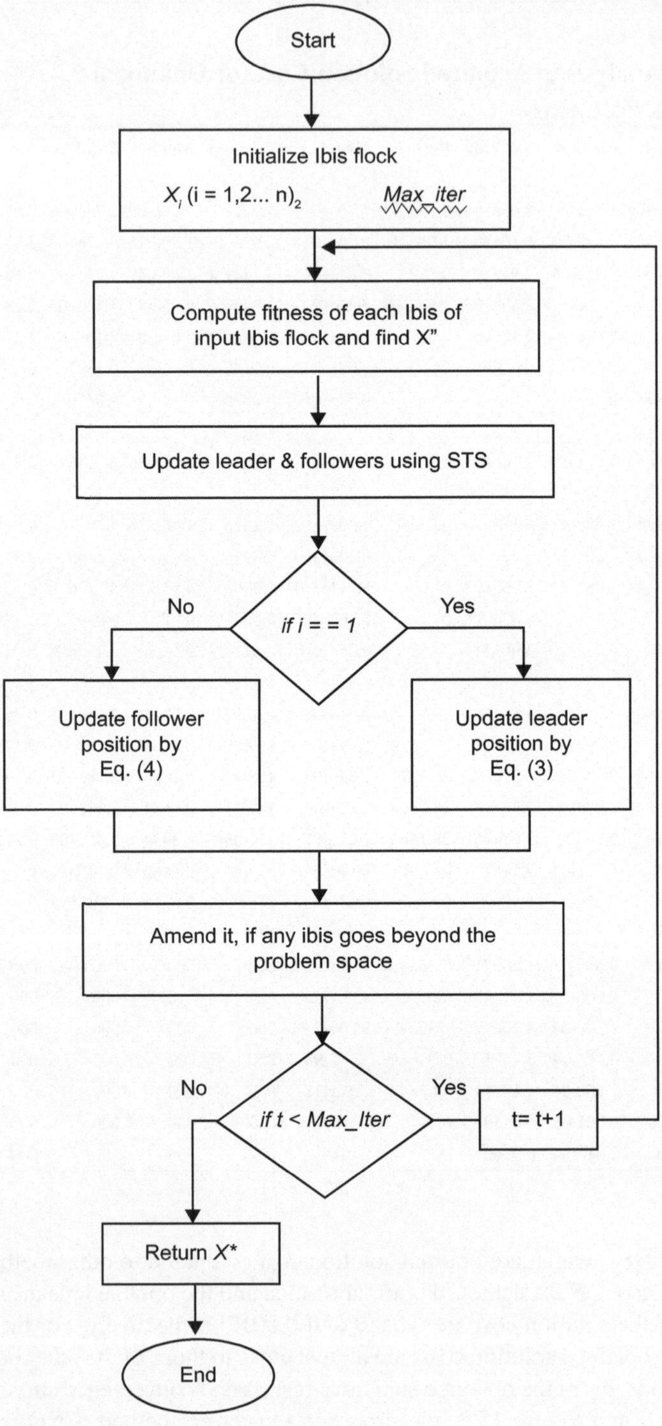

FIGURE 15.2 Flowchart of NOA.

TABLE 15.5
Statistical Analysis of Acquired Colution cost for Multimodal Benchmark Problems

Alg.	Average	Median	Std. Dev.	Best	Worst	Average	Median	Std. Dev.	Best	Worst
	F8					*F9*				
PSO	-1.25e+04	-1.42e+04	1.79e+03	-1.82e+04	-3.49e+03	30.2458	0.1762	86.1954	0.0361	485.1668
WOA	-1.15e+04	-1.23e+04	1.75e+03	-1.23e+04	-2.44e+03	27.1213	0	76.7544	0	450.6246
HWOA	-1.17e+04	-1.22e+04	1.37e+03	-1.22e+04	-1.54e+03	1.4091	0	22.7304	0	425.1632
TOA	-3.17e+03	-2.86e+03	635.6914	-4.37e+03	-1.79e+03	9.1750	0	48.0624	0	435.2462
HTOA	-2.65e+04	-1.65e+04	1.25e+03	-1.25e+04	-3.11e+03	9.0452	0	24.6906	0	425.3651
NOA	-1.55e+03	-1.55e+03	49.0901	-1.55e+03	0	0.2998	0	3.6485	0	91.1446
CNOA	-1.84e+03	-1.84e+03	82.6653	-1.85e+03	0	8.44e-05	0	5.12e-04	0	0.0038
	F10					*F11*				
PSO	0.6583	6.60e-15	3.0201	4.62e-15	20.6903	6.9046	0	40.6608	0	614.7406
WOA	0.4300	6.22e-15	2.5248	4.44e-15	20.4593	6.8496	0	51.1150	0	573.1051
HWOA	0.0448	8.88e-16	0.9233	8.88e-16	20.5905	1.1028	0	24.3672	0	501.3912
TOA	0.1464	8.88e-16	1.4175	8.88e-16	20.2126	1.9288	0	27.1083	0	589.0046
HTOA	0.1352	8.88e-16	0.9811	8.88e-16	20.2281	1.5153	0	27.0531	0	581.8904
NOA	0.0277	8.88e-16	0.5347	0	16.2321	1.0034	0.0002	24.0612	0	540.5046
CNOA	0.0232	8.88e-16	0.1242	0	1.1052	0.1809	0	4.0251	0	90.0045
	F12					*F13*				
PSO	6.81e+06	0.0465	5.62e+07	0.0465	7.83e+08	9.51e+06	0.3151	8.34e+07	0.3151	3.49e+08
WOA	6.54e+06	0.0472	5.60e+07	0.0472	7.44e+08	9.61e+06	0.2720	8.26e+07	0.2720	9.54e+08
HWOA	1.29e+06	0.0130	2.89e+07	0.0130	6.46e+08	2.16e+06	0.2324	4.84e+07	0.2323	1.08e+09
TOA	1.02e+06	0.0023	2.24e+07	0.7495	5.00e+08	2.70e+06	2.9942	4.67e+07	2.9927	1.04e+09
HTOA	1.05e+06	0.0151	2.17e+07	0.0451	4.00e+08	2.05e+06	0.3451	4.82e+07	0.8245	2.55e+09
NOA	0.0177	0.0043	0.1747	0	2.65	0.0015	1.92e-06	0.0357	0	1.1016
CNOA	0.0326	0.0077	0.2564	0	2.8230	4.38e+05	6.65e-07	9.79e+06	0	2.19e+08

TABLE 15.6
Mean and Standard Deviation (Std_Dev.) for Ten Well-Known Datasets over 30 Independent Runs for 500 Iterations

Alg.	*Iris_Dataset*			*Glass_Dataset*		
	Solution_cost	*Intracluster*	*Intercluster*	*Solution_cost*	*Intracluster*	*Intercluster*
DE	0.2787 ± 0.0157	2.4925 ± 0.3709	1.9247 ± 0.2746	0.0443 ± 0.0049	5.2019 ± 0.8524	5.2634 ± 6.4387
WOA	0.2672 ± 0.0141	2.3902 ± 0.3562	1.9291 ± 0.2789	0.0443 ± 0.0042	5.0199 ± 0.7014	4.7812 ± 9.2109
RWOA	0.2689 ± 9.5741e-004	2.1222 ± 0.3522	1.9802 ± 0.2189	0.0433 ± 0.0017	4.8745 ± 0.8957	6.4189 ± 6.9702
MSWOA	0.2655 ± 0.0067	2.0641 ± 0.3434	1.9291 ± 0.2789	0.0429 ± 0.0031	4.0934 ± 0.7134	6.7204 ± 9.0704
HWOA	0.2645 ± 0.0064	2.0176 ± 0.3427	1.9802 ± 0.2189	0.0414 ± 0.0014	4.3705 ± 0.7447	6.7612 ± 6.4308
TOA	0.2628 ± 9.5741e-004	2.1920 ± 0.0512	1.9802 ± 0.2189	0.0412 ± 0.0017	4.8745 ± 0.8957	6.4189 ± 6.9702
CTOA	0.2604 ± 8.6802e-004	2.1281 ± 0.0503	1.9237 ± 0.2403	0.0411 ± 0.0021	4.8682 ± 0.4907	5.9403 ± 2.1643
HTOA	0.2600 ± 0.00054	2.1164 ± 0.0442	1.9735 ± 0.2325	0.0400 ± 0.0021	4.3601 ± 0.4291	6.4518 ± 3.1063
NOA	0.2516 ± 0.00414	2.1164 ± 0.0127	1.9085 ± 0.0325	0.0408 ± 0.0021	4.3271 ± 0.1391	5.0318 ± 3.1003
ENOA	0.2590 ± 8.3412e-004	2.1164 ± 0.0311	1.9103 ± 0.1275	0.0401 ± 0.0018	4.3134 ± 0.1307	6.5348 ± 4.1338
CNOA	0.2503 ± 8.2716e-004	2.1117 ± 0.0243	1.9897 ± 0.0391	0.0400 ± 0.0015	4.3118 ± 0.1308	6.1108 ± 3.1105
Alg.	*Wine_Dataset*			*Breast_Dataset*		
	Solution_cost	*Intra-cluster*	*Inter-cluster*	*Solution_cost*	*Intra-cluster*	*Inter-cluster*
DE	340.8613 ± 1.8143	364.6584 ± 8.6347	328.5411 ± 3.4305	0.0647 ± 0.0214	17.1289 ± 3.72e-15	13.4628 ± 0
WOA	340.1125 ± 10.9298	350.1951 ± 7.5439	329.2461 ± 71.9199	0.0561 ± 0.0285	17.0246 ± 7.95e-15	13.4529 ± 0
RWOA	341.7219 ± 6.80e-014	332.7921 ± 2.2318	359.8712 ± 52.3245	0.0551 ± 6.79e-04	16.9804 ± 7.45e-15	13.5521 ± 0
MSWOA	341.6293 ± 1.6702	330.1041 ± 7.4414	345.2121 ± 47.3042	0.0547 ± 0.1385	16.0426 ± 7.75e-15	13.4521 ± 0
HWOA	339.6378 ± 1.6716	330.1181 ± 7.4034	329.2711 ± 47.9159	0.0534 ± 0.0234	16.4036 ± 7.71e-15	13.4520 ± 0
TOA	336.7219 ± 6.80e-014	332.7921 ± 2.2318	359.8712 ± 52.3245	0.0517 ± 6.79e-04	16.9804 ± 7.45e-15	13.5521 ± 0
CTOA	335.2041 ± 5.91e-014	330.3072 ± 8.1084	345.1492 ± 14.3351	0.0512 ± 0.0248	16.6703 ± 7.31e-15	13.3417 ± 0

HTOA	340.2041 ± 4.53e-014	330.2152 ± 6.5457	345.0544 ± 35.7405	0.0512 ± 0.0182	16.6402 ± 7.35e-15	13.4055 ± 0
NOA	333.7219 ± 2.31e-014	330.0371 ± 6.0654	345.1064 ± 15.3504	0.0518 ± 0.0304	16.0371 ± 7.35e-15	13.4154 ± 0
ENOA	333.1900 ± 3.34e-014	330.0324 ± 7.0008	346.3008 ± 14.7205	0.0519 ± 0.0712	16.0369 ± 7.35e-15	13.4073 ± 0
CNOA	333.2099 ± 3.02e-014	330.0398 ± 6.8399	349.3004 ± 22.0199	0.0509 ± 0.0294	16.0304 ±7.35e-15	13.4002 ± 0
Alg.	***Diabetes_Dataset***			***Haberman_Dataset***		
	Solution_cost	***Intra-cluster***	***Inter-cluster***	***Solution_cost***	***Intra-cluster***	***Inter-cluster***
DE	14.4641 ±0.1048	730.0298 ± 0.15	30.4512 ± 0.4152	6.2107 ± 0.3273	40.3214 ± 2.11e-14	5.2181 ± 8.8e-16
WOA	14.2800 ± 0.1654	708.0127± 4.4521	36.2301 ± 14.7728	5.7700 ± 0.2845	35.3026 ± 3.6534	7.6424 ± 9.8043
RWOA	13.7689± 0.0025	705.9802± 4.3512	36.8425 ± 14.4512	5.5689± 3.61e-06	34.9885 ± 3.6875	8.1285 ± 10.4219
MSWOA	13.7741 ±0.0454	705.0164± 4.6151	36.3051 ± 14.3508	5.3655 ±0.3065	34.3721 ± 3.5504	7.3047 ± 9.8042
HWOA	13.7453 ±0.0656	705.0045± 4.3402	36.3801 ± 14.3461	5.2814 ±0.3061	34.3125 ± 3.5530	7.3607 ± 9.8342
TOA	14.4523 ± 0.0025	705.9802± 4.3512	36.8425 ± 14.4512	4.9842 ± 3.61e-06	34.9885 ± 3.6875	8.1285 ± 10.4219
CTOA	14.0013 ± 0.1501	705.1504 ±4.3501	36.8183 ±14.4501	5.3901 ± 3.63e-06	34.8161 ± 3.6002	8.0812 ± 9.9723
HTOA	13.5131 ± 0.4228	705.3518± 6.4868	36.6105 ± 14.6405	5.1901 ± 3.64e-06	34.8015 ± 3.6812	7.8145 ± 9.5508
NOA	13.8162 ± 0.1056	705.3518± 6.0054	36.3541 ± 14.6074	5.2001 ± 0.5219	34.8513 ± 2.9145	7.3804 ± 9.3604
ENOA	13.7351 ± 0.2813	705.3504± 1.7113	36.6035 ± 14.3441	5.0812 ± 0.1305	34.8003 ± 3.1648	8.6725 ± 10.0034
CNOA	13.4909 ± 0.0981	705.2401± 5.6009	36.7081 ± 14.3908	5.2772 ± 0.0924	34.8491 ± 2.9504	8.6009 ± 10.5083
Alg.	***Hayes-Roth_Dataset***			***E. Coli_Dataset***		
	Solution_cost	***Intra-cluster***	***Inter-cluster***	***Solution_cost***	***Intra-cluster***	***Inter-cluster***
DE	0.2861 ± 0.0324	3.5643 ± 0.4428	1.3915 ± 0.3541	0.0642 ± 0.037	0.6648 ± 0.0834	0.2445 ± 0.0598
WOA	0.3672 ± 0.0284	3.5492 ± 0.4865	1.4256 ± 0.6845	0.0641 ± 0.0028	0.6745 ± 0.0825	0.3176 ± 0.1517
RWOA	0.2689± 0.0129	3.2508 ± 0.3512	1.8251 ± 0.6702	0.0641± 0.0012	0.6012 ± 0.0542	0.4643 ± 0.1728
MSWOA	0.2655 ±0.0105	3.3415 ± 0.4181	1.4046 ± 0.6451	0.0637 ±0.00124	0.6015 ± 0.0621	0.3206 ± 0.1410
HWOA	0.2645 ±0.0108	3.5472 ± 0.4160	1.4053 ± 0.6415	0.0624 ±0.00113	0.6018 ± 0.0721	0.3603 ± 0.1421
TOA	0.2418 ± 0.0129	3.2508 ± 0.3512	1.8251 ± 0.6702	0.0558 ± 0.0012	0.6012 ± 0.0542	0.4643 ± 0.1728

(Continued)

TABLE 15.6 Continued
Mean and Standard Deviation (Std_Dev.) for Ten Well-Known Datasets over 30 Independent Runs for 500 Iterations

Alg.	*Iris_Dataset*			*Glass_Dataset*		
	Solution_cost	*Intracluster*	*Intercluster*	*Solution_cost*	*Intracluster*	*Intercluster*
CTOA	0.2381 ± 0.0148	3.2518 ± 0.3414	1.4108 ± 0.5168	0.0553 ± 0.0031	0.5981 ± 0.0601	0.3501 ± 0.0918
HTOA	0.2514 ± 0.0134	3.2415 ± 0.3251	1.4206 ± 0.6105	0.0545 ± 0.0018	0.6003 ± 0.0567	0.3164 ± 0.1306
NOA	0.3201 ± 0.0134	3.2415 ± 0.3251	1.4206 ± 0.6105	0.0581 ± 0.0234	0.5191 ± 0.0725	0.3234 ± 0.1284
ENOA	0.2505 ± 0.0203	3.2207 ± 0.4028	1.4053 ± 0.5061	0.0565 ± 0.0012	0.6061 ± 0.0904	0.3537 ± 0.1076
CNOA	0.2601 ± 0.2804	3.2376 ± 0.3904	1.4299 ± 0.6811	0.0519 ± 0.0012	0.6992 ± 0.0944	0.3801 ± 0.1435
Alg.	*Zoo_Dataset*			*Vowel_Dataset*		
	Solution_cost	*Intra-cluster*	*Inter-cluster*	*Solutioncost*	*Intra-cluster*	*Inter-cluster*
DE	0.0029 ± 0.0026	1.6378 ± 1.15e-15	2.2315 ± 5.021e-16	32.9214 ± 1.3465	901.325 ± 108.8745	348.3268 ± 230.352
WOA	0.0120 ± 0.0022	1.6401 ± 1.13e-15	2.6471 ± 2.274e-15	38.4012 ± 2.2314	899.246±101.8705	350.3548± 15.655
CSA-WOA	0.0090 ± 0.0019	1.6381 ± 1.12e-15	2.6487 ± 2.258e-15	36.8140 ± 2.3456	899.228± 02.3815	350.8086 ± 215.95
RWOA	0.0094± 8.3665–04	1.5254 ± 4.54e-16	2.6502 ± 2.256e-15	35.8941± 2.2145	898.145± 02.9821	353.3282± 18.254
MSWOA	0.0100±0.0012	1.5332 ± 1.13e-14	2.6451 ± 2.264e-15	34.1553± 2.2645	898.224± 02.3142	350.7301± 15.961
HWOA	0.0053 ±0.0024	1.4301 ± 1.13e-14	2.6432 ± 2.261e-15	32.4501± 2.2416	898.138± 02.3107	350.6126± 15.941
TOA	0.0102 ± 8.3665–04	1.5254 ± 4.55e-16	2.6502 ± 2.256e-15	30.2145 ± 2.2145	898.145± 02.9821	353.3282± 18.254
CTOA	0.0113 ± 0.0023	1.6506 ± 1.13e-15	2.6142 ± 2.278e-15	30.0018 ± 2.0604	894.627± 01.6921	353.1527± 18.044
NOA	0.0079 ± 0.3421	1.3418 ± 2.04e-16	2.6685 ± 2.216e-15	32.4000 ± 3.1491	896.407± 101.5401	353.6404± 218.004
ENOA	0.0040 ± 0.3901	1.5248 ± 4.36e-16	2.6702 ± 2.268e-15	29.0041 ± 3.4368	894.919 ± 101.0371	353.2904± 218.380
CNOA	0.0042 ± 6.164e-04	1.4608 ± 3.61e-16	2.6108 ±2.261e-15	29.2739 ± 2.3699	895.558 ± 101.8255	353.8332± 218.661

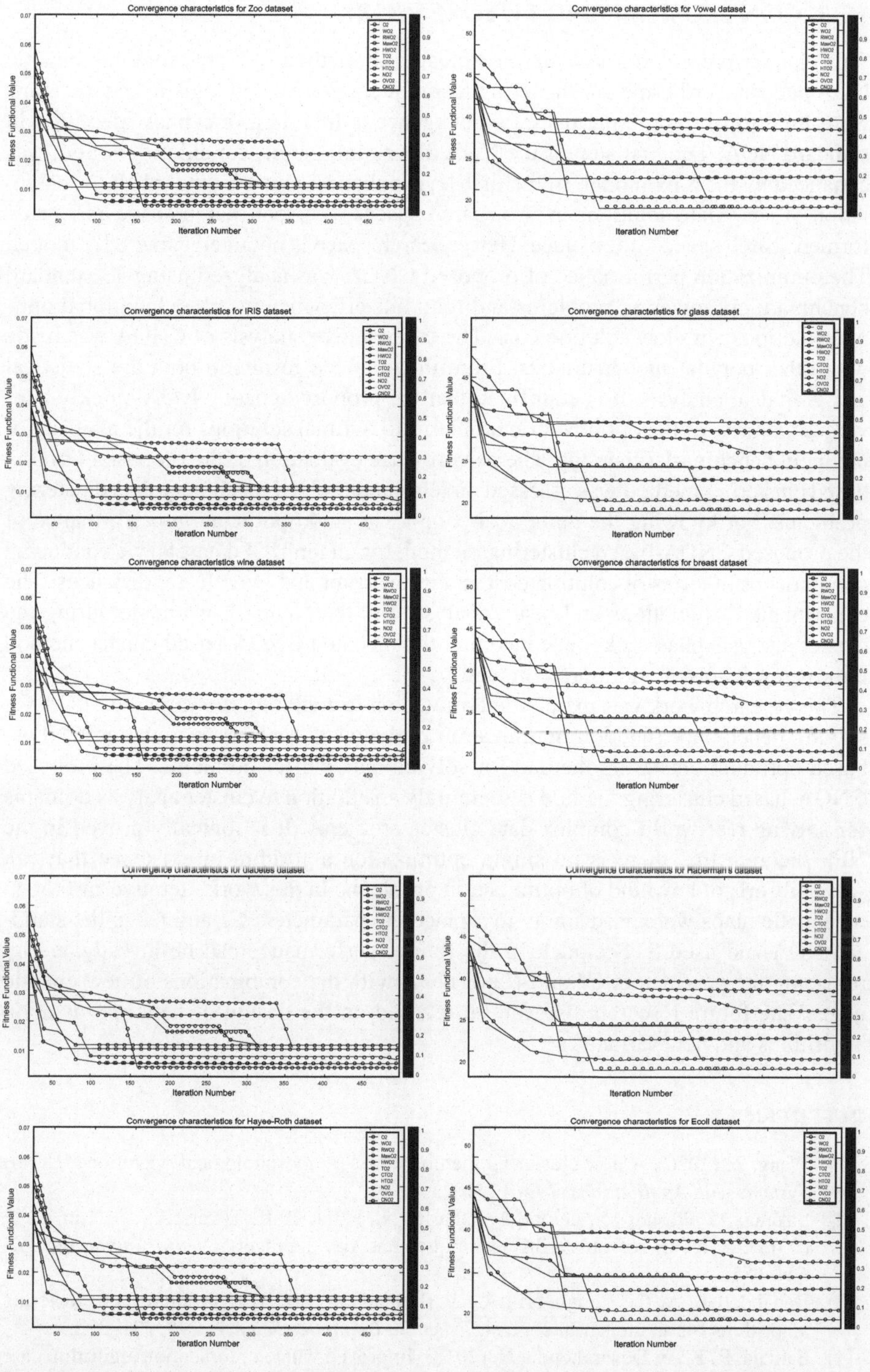

FIGURE 15.3 (a), (b), (c): Convergence of UCI machinery data clustering problems.

15.5 CONCLUSION AND FUTURE WORK

This chapter presented a new optimal clustering method by integrating the chaotic NOA and standard k-means clustering technique. Iterative and logistic chaotic maps were used to overcome premature convergence, which is a drawback identified in standard NOA. The first step of the NOA algorithm, that is, candidate solution, was enhanced by the STS model. In the first-level enhancement of standard NOA, calculation of candidate solutions in two search spaces—current search space and transformed search space—took place. Newer search space is obtained using STA model. The optimization performance of proposed CNOA was analyzed using 13 standard benchmark optimization problems and found its efficiency in achieving global optimal solutions with low solution cost. The comparative analysis of CNOA was made with other popular metaheuristics. From the obtained results in both the statistical and graphical analysis, it is confirmed that the proposed chaotic NOA quickly converges majority of the optimization problems to optimal solutions for the majority of the input benchmark functions. The performance evaluation of the proposed CNOA-based clustering method was assessed on ten standard UCI machinery data clustering problems. For knowing the numerical complexity and optimization performance of the proposed CNOA-based clustering method, for all ten UCI datasets, we conducted comparisons in terms of solution cost and intracluster and intercluster distances. The experimental simulations and result analysis and the comparisons made during the cluster analysis phase acknowledged that the invented CNOA-based cluster method returns results faster and more efficiently.

This research work was mainly determined on formulating the variant of NOA for solving benchmark function optimization problems as well as formulating CNOA-based optimum clustering method for solving data cluster problems. The proposed CNOA-based clustering method is seemingly an addition to cluster analysis domains for solving real-world complex data cluster problems. It is logically proved in the NFL theorem that there is no single optimization algorithm in existence that can solve all sorts of any kind of optimization problems. In this work, iterative and logistic chaotic maps were used singly to replace the parameters *C1* and *C2* in the standard NOA and used STS model. In the future, performance can be investigated by replacing the parameters *C1* and *C2* of NOA with the combinations of new chaotic maps. The future research also can be focused on the design of new optimization algorithms and new variants.

REFERENCES

[1] Wang, Z. (2020). A new clustering method based on morphological operations. *Expert Systems with Applications*, *145*, 113102.

[2] Saxena, A., Prasad, M., Gupta, A., Bharill, N., Patel, O. P., Tiwari, A., . . . Lin, C. T. (2017). A review of clustering techniques and developments. *Neurocomputing*, *267*, 664–681.

[3] Benabdellah, A. C., Benghabrit, A., & Bouhaddou, I. (2019). A survey of clustering algorithms for an industrial context. *Procedia Computer Science*, *148*, 291–302.

[4] Saidala, R. K., & Devarakonda, N. (2018). Improved whale optimization algorithm case study: Clinical data of anaemic pregnant woman. In *Data engineering and intelligent computing* (pp. 271–281). Springer, Singapore.

[5] Saidala, R. K., & Devarakonda, N. (2017, July). The tornadogenesis optimization algorithm. In *2017 IEEE 16th International Conference on Cognitive Informatics & Cognitive Computing (ICCI* CC)* (pp. 378–387). IEEE, Oxford University, Oxford, London, UK.

[6] Mirjalili, S., & Lewis, A. (2016). The whale optimization algorithm. *Advances in Engineering Software*, *95*, 51–67.

[7] Saidala, R. K., & Devarakonda, N. (2018). Multi-swarm whale optimization algorithm for data clustering problems using multiple cooperative strategies. *International Journal of Intelligent Systems and Applications*, *10*(8), 36.

[8] Saidala, R. K., & Devarakonda, N. R. (2017, April). A new parallel metaheuristic optimization algorithm and it's application in CDM. In *2017 2nd International Conference for Convergence in Technology (I2CT)* (pp. 667–674). IEEE, Pune, India.

[9] Saidala, R. K., & Devarakonda, N. R. (2017, April). Bubble-net hunting strategy of whales based optimized feature selection for e-mail classification. In *2017 2nd international conference for convergence in technology (I2CT)* (pp. 626–631). IEEE, Pune, India.

[10] Wolpert, D. H., & Macready, W. G. (1997). No free lunch theorems for optimization. *IEEE Transactions on Evolutionary Computation*, *1*(1), 67–82.

[11] Lin, L., & Gen, M. (2009). Auto-tuning strategy for evolutionary algorithms: Balancing between exploration and exploitation. *Soft Computing*, *13*(2), 157–168.

[12] Nanda, S. J., & Panda, G. (2014). A survey on nature inspired metaheuristic algorithms for partitional clustering. *Swarm and Evolutionary Computation*, *16*, 1–18.

[13] Saidala, R. K., & Devarakonda, N. (2018). Chaotic tornadogenesis optimization algorithm for data clustering problems. *International Journal of Software Science and Computational Intelligence (IJSSCI)*, *10*(1), 38–64.

[14] Saidala, R. K., & Devarakonda, N. (2018, July). Northern bald ibis optimization algorithm: Theory and application. In *2018 IEEE 17th International Conference on Cognitive Informatics & Cognitive Computing (ICCI* CC)* (pp. 541–551). IEEE, University of California, Berkeley, USA.

[15] Masdari, M., & Barshandeh, S. (2020). Discrete teaching—learning-based optimization algorithm for clustering in wireless sensor networks. *Journal of Ambient Intelligence and Humanized Computing*, *11*(11), 5459–5476.

[16] Saidala, R. K., Devarakonda, N., & Edara, S. R. (2018). Enhanced northern bald ibis optimization algorithm based clustering method for web document clustering. *Journal of Advanced Research in Dynamical and Control Systems*, (9), 2081–2094.

[17] Saidala, R. K., & Devarakonda, N. (2019). A novel chaotic northern bald Ibis optimization algorithm for solving different cluster problems [ICCICC18# 155]. *International Journal of Software Science and Computational Intelligence (IJSSCI)*, *11*(2), 1–25.

[18] Saidala, R. K. (2020). Variant of northern bald ibis algorithm for unmasking outliers. *International Journal of Software Science and Computational Intelligence (IJSSCI)*, *12*(1), 15–29.

[19] Saidala, R. K. (2019). Northern bald ibis algorithm-based novel feature selection approach. *International Journal of Software Science and Computational Intelligence (IJSSCI)*, *11*(4), 17–30.

16 Role of Artificial Intelligence and Neural Network in the Health-Care Sector

An Important Guide for Health Prominence

Ankur Narendra Bhai Shah, Nimisha Patel, Jay A. Dave, and Rajanikanth Aluvalu

CONTENTS

Nomenclature

AI: Artificial Intelligence
NN: Neural Network

DOI: 10.1201/9781003309451-16

ANN: Artificial Neural Network
AGI: Artificial General Intelligence
VLSI: Very Large Scale Intergration
ML: Machine Learning
NLP: Natural Language Processing
XAI: Explainable AI
SARSA: State Action Reward State Action
DQN: Deep Q Network
DDPG: Deep Deterministic Policy Gradient
MRI: Magnetic Resonance Imaging
Bi-ANN: Bidirectional Artificial Neural Networks
SOM: Self-Organizing Map

16.1 ARTIFICIAL INTELLIGENCE (AI)

16.1.1 Introduction

The term *artificial intelligence* (AI) was coined by John McCarthy in 1956 during a conference held on this subject [5]. Artificial intelligence is the branch of computer which is able to make a machine think the same as an ordinary human. It is also true that not all the tasks or jobs performed by humans can be perform by artificial intelligence. In artificial intelligence, we need to train the machine as per our wish, that is, for the task that we want to be performed by the machine. We can define *artificial intelligence* as "a system's ability to interpret external data correctly to learn from such data and to use those learning to achieve specific goals and tasks through flexible adaption" [12]. The definition of *artificial intelligence* mainly has four different approaches: think like human, think wisely, act like human, act wisely. The first two ideas are relevant to thought processes, and the remaining two are relevant to behavior. [37] Improvement of computer functions that are relevant to human knowledge like reasoning, learning, and problem-solving is the main aim of artificial intelligence. [26] We can produce a computer, a robot, or a product that can think like how smart humans think using artificial intelligence. [26] "AI is a study of how human brain think, learn, decide and work, when it tries to solve problems. And finally this study outputs intelligent software systems." [26] It is interesting to know why artificial intelligence is to be used. By using artificial intelligence, we can create software or device which can solve real-world problems like health issues, traffic issues, etc. [27] By using artificial intelligence, we can create personal virtual assistants, like Cortana, Google Assistant, Siri, etc. [27] By using artificial intelligence, we can create a robots which can work in an environment where the survival of humans can be at risk. [27] "AI opens a path for other new technologies, new devices and new opportunities." [27] Artificial intelligence has developed a variety of tools to solve many complex problems of the computer society, like search and optimization, logic, probabilistic methods for uncertain reasoning, classifiers and statistical learning methods, control theory, languages, etc. [28] That is the reason that people resort to using artificial intelligence widely.

16.1.2 Type

The different types of artificial intelligence are artificial intelligence, machine learning, and deep learning. [29] In machine learning, algorithms learn by examples and experience. [29] "Machine Learning focuses on development of computer programs that can access data and use it to learn for themselves." [30] Some patterns exist in data machine learning that are used for future predictions. [29] In machine learning, a computer learns by itself without much human involvement and support and changes accordingly, which is the main chief endeavor of machine learning. [30] Machine learning algorithms are mainly classified as supervised algorithms and unsupervised algorithms. [30] In supervised learning, algorithms use past data to label the new data to predict future events. [30] The algorithm produces contingent function to forecast about the output values and which start with training datasets. [30] After adequate training, the system is able to grant targets for any new input. [30] The output can also be compared with correct intended output by learning algorithms in order to modify the models if error is found. [30] The supervised learning problems can be further classified as classification problem and regression problem. [31] When output variables have categoiezy like "pink," "white," "table," etc., then the problem is known as classification problem. [31] When output variables have real values like "height," "rupees," etc., then the problem is known as regression problem. [31] Supervised machine learning algorithm examples are linear regression for regression problems, random forest for classification and regression problems, support vector machines for classification problems, etc. [31] When information used to train is not classified or labeled, then we use unsupervised learning algorithms. [30] In unsupervised learning algorithm, the system can infer a function using unlabeled data to describe hidden structures. [30] "The system doesn't figure out the right output, but it explores the data and can draw inferences from datasets to describe hidden structures from unlabeled data." [30] The main goal of unsupervised learning algorithms is to learn more about data. [31] In unsupervised learning, there is no teacher and there is no correct answer. [31] Algorithms learn by themselves and find interesting structure in a data. [31] Unsupervised learning problems can be further classified as clustering problems and association rule learning problems. [31] A problem in which you want to discover the inherent grouping in a data like grouping patients by their behavior is known as clustering problem. [31] A problem in which you want to discover rules that describe large portions of your data, like patients that have disease X also have disease Y, is known as association rule learning problem. [31] Unsupervised learning algorithms examples are k-means for clustering problems, Apriori algorithm for association rule learning problems, etc. [31] Algorithms that use both techniques, that is, supervised algorithms and unsupervised algorithms, are classified as semisupervised machine learning algorithms. [30] In semisupervised machine learning algorithms for training purposes, they use both labeled and unlabeled data, among which is a small amount of labeled data and a large amount of unlabeled data. [30] Learning accuracy is more when the system uses semisupervised machine learning algorithms. [30] "Usually, semi-supervised learning is chosen when the acquired labeled data requires skilled and relevant resources in order to train it/learn from it. Otherwise, acquiring unlabeled data generally doesn't require additional resources." [30] One of the "good examples is a photo archive where only some of the images are labeled (e.g., dog, cat, person) and the majority are unlabeled." [31] Many real-world applications use semisupervised learning algorithms because it is cheaper than supervised learning algorithms. [31] As domain experts are required in supervised learning algorithms, that makes them expensive. [31] Also, data labeling task is expensive and time-consuming. [31] "The unlabeled data is cheap and easy to collect and store." [31] All these makes semisupervised learning algorithms the more preferred choice for users. Text document classifier is one example of semisupervised learning algorithms. [32]

> This is the type of situation where semi-supervised learning is ideal because it would be nearly impossible to find a large amount of labeled text documents. This is simply because it is not time efficient to have a person read through entire text documents just to assign it a simple classification.
>
> [32]

When learning methods produce actions by interacting with its environment and discover errors or rewards, that type of algorithm is known as reinforcement machine learning algorithm. [30] The characteristics of reinforcement machine learning algorithms are trial and error, search and delayed. [30] A simple reward feedback is known as reinforcement signal, which is required for an agent to learn which action is best. [30] "This method allows machines and software agents to automatically determine the ideal behavior within a specific context in order to maximize its performance." [30] Types of reinforcement learning may be positive or negative. Examples of reinforcement learning algorithms are Monte Carlo, Q-learning, SARSA, DQN, DDPG, etc. [33] A subfield of machine learning is deep learning. [29] "Deep learning does not mean machine learns more in depth knowledge." [29] When a machine uses different layers to learn from the data, the technique is known as deep learning. [29] The number of layers in the model represents the depth of the model. For example, for image reorganization, Google LeNet model has 22 layers. [29] A neural network is used for the learning phase in deep learning. [29]

16.1.3 Category

Artificial intelligence mainly falls under two categories. First is narrow AI, which is also known as "weak AI." This type of AI mainly operates within a limited context and is a simulation of human intelligence. It is mainly focused on performing a single task extremely well. [37] For example, "[a] weak AI system designed to identify cancer from X-ray or ultrasound images, for example, might be able to spot a cancerous mass in images faster and more accurately than a trained radiologist." [34] The main pitfall of narrow AI is that systems can only do what they are designed to do and can only make decisions based on their training data. [34] For example, a retailer's customer service chatbot can give answers to queries like working hours of the store, item price, item return policy, etc., but if it is not trained for it, then it cannot give answers to other questions, like why a certain product is better than other similar products. [34] Examples of narrow AI are image and facial recognition systems, chatbots and conversational assistants, predictive maintenance models, self-driving vehicles, recommendation engines, etc. [34] The other type of AI is artificial general intelligence (AGI), which is also known as "strong AI," "general AI," or "superintelligence." [35] This type of AI, we see in movies like *Robots*. "The next generation AGI is a machine with general intelligence and much like human being, it can apply that intelligence to solve any problem." [37] Strong AI is more like a brain; it does not classify but uses clustering and associations to process data. [35]

> In short, it means there isn't a set answer to your keywords. The function will mimic the result, but in this case, we aren't certain of the result. Like talking to a human, you can assume what someone would reply to a question with, but you don't know.
>
> [35]

Examples of strong AI are machines that start the coffee maker when it hears "Good morning," AI in games like 49 classic Atari games, etc. [35]

16.1.4 Applications

Nowadays, due to the growing use of big data and cloud computing, AI has been strengthened, and there are huge numbers of applications of AI in various fields. AI can be used in marketing, banking, finance, agriculture, healthcare, gaming, space exploration, autonomous vehicles, social media, chatbots, artificial creativity, etc. When we talk about health of the people, many organization rely on AI.

> An organization called Cambio Health Care developed a clinical decision support system for stroke prevention that can give the physician a warning when there's a patient at risk of having a heart stroke. Another such example is Coala life which is a company that has a digitalized device that can find cardiac diseases. Similarly, Aifloo is developing a system for keeping track of how people are doing in nursing homes, home care, etc. The best thing about AI in healthcare is that you don't even need to develop a new medication. Just by using an existing medication in the right way, you can also save lives.
>
> [39]

16.1.5 Applications of AI in Health Care

Accurate cancer diagnosis with AI. To assist pathologists to make more accurate diagnosis, the PathAI machine learning technology was developed. [36] It was developed at Cambridge. The main aim of the company is to reduce error in cancer diagnosis, and they also want to develop a method for individualized medical treatment. [36] "PathAI has worked with drug developers like Bristol-Myers Squibb and organizations like the Bill & Melinda Gates Foundation to expand its AI technology into other healthcare industries." [36]

An intelligent symptom checker. Buoy Health is used to diagnose and treat illness using an AI-based symptom and cure checker that uses algorithms. [36] It uses the chatbot for providing correct care based on a patient's diagnosis. The patient's symptoms and health problems are first "heard" by the chatbot, and then it guides regarding the same. [36] Buoy's AI, which helps in the diagnosis and treatment of patients more quickly, is used in Harvard Medical School, which is one of the many hospitals and healthcare settings that use it. [36]

Streamline radiology diagnoses. Enlitic is used to streamline radiology diagnoses. [36] It was develop using deep learning medical tools. [36] Unstructured medical data like EKGs, genomics, patient medical history, radiology images, blood tests, etc. are analyzed using the company's deep learning platform. [36] This will provide better insight into patients' real-time need to doctors. [36] The fifth smartest artificial intelligence company in the world is MIT, which is ranked above Facebook and Microsoft, who named Enlitic. [36]

Earlier cancer detection with AI. Freenome is used for earlier cancer detection using AI in screenings, diagnostic tests, and blood work for cancer tests. Freenome aims to detect cancer in its earliest stages and subsequently develop new treatments by deploying AI at general screenings. [36]

Diagnosing deadly blood diseases faster using AI. "Harvard University's teaching hospital, Beth Israel Deaconess Medical Center, is using artificial intelligence to diagnose potentially deadly blood diseases at a very early stage." [36] Doctors scan for harmful bacteria like *E. coli* and *staphylococcus* in blood samples at a faster rate by using AI-enhanced microscopes than is possible using manual scanning. [36] To teach the machines how to search for bacteria in blood sample, scientists used 25,000 images of blood samples. [36] After learning, the machines are able to identify and predict harmful bacteria in blood with 95% accuracy. [36]

AI-powered radiology assistant. Zebra Medical Vision provides radiologists with an AI-enabled assistant. [36] It receives imaging scans and automatically analyzes them for various clinical findings it has studied. [36] "The findings are passed onto radiologists, who take the assistant's reports into consideration when making a diagnosis." [36]

New medicines development using AI. In the drug development industry, the cost for development is high, and research as well takes thousands of human hours. [36] Only 10% of drugs are successfully brought to market, while they spend $2.6 billion for each drugs through clinical trials. [36] Due to advancement in technology, pharma companies take a note about efficiency, accuracy, and knowledge that artificial intelligence can provide. [36]

> One of the biggest AI breakthroughs in drug development came in 2007 when researchers tasked a robot named Adam with researching functions of yeast. Adam scoured billions of data points in public databases to hypothesize about the functions of 19 genes within yeast, predicting 9 new and accurate hypotheses. Adam's robot friend, Eve, discovered that triclosan, a common ingredient in toothpaste, can combat malaria-based parasites.
>
> [36]

In the fields of immuno-oncology and neuroscience, to identify and develop new medicines, BioXcel Therapeutics uses AI. [36] For finding new applications for existing drugs or to identify new patients, the company's drug reinnovation program employs AI. [36] "BioXcel Therapeutics' work in AI-based drug development was named as one of the Most Innovative Healthcare AI Developments of 2019." [36] A clinical-stage, AI-based biotech platform that maps diseases to accelerate the discovery and development of breakthrough medicines is BERG. [36] BERG can develop more robust product candidates that fight rare diseases by combining its "interrogative biology" approach with traditional R&D. [36] "BERG recently presented its findings on Parkinson's Disease treatment—they used AI to find links between chemicals in the human body that were previously unknown—at the Neuroscience 2018 conference." [36] Some of the serious diseases nowadays, like Ebola and multiple sclerosis, can be tackled by Atomwise using artificial intelligence. AtomNet is the company's neural network, which helps predict bioactivity and identify patient characteristics for clinical trials. [36] "Atomwise's AI technology screens between 10 and 20 million genetic compounds each day and can reportedly deliver results 100 times faster than traditional pharmaceutical companies." [36] Deep Genomic's AI platform helps researchers find better candidates for development of drugs related to neuromuscular and neurodegenerative disorders. [36] This will statistically raise the chances of successfully passing clinical trials while also decreasing time and cost to market due to finding the right candidates during a drug's development. [36] "Deep Genomics is also working on Project Saturn, which analyzes over 69 billion different cell compounds and provides researchers with feedback" [36]

> Combining AI, the cloud and quantum physics, XtalPi's ID4 platform predicts the chemical and pharmaceutical properties of small-molecule candidates for drug design and development. Additionally, the company claims its crystal structure prediction technology (aka polymorph prediction) predicts complex molecular systems within days rather than weeks or months. XtalPi's big-name investors include Google, Tencent and Sequoia Capital.
>
> [36]

Deep learning for targeted treatment. BenevolentAI uses deep learning for targeted treatment. [36] "The primary goal of BenevolentAI is to get the right treatment to the right patients at the right time by using artificial intelligence to produce a better target selection and provide previously undiscovered insights through deep learning." [36] "BenevolentAI is working with major pharmaceutical groups to license drugs, while also partnering with charities to develop easily transportable medicines for rare diseases." [36]

Streamlining patient experiences with artificial intelligence. Time is money in the health-care industry. [36] "Efficiently providing a seamless patient experience allows hospitals, clinics and physicians treat more patients on a daily basis." [36]

> US hospitals saw more than 35 million patients in 2016, each with different ailments, insurance coverage and condition that factor into providing service. A 2016 study of 35,000 physician reviews revealed 96% of patient complaints are about lack of customer service, confusion over paperwork and negative front desk experiences. New innovations in AI healthcare technology are streamlining the patient experience, helping hospital staff process millions, if not billions of data points, faster and more efficiently.
>
> [36]

> Babylon Health is on a mission to reengineer healthcare by shifting the focus away from simply caring for the sick but helping prevent sickness in the first place, leading to better health and less health-related expenses. The platform features an AI engine created by experienced doctors and skilled deep-learning scientists that operates an interactive symptom checker, assisting known symptoms and risk factors to provide the most informed and up-to-date medical information possible. Additionally, the Babylon Health platform features a health monitoring system to help people stay healthy for longer periods of time.
>
> [36]

For the betterment of mental health, Spring Health is used.

> It offers a powerful mental health benefit solution that employers can adapt to provide their employees with the resources they need to best keep their mental health in check. The powerful, clinically validated technology works by collecting a comprehensive dataset from each individual and comparing that against hundreds of thousands of other data points. The platform then uses a machine-learning model to match them with the right specialist for either in-person care or telehealth appointments.
>
> [36]

> One Drop provides a complete, discreet solution for managing chronic conditions like diabetes, heart health and blood pressure, as well as weight management. The One Drop Premium app allows people to manage their conditions head-first, offering interactive coaching from real-world professionals, predictive glucose readings powered by artificial intelligence and data science, learning resources and daily tracking of readings taken from One Drop's Bluetooth-enabled glucose reader and other devices.
>
> [36]

> Kaia Health operates a digital therapeutics platform that features live physical therapists to provide people with the physical care they need within the boundaries of their own schedule. The platform includes personalized programs with case reviews, exercise routines, relaxation activities and learning resources all contained within, perfect for treating chronic back, neck, shoulder, elbow, hip, knee and additional pain in the moments when it's most needed. In addition to Kaia Health's coaches and professionals, Kaia Health also features a PT-grade automated feedback coach that uses AI technology to ensure patients receive the best care possible. Kaia Health is available as integration with leading medical professionals or as an employer-offered benefit.
>
> [36]

Twin Health is pioneering a holistic method of addressing and potentially reversing chronic conditions like Type 2 Diabetes through a mixture of IoT tech, AI, data science, medical science and healthcare. The company has created the Whole Body Digital Twin—a dynamic, digital representation of human metabolic function built around thousands of health data points, daily activities and personal preference. The company's Twin Service is dedicated to providing personalized and precise nutrition, sleep, activity and breathing guidance to each member along with medication management guidance to their physicians, creating a goal of providing healthcare with enough accuracy to potentially reverse chronic ailments.

[36]

Olive's AI platform is designed to automate the healthcare industry' most repetitive tasks, freeing up administrators to work on higher-level ones. The platform automates everything from eligibility checks to un-adjudicated claims and data migrations so staffers can focus on providing better patient service. Olive's AI-as-a-Service easily integrates within a hospital's existing software and tools, eliminating the need for costly integrations or downtimes.

[36]

Qventus is an AI-based software platform that solves operational challenges, including those related to emergency rooms and patient safety. The company's automated platform prioritizes patient illness/injury; tracks hospital waiting times and can even chart the fastest ambulance routes. CB Insights named Qventus one of its 100 Most Innovative AI Startups for 2019 based on the company's work in automating and prioritizing patient safety.

[36]

Babylon uses AI to provide personalized and interactive healthcare, including anytime face-to-face appointments with doctors. The company's AI-powered chatbot streamlines the review of a patient's symptoms, and then recommends either a virtual check-in or a face-to-face visit with a healthcare professional. Babylon and Canada's Telus Health teamed up to develop a Canada-specific AI app that scans a patient's survey answers, then connects them via video with the right healthcare provider or professional.

[36]

CloudMedX uses machine learning to generate insights for improving patient journeys throughout the healthcare system. The company's technology helps hospitals and clinics manage patient data, clinical history and payment information by using predictive analytics to intervene at critical junctures in the patient care experience. Healthcare providers can use these insights to efficiently move patients through the system without any of the traditional confusion.

[36]

The Cleveland Clinic teamed up with IBM to infuse its IT capabilities with artificial intelligence. The world-renowned hospital is using AI to gather information on trillions of administrative and health record data points to streamline the patient experience. This marriage of AI and data is helping the Cleveland Clinic personalize healthcare plans on an individual basis.

[36]

Johns Hopkins Hospital recently announced a partnership with GE to use predictive AI techniques to improve the efficiency of patient operational flow. A task force, augmented with artificial intelligence, quickly prioritized hospital activity for the benefit of all patients. Since implementing the program, the facility has seen a 60% improvement in its ability to admit patients and a 21% increase in patient discharges before noon, resulting in a faster, more positive patient experience.

[36]

Mining and managing medical data with AI.

Healthcare is widely considered one of the next big data frontiers to tame. Highly valuable information can sometimes get lost among the forest of trillions of data points, losing the industry around $100 billion a year. Additionally, the inability to connect important data points are slows the development of new drugs, preventative medicine and proper diagnosis. Many in healthcare are turning to artificial intelligence as way to stop the data hemorrhaging. The technology breaks down data silos and connects in minute's information that used to take years to process.

[36]

Tempus is using AI to sift through the world's largest collection of clinical and molecular data in order to personalize healthcare treatments. The company is developing AI tools that collect and analyze data in everything from genetic sequencing to image recognition, that can give physicians better insights into treatments and cures. Tempus is currently using its AI-driven data to tackle cancer research and treatment.

[36]

KenSci combines big data and artificial intelligence to predict clinical, financial and operational risk by taking data from existing sources to foretell everything from who might get sick to what's driving up a hospital's healthcare costs. KenSci has partnered with some of the biggest names in tech and data science, including GE, KPMG, Allscripts and Microsoft.

[36]

Proscia is a digital pathology platform that uses AI to detect patterns in cancer cells. The company's software helps pathology labs eliminate bottlenecks in data management and uses AI-powered image analysis to connect data points that support cancer discovery and treatment. Proscia recently raised $8.3M in Series A funding that will be used to expand deployment of the company's digital pathology software and AI tools.

[36]

H2O.ai's AI analyzes data throughout a healthcare system to mine, automate and predict processes. It has been used to predict ICU transfers, improve clinical workflows and even pinpoint a patient's risk of hospital-acquired infections. Using the company's artificial intelligence to mine health data, hospitals can predict and detect sepsis, which ultimately reduces death rates.

[36]

When IBM's Watson isn't competing on *Jeopardy!* It's helping healthcare professionals harness their data to optimize hospital efficiency, better engage with patients and improve treatment. Watson is currently applying its skills to everything from developing personalized health plans to interpreting genetic testing results and catching early signs of disease.

[36]

Google's DeepMind Health AI software is being used by hospitals all over the world to help move patients from testing to treatment more efficiently. The DeepMind Health program notifies doctors when a patient's health deteriorates and can even help in the diagnosis of ailments by combing its massive dataset for comparable symptoms. By collecting symptoms of a patient and inputting them into the DeepMind platform, doctors can diagnose quickly and more effectively.

[36]

ICarbonX is using AI and big data to look more closely at human life characteristics in a way they describe as digital life. By analyzing the health and actions of human beings in a carbon cloud the company hopes its big data will become so powerful that it can manage all aspects of health. ICarbonX believes its technology can gather enough data to better classify symptoms, develop treatment options and get people healthier.

[36]

AI robot assistant surgery.

> Popularity in robot-assisted surgery is skyrocketing. Hospitals are using robots to help with everything from minimally-invasive procedures to open heart surgery. According to the Mayo Clinic, robots help doctors perform complex procedures with a precision, flexibility and control that go beyond human capabilities. Robots equipped with cameras, mechanical arms and surgical instruments augment the experience, skill and knowledge of doctors to create a new kind of surgery. Surgeons control the mechanical arms while seated at a computer console while the robot gives the doctor a three dimensional, magnified view of the surgical site that surgeons could not get from relying on their eyes alone. The surgeon then leads other team members who work closely with the robot through the entire operation. Robot-assisted surgeries have led to fewer surgery-related complications, less pain and a quicker recovery time.
>
> [36]

> Vicarious Surgical combines virtual reality with AI-enabled robots so surgeons can perform minimally invasive operations. Using the company's technology, surgeons can virtually shrink and explore the inside of a patient's body in much more detail. Vicarious Surgical's technology impressed former Microsoft chief Bill Gates, who invested in the company.
>
> [36]

> Auris Health develops a variety of robots designed to improve endoscopies by employing the latest in micro-instrumentation, endoscope design, data science and AI. Consequently, doctors get a clearer view of a patient's illness from both a physical and data perspective. The company is developing AI robots to study lung cancer, with the goal of curing it someday.
>
> [36]

> The Accuray CyberKnife System uses robotic arms to precisely treat cancerous tumors all over the body. Using the robot's real-time tumor tracking capabilities, doctors and surgeons are able to treat only affected areas rather than the whole body. The Accuray CyberKnife robot uses 6D motion-sensing technologies to aggressively track and attack cancerous tumors while saving healthy tissue.
>
> [36]

> Intuitive's da Vinci platforms have pioneered the robotic surgery industry. Being the first robotic surgery assistant approved by the FDA over 18 years ago, the surgical machines feature cameras, robotic arms and surgical tools to aide in minimally invasive procedures. The da Vinci platform is constantly

> taking in information and providing analytics to surgeons to improve future surgeries. So far, da Vinci has assisted in over five million operations.
>
> [36]

> The robotics department at Carnegie Mellon University developed Heartlander, a miniature mobile robot designed to facilitate therapy on the heart. Under a physician's control, the tiny robot enters the chest through a small incision, navigates to certain locations of the heart by itself, adheres to the surface of the heart and administers therapy.
>
> [36]

"MicroSure's robots help surgeons overcome their human physical limitations. The company's motion stabilizer system reportedly improves performance and precision during surgical procedures. Currently, eight of MicroSure's micro-surgical operations are approved for lymphatic system procedures." [36] "Surgeons use the Mazor Robotics' 3D tools to visualize their surgical plans, read images with AI that recognizes anatomical features and perform a more stable and precise spinal operation." [36]

16.1.6 Advantages of AI in Health Care

By integrating AI with the health-care system, it gives a number of benefits.

Low cost. By using AI, it makes various administration tasks easy, and thus, it is possible to have low-cost solution. [40]

> According to Insider Intelligence, 30% of healthcare costs are associated with administrative tasks. AI can automate some of these tasks, like pre-authorizing insurance, following-up on unpaid bills, and maintaining records, to ease the workload of healthcare professionals and ultimately save them money.
>
> [40]

Analyzing big datasets. Analyzing big datasets of patients helps improve key areas of patient health. [40]

Wearable health-care technology. AI uses wearable health-care technology to serve patients better. One of the examples of this is the smart watch. [40]

Clinical decision-making. AI can be helpful in making plans for patients' treatment, thus making better decision-making. [41]

Streamlining processes.

> There are solutions that are intelligent enough to identify possible markers on radiology images and there are solutions that ease the physician admin burden by translating clinical notes or streamlining appointments or tracking patient notes and care recommendations. In essence, the benefits of AI in healthcare are as numerous as the applications for which it is invented and applied.
>
> [41]

Information sharing.

> In addition to physician support, the benefits of AI in healthcare extend to information sharing and precision medicine. AI can be used to track specific patient data more accurately—an essential tool in weighty healthcare institutions such as the NHS—and thereby allow for more accurate patient care and improved doctor time to patient ratios. In 2019, the NHS established a national AI lab with health secretary Matt Hancock pledging £250 million to increase its future role within the healthcare sector. In terms of precision medicine, according to the U.S National Institute of Health, AI can be used to increase outcome precision and accuracy and it can be used to potentially identify patterns within patient data to determine their probability of getting a specific disease or illness. This level of insight can have immense value to the medical profession as it can fully streamline patient care and reduce potential risks by addressing their root causes earlier. AI's ability to read and analyze vast quantities of information is the key to unlocking the full potential of precision medicine.
>
> [41]

16.1.7 Disadvantages of AI in Health Care

Though there are various advantages of integrating AI within the health-care system, there are some disadvantages or dangers in using AI in healthcare.

Implementation cost. Implementation of AI machines, computers, etc. increases cost. [17] Also, a skilled man is required to handle its increased cost.

Unsafe failure mode. "Unlike a human doctor, an AI system can diagnose patients without having confidence in its prediction, especially when working with insufficient information." [18]

Reinforcement of outmoded practice. "AI can't adapt when developments or changes in medical policy are implemented, as these systems are trained using historical data." [18]

Negative side effects. "AI systems may suggest a treatment but fail to consider any potential unintended consequences." [18]

Unemployment. The ratio of unemployment is increased in the health sector if it is successfully implemented.

Implementing the right AI platform.

> The technology isn't completely in its infancy right now, but it does need to be refined and adapted consistently as lessons are learned and algorithms become more adept. Medical institutions need to invest into AI carefully, strategically, and with the right partners. There are plenty of organizations claiming that their platform and solution is the right one, but not all AI solutions are created equal. The risk is that a healthcare practice implements an AI platform that doesn't have rigorous controls or accreditations.
>
> [41]

We organize this paper as follows section 2 presents basics about Artificial Neural Network. Section 3 presents various techniques of AI used in healthcare. Section 4 presents literature survey about various papers. Section 5 is about examples of AI in healthcare and then final conclusion.

16.2 ARTIFICIAL NEURAL NETWORK (ANN)

16.2.1 Introduction

Artificial neural network is also known as neural network.

> Neural networks represent a class of algorithms that are designed to recognize patterns, and which can help in clustering or classifying input data. Their design follows a structure which is loosely inspired by building blocks of information processing in the human brain, hence their name.
>
> [3]

In short, we can say a neural network is a machine that is designed to model the way in which the brain performs a particular task or function of interest. The network is usually implemented by using electronics components or is simulated in a software on a digital computer. [4] "Artificial neural networks may be thought of as simplified models of the networks of neurons that occur naturally in the animal brain." [2] Neural network is also referred to as neurocomputers, connectionist networks, parallel distributed processors, etc. [1]

> Neural networks represent deep learning using artificial intelligence. Certain application scenarios are too heavy or out of scope for traditional machine learning algorithms to handle. As they are commonly known, Neural Network pitches in such scenarios and fills the gap. Artificial neural networks are inspired from the biological neurons within the human body which activate under certain circumstances resulting in a related action performed by the body in response. Artificial neural nets consist of various layers of interconnected artificial neurons powered by activation functions which help in switching them ON/OFF. Like traditional machine algorithms, here too, there are certain values that neural nets learn in the training phase.
>
> [42]

16.2.2 Features

Following are various features of the neural network. *Nonlinearity*. An artificial neuron can be linear or nonlinear. A neural network is made up of an interconnection of nonlinear neurons, so the neural network itself is a nonlinear. [1] Nonlinearity is a special kind of sense that is distributed throughout the network. [1] Nonlinearity is a highly important property, particularly if the underlying physical mechanism responsible for generation of the input signal (e.g., speech signal) is inherently nonlinear. [1]

Input–output mapping. Supervised learning is learning with a teacher. It involves modification of the synaptic weight of a neural network. It's done by applying a set of labeled training samples or task examples. Each example consists of a unique input signal and a corresponding desired response. [1] The network is presented with an example picked at random from the set. [1] The synaptic weights of the network are modified to minimize the difference between the desired response and the actual response of the network. [1] The training of the network is repeated until it reaches a steady state; it means, there are no significant changes in the synaptic weights. [1] The previously applied training examples may be reapplied during the training session, but in a different order. [1] So the network learns from the examples by constructing an input–output mapping for the problem. [1]

Adaptivity. Neural networks have a built-in capability to adapt change in synaptic weights according to the surrounding environment. [1] In general, a neural network trained to operate in a specific environment can be easily retrained to deal with minor changes in the operating environment conditions. [1] Moreover, when it is operating in a nonstationary environment (i.e., one where statistics change with time), a neural network can be designed to change its synaptic weights in real time. [1] To get full benefit of adaptivity, principal time constants of the system should be long enough to ignore disturbances and yet short enough to respond to meaningful changes in the environment. [1] The problem described here is known as the stability plasticity dilemma. [1]

Evidential response. A neural network can be designed to provide information not only about which particular pattern to select but also about confidence in the decision made. This information may be used to reject ambiguous patterns and improve the classification performance of the network. [1]

Contextual information. Contextual information is dealt with naturally by a neural network. Knowledge is represented by a varying structure and activation state of a neural network. Every neuron in the network is potentially affected by the global activity of all other neurons in the network. [1]

VLSI implementability. The massively parallel nature of neural network makes it potentially fast for the computation of certain tasks. So it is well suited for VLSI technology. [1]

Uniformity of analysis and design. In neural network, all domains of the applications use the same notations. So it is easy to share theories and learning algorithms between different applications of neural network. It provides a seamless integration of modules. [1]

Fault tolerance. A neural network implemented in hardware form has the potential to be inherently fault-tolerant. In order to make neural network fault-tolerant, it is necessary to train the network in such a way. [1]

Neurobiological analogy. The design of a neural network is motivated by analogy of the brain. It gives proof that fault tolerance is fast and powerful. [1]

16.2.3 Types of ANN

"There are many types of neural networks available or that might be in the development stage. They can be classified depending on their: Structure, Data flow, Neurons used and their density, Layers and their depth activation filters etc." [42]

Single-layer feed-forward networks. In layered neuron network, the neurons are organized in the form of layers. In the simplest form of a layered network, we have an input layer of source nodes that projects onto an output layer of neurons, but not vice versa. This network is strictly a feed-forward or acyclic type. This network is called a single-layer network, where *single layer* refers to output layers of neurons. We do not count the input layer performed there. [1]

Multilayered feed-forward networks. In this class of neural network, there may be the presence of one or more hidden layers. Hidden layers neurons (computation nodes) are called hidden neurons or hidden units. Hidden neurons remain between the input layer and the output layer of a network and provide some useful functionality. By adding one or more hidden layers, the network is enabled to extract higher-order statistics. The ability of hidden neurons to extract higher-order statistics is particularly valuable when the size of the input layer is large. The output of the input layer is supplied to the second layer (i.e., first hidden layer) as input, and the output of the second layer is supplied to the third, and so on. The neural network is said to be fully connected in the sense that every node in each layer of the network is connected to every other node in the adjacent forward layer. If some of the communication links are missing, then the network is said to be partially connected. [1]

Recurrent networks. This network has at least one feedback loop. For example, a recurrent network may consist of a single layer of neurons where each neurons feed its output signal back to the input of all other neurons. This type of structure does not have self-feedback loops in the network. The *self-feedback* refers to a situation where the output of neurons is fedback into its own input. The recurrent network may have some hidden neurons, or the recurrent network may not have some hidden neurons. [1]

Input-delay feed-forward back-propagation neural network. Input-delay feed-forward back-propagation neural network is a time-delay neural network. In this type of neural network, hidden neurons and output neurons are replicated across the time. Generally, inside this type of neural network, delay is taken from top to bottom. Here, the network traps delay lines that sense the current signal, previous signal, and delay signal before it connects to the network weight matrix through the delay-time unit. These are added to the corresponding weight matrix from left to right in ascending order. All other features are the same as the feed-forward neural network, except input given to the neural network is delayed. In this neural network, memory is limited by the length of the tapped delay line. [1]

Hopfield artificial neural network.

> A Hopfield artificial neural network is a kind of repetitive artificial neural network that utilized to store at least one stable objective vectors. These steady vectors can be shown as recollections that the network reviews when furnished with comparative vectors that go about as a prompt to the network memory. These paired units stake two distinct qualities for their states that are controlled by whether the units' information surpasses their limit. Paired units can take either estimation of 1 or—1 or estimations of 1 or 0.
>
> [16]

Elman and Jordan artificial neural network.

Elman network likewise alluded as Simple Recurrent Network is an extraordinary instance of intermittent artificial neural networks. It varies from customary two-layer networks in that the primary layer has a repetitive association. It is a primary three-layer artificial neural network that has back-circle from shrouded layer to include layer trough alleged setting unit. This kind of artificial neural network has a memory that is permitting it to both identify and produce time-shifting examples. The Elman artificial neural network has commonly artificial sigmoid neurons in its concealed layer and straight artificial neurons in its yield layer. This mix of artificial neurons moves capacities can make inexact any capacity with discretionary exactness if there are sufficient artificial neurons in concealed layer. Having the option to store data Elman artificial neural network is fit for producing fleeting examples just as spatial examples and reacting on them. The main distinction is that setting units are taken care of from the yield layer rather than the shrouded layer.

[16]

Long short-term memory.

Long Short Term Memory is one of the intermittent artificial neural networks geographies. Conversely, with virtual intermittent artificial neural networks, it can gain from its experience to process, group and anticipate time arrangement with long delays of obscure size between significant occasions. This makes Long Short Term Memory to outflank other intermittent artificial neural networks, Hidden Markov Models and other grouping learning techniques. Long Short Term Memory artificial neural network is work from Long Short Term Memory hinders that equipped for recalling an incentive for any periods.

[16]

Bidirectional artificial neural networks (Bi-ANN).

Bidirectional artificial neural networks intended to foresee complex time arrangement. They comprise of two individual interconnected artificial neural (sub) networks that perform immediately and converse (bidirectional) change. Interconnection of artificial neural is sub-networks done through two unique artificial neurons that equipped for recollecting their interior states. This sort of interconnection among future and past estimations of the handled signs increment time arrangement expectation capacities. As such, these artificial neural networks anticipate future estimations of information as well as past qualities. That brings a requirement for two-stage learning; in the first stage, we show one artificial neural sub-network for anticipating future, and in the second stage, we show a second artificial neural sub-network for foreseeing past.

[16]

Self-organizing map (SOM).

> Self-sorting out guide is an artificial neural network that identified with feed-forward networks, yet it should inform that this kind of engineering is in a general sense distinctive in the game plan of neurons and inspiration. Regular course of action of neurons is in a hexagonal or rectangular matrix. Self-sorting out guide is distinctive in contrast with other artificial neural networks as in utilize a local capacity to protect the topological properties of the information space. They utilize unaided learning worldview to deliver a low-dimensional, discrete portrayal of the info space of the preparation tests, called a guide what makes them incredibly helpful for envisioning low-dimensional perspectives on high-dimensional information. Such networks can figure out how to recognize regularities and connections in their information and adjust their future reactions to that input as needs be.
>
> [16]

Stochastic artificial neural network.

> Stochastic artificial neural networks are a kind of an artificial knowledge apparatus. They worked by bringing arbitrary varieties into the network, either by giving the network's neurons stochastic exchange capacities or by giving them stochastic loads. This makes them valuable devices for improvement issues since the irregular changes assist it with getting away from nearby minima. Stochastic neural networks that worked by utilizing stochastic exchange capacities are frequently called Boltzmann machine.
>
> [16]

Physical artificial neural network.

> The more significant part of the artificial neural networks today is programming based yet that does not prohibit the likelihood from making them with physical components which based on customizable electrical flow opposition materials. History of physical artificial neural networks returned in the 1960s when first physical artificial neural networks made with memory semiconductors called memristors. Memristorsimitate and neurotransmitters are of artificial neurons. Even though these artificial neural networks marketed, they did not keep going for long because of their ineptitude for adaptability. After this endeavor a few others followed, for example, endeavor to make physical artificial neural network dependent on nanotechnology or stage change material.
>
> [16]

16.2.4 Advantages of ANN in Health Care

Artificial neural networks are widely used in various medical applications, like in cardiology, diagnosis, electronic signal analysis, medical image analysis, radiology etc. [6], and gain all the advantages of artificial neural network, like nonlinearity, uniformity of analysis and design, adaptivity, etc. "The science of artificial neural networks has been applied in efforts to solve several problems in four general areas of cardiovascular medicine. These areas are: coronary artery disease, electrocardiography, cardiac image analysis and cardiovascular drug dosing." [13] "Artificial neural network is used in the diagnosis of cancer, sclerosis, diabetes, heart diseases, etc. An adaptive algorithm is developed and applied to yield maximum accuracy in outputs with the statistics in clinical trials." [38]

> Algorithms in the framework of Neural Networks in Signal processing have found new applications potentials in the field of Nuclear Engineering. Nuclear Engineering has matured during the last decade. In research & design, control, supervision, maintenance and production, mathematical models and theories are used extensively. In all such applications signal processing is embedded in the process. Artificial Neural Networks (ANN), because of their nonlinear, adaptive nature is well suited to such applications where the classical assumptions of linearity and second order Gaussian noise statistics cannot be made. ANN's can be treated as nonparametric techniques, which can model an underlying process from example data. They can also adopt their model parameters to statistical change with time.
>
> [14]

> Artificial neural network is used in medical image segmentation and edge detection toward visual content analysis, and medical image registration for its preprocessing and post processing. Artificial neural network applications in computer-aided diagnosis represent the main stream of computational intelligence in medical imaging. Other applications of artificial neural network include data compression, image enhancement and noise suppression, and disease prediction etc. Moreover application of artificial neural network for functional magnetic resonance imaging (MRI) simulation becomes a research hotspot, where certain structured artificial neural networks are employed to simulate the functional connectivity of brain networks. Due to the similar nature of artificial neural network and human neurons, artificial neural network has been proved to be a very useful for this new task.
>
> [15]

> In radiology artificial neural network can be used in automated radiation dose estimation, image quality estimation, radiology reporting analytics, post processing of image segmentation, registration and quantification. By applying machine learning techniques to natural language processing, researchers can extract data from free text radiology reports, extra findings and measurements from narrative radiology reports and even track recommendations made by radiologists to referring physicians. This segment presents late applications inside radiology, which partitioned into the accompanying classes: characterization, division, discovery, and others. Automated radiation dose estimation

> calculations by artificial intelligence could support radiologists and technologists with making portion gauges before tests, the creators noted. This comes while presenting patients to the most reduced portion conceivable is all the more a concentration in clinical imaging than any time in recent memory.
>
> [16]

16.2.5 Disadvantages of ANN in Health Care

Along with the various advantages of artificial neural network in healthcare, it has some disadvantages, like *black box nature*. Due to black box nature, you do not know how and why your artificial neural network will come with a certain output. [19] Next is *duration of development*. Sometimes we develop an application that is not yet developed, then we use Tensorflow, which provides more opportunity, but it is more complicated and development takes more time. [19] Then there is *requirement of data*. Data requirements for neural network are more when compared to other traditional machine learning algorithms. That is not the easy problem to deal with. [19] Another disadvantage is its being *computationally expensive*. Neural networks are expensive when compared to traditional algorithms. The amount of computational power needed for a neural network depends heavily on the size of data but also on the depth and complexity of the network. [19] Last is *performance of network*. "Artificial neural networks can work with numerical information. Problems have to be translated into numerical values before being introduced to artificial neural network. The display mechanism to be determined here will directly influence the performance of the network." [20]

16.3 VARIOUS TECHNIQUES OF AI USED IN HEALTH CARE

Artificial intelligence has become more famous and is used in various fields, even in the health-care sector nowadays. The various techniques of AI used in healthcare are: (i) *Machine learning (ML)*. It is the most widely used technique in AI for healthcare. It uses the concept of supervised learning to predict accurate medicine and treatment. [21] (ii) *Natural language processing (NLP)*. Mainly, NLP is used for speech reorganization, text analysis, and translation. In healthcare, NLP is used to understand and classify clinical documents. [21] (iii) *Rule-based expert systems*. In rule-based expert systems, it make software based on "if–then" rules, but as the number of rules grow, system complexity is also increased. In recent times, machine learning has tried to replace rule-based system by using propriety medical algorithms. [21] (iv) *Diagnosis and treatment applications*. Earlier rule-based system is used for diagnosis and treatment purpose in healthcare. It identifies diseases, but it is not fully accepted for clinical practice. It is also complex. [21] (v) *Administrative applications*.

> There are a number of administrative applications for artificial intelligence in healthcare. AI in healthcare can be used for a variety of applications, including claims processing, clinical documentation, revenue cycle management and medical records management. Another use of artificial intelligence in healthcare

applicable to claims and payment administration is machine learning, which can be used for pairing data across different databases. Insurers and providers must verify whether the millions of claims submitted daily are correct. Identifying and correcting coding issues and incorrect claims saves all parties time, money and resources.

[21]

16.4 LITERATURE SURVEY

In this section, we will try to show the use of AI in health-care sectors using literature survey. For this, we refer to various IEEE, Springer, Elsevier, and other journals and papers. In this reference, authors Urja Pawar, Donna O'Shea, Susan Rea, and Ruairi O'Reilly (2020) discuss explainable AI in healthcare.

> Explainable AI (XAI) is a domain in which techniques are developed to explain predictions made by AI systems. In this paper, XAI is discussed as a technique that can use in the analysis and diagnosis of health data by AI based systems and a proposed approach presented with the aim of achieving accountability, transparency, result tracing, and model improvement in the domain of healthcare.
>
> [8]

Authors Luca Greco, Gennaro Percannella, Pierluigi Ritrovato, Francesco Tortorella, and Mario Vento (2020), meanwhile, discuss trends in IoT-based solutions for health care.

> Moving AI to the edge. In this paper authors proposed a short review about the general use of IoT solutions in health care, starting from early health monitoring solutions from wearable sensors up to a discussion about the latest trends in fog/edge computing for smart health.
>
> [9]

Mr. Ankur Narendra Bhai Shah and Dr. Jay A. Dave (2020), authors, discuss especially about the security of health records and its classification. "In this paper authors plan to design and develop an algorithm to secure health record for big data in cloud computing environment." [10]

Lo'ai A. Tawalbeh, Rashid Mehmood, Elhadj Benkhlifa, and Houbing Song (2016), authors, discuss mobile cloud computing model and big data analysis for health-care applications. They also discussed about networked health-care system, which is the integration of networking, healthcare, and other smart city systems. [7]

Shilpa Srivastava, Millie Pant, and Ritu Agarwal (2019), authors, discuss role of AI techniques and deep learning in analyzing critical health conditions.

> The purpose of this article is to find the relevance of various techniques of AI in different critical health scenarios. A comparative analysis is done based on

> the publications since 1995. The challenges and risks associated with the usage of AI in healthcare have been analyzed and suggestions made for making the analysis in the health domain more accurate and effective. Further the concept of deep learning has also been explained and its inculcation with the medical domain is discussed.
>
> [11]

16.5 EXAMPLES OF AI IN HEALTH CARE

In this section, we will see some of the examples of AI in health-care systems. There are a large number of examples of AI. Here, we have listed a few of them.

PathAI is used for more accurate cancer diagnosis with AI. Its location is Cambridge, Massachusetts.

> PathAI is developing machine learning technology to assist pathologists in making more accurate diagnoses. The company's current goals include reducing error in cancer diagnosis and developing methods for individualized medical treatment. PathAI has worked with drug developers like Bristol-Myers Squibb and organizations like the Bill & Melinda Gates Foundation to expand its AI technology into other healthcare industries.
>
> [22]

Buoy Health is an intelligent symptom checker. Its location is Boston, Massachusetts.

> Buoy Health is an AI-based symptom and cure checker that uses algorithms to diagnose and treat illness. Here's how it works: a chatbot listens to a patient's symptoms and health concerns, then guides that patient to the correct care based on its diagnosis. Harvard Medical School is just one of the many hospitals and healthcare providers that use Buoy's AI to help diagnose and treat patients more quickly.
>
> [22]

Freenome is used for earlier cancer detection with AI. Its location is San Francisco, California. "Freenome uses AI in screenings, diagnostic tests and blood work to test for cancer. By deploying AI at general screenings, FREENOME aims to detect cancer in its earliest stages and subsequently develop new treatments." [22]

16.6 CONCLUSION

In today's era, the use of artificial intelligence in the medical field is not just in demand but essential as well. A number of research has already been done to integrate AI into the medical field, and still many are ongoing. In this chapter, we tried to cover all the aspects of AI in the medical field and some survey part as well. From all these, we can concluded that, still, AI is very new to the medical field and much more research is required to use it efficiently.

REFERENCES

Book

1. Simon Haykin: "Neural networks and learning machines, 3/e", Pearson (2010).
2. Kevin Gurney: "An introduction to neural networks", UCL Press (1997).

Journal Article

3. Benedikt Wiestler, Bjoern Menze: "Deep learning for medical image analysis: A brief introduction", Oxford University Press, iv35–iv41 (2020).
4. Oludare Isaac Abiodun, Aman Jantan, Abiodun Esther Omolara, Kemi Victoria Dada, Nachaat AbdElatif Mohamed, Humaira Arshad: "Review article state-of-the-art in artificial neural network applications: A survey", Elsevier, 1–41 (2018).
5. Yoav Mintz, Ronit Brodie: "Introduction to artificial intelligence in medicine", Taylor & Francis Group, 1–9 (2019).
6. Jignesh Kumar L. Patel, Ramesh K. Goyal: "Applications of Artificial Neural Networks in Medical Science", *Current Clinical Pharmacology*, Vol. 2, No. 3, 217–226 (2007).
7. Lo'ai A. Tawalbeh, Rashid Mehmood, Elhadj Benkhlifa, Houbing Song: "Mobile cloud computing model and big data analysis for healthcare applications", IEEE (2016).
8. Urja Pawar, Donna O'Shea, Susan Rea, Ruairi O'Reilly: "Explainable AI in healthcare", IEEE (2020).
9. Luca Greco, Gennaro Percannella, Pierluigi Ritrovato, Francesco Tortorella, Mario Vento: "Trends in IoT based solutions for health care: Moving AI to the edge", Elsevier, 346–353 (2020).
10. Ankur N. Shah, Jay A. Dave: "Classification and security enforcement of the semi-structured health records on MongoDB leveraging hybridization of Modified AES algorithm with Apriory algorithm", *Juni Khyat*, Vol.10, No.1, 91–95 (2020).
11. Shilpa Srivastava, Millie Pant, Ritu Agarwal: "Role of AI techniques and deep learning in analyzing the critical health conditions", Springer (2019).
12. Michael Haenlein, Andreas Kaplan: "A brief history of artificial intelligence: On the past, present, and future of artificial intelligence", Sage, 1–10 (2019).
13. Dipti Itchhaporia, Peter B. Snow, Robert J. Almassy, William J. Oetgen: "Artificial neural networks: Current status in cardiovascular medicine", *JACC*, 515–521 (1996).
14. Rekha Govil: "Neural networks in signal processing", Springer (2000).
15. J. Jiang, P. Trundle, J. Ren: "Medical imaging analysis with artificial neural networks", Elsevier (2010).
16. Mayur Pankhania: "Artificial neural network: A primer for radiologists", IJCMR (2020).

Website

17. www.proschoolonline.com/blog/what-are-the-disadvantages-of-ai, last accessed on 13 February 2021.
18. www.thomasnet.com/insights/the-challenges-and-dangers-of-ai-in-the-health-care-industry-report/, last accessed on 13 February 2021.
19. https://builtin.com/data-science/disadvantages-neural-networks, last accessed on 13 February 2021.
20. www.linkedin.com/pulse/artificial-neural-networks-advantages-disadvantages-maad-m-mijwel, last accessed on 13 February 2021.

21. www.foreseemed.com/artificial-intelligence-in-healthcare, last accessed on 13 February 2021.
22. https://builtin.com/artificial-intelligence/artificial-intelligence-healthcare, last accessed on 13 February 2021.
23. www.byteant.com/blog/ai-adoption-in-healthcare-10-pros-and-cons/, last accessed on 13 February 2021.
24. https://royaljay.com/healthcare/neural-networks-in-healthcare/, last accessed on 13 February 2021.
25. www.businessinsider.com/artificial-intelligence-healthcare?IR=T#:~:text=Integrating%20AI%20into%20the%20healthcare,are%20associated%20with%20administrative%20tasks., last accessed on 13 February 2021.
26. https://becominghuman.ai/introduction-to-artificial-intelligence-5fba0148ec99, last accessed on 18 August 2021.
27. www.javatpoint.com/artificial-intelligence-tutorial, last accessed on 18 August 2021.
28. www.geeksforgeeks.org/artificial-intelligence-an-introduction/, last accessed on 18 August 2021.
29. www.guru99.com/artificial-intelligence-tutorial.html, last accessed on 18 August 2021.
30. www.expert.ai/blog/machine-learning-definition/, last accessed on 18 August 2021.
31. https://machinelearningmastery.com/supervised-and-unsupervised-machine-learning-algorithms/, last accessed on 18 August 2021.
32. https://algorithmia.com/blog/semi-supervised-learning, last accessed on 18 August 2021.
33. https://en.wikipedia.org/wiki/Reinforcement_learning, last accessed on 18 August 2021.
34. https://searchenterpriseai.techtarget.com/definition/narrow-AI-weak-AI, last accessed on 18 August 2021.
35. https://spotle.ai/feeddetails/What-Is-AI-Weak-AI-Strong-AI-with-examples-Key-disciplines-and-applications-Of-AI-/3173, last accessed on 18 August 2021.
36. https://builtin.com/artificial-intelligence/artificial-intelligence-healthcare, last accessed on 18 August 2021.
37. https://builtin.com/artificial-intelligence, last accessed on 13 February 2021.
38. https://pubrica.com/academy/medical-writing/overview-of-artificial-neural-network-in-medical-diagnosis/, last accessed on 18 August 2021.
39. www.edureka.co/blog/artificial-intelligence-applications/, last accessed on 13 February 2021.
40. www.businessinsider.com/artificial-intelligence-healthcare?IR=T, last accessed on 13 February 2021.
41. www.aidoc.com/blog/pros-cons-artificial-intelligence-in-healthcare/, last accessed on 13 February 2021.
42. www.mygreatlearning.com/blog/types-of-neural-networks/, last accessed on 18 August 2021.

Index

T

X

Y